✳ *Fire from Ice*

Searching for the Truth Behind the Cold Fusion Furor

by Eugene F. Mallove

Wiley Science Editions

John Wiley & Sons, Inc.

New York Chichester Brisbane Toronto Singapore

In recognition of the importance of preserving what has been written, it is a policy of John Wiley & Sons, Inc. to have books of enduring value published in the United States printed on acid-free paper, and we exert our best efforts to that end.

Published by John Wiley & Sons, Inc.

Library of Congress Cataloging-in-Publication Data

Mallove, Eugene F.
Fire from Ice : searching for the truth behind the cold fusion furor / by Eugene F. Mallove.
p. cm. -- (Wiley science editions)
Includes index.
ISBN 0-471-53139-1
1. Controlled fusion. I. Title. II. Title: Title: Cold fusion furor. III. Series.
QC791.73M35 1990
621.48′4--dc20 91-8036
CIP

Printed in the United States of America

10 9 8 7 6 5 4 3 2 1

Science is magic that works.

Kurt Vonnegut

To all who have struggled to bring
the fire of stars down to Earth.
To seekers of Truth, everywhere.

Great is truth. Fire cannot burn,
nor water drown it.

Alexander Dumas the Elder,
The Count of Monte Cristo, 1841–45

Contents

Preface

> It is really quite amazing by what margins competent but conservative scientists and engineers can miss the mark, when they start with the preconceived idea that what they are investigating is impossible. When this happens, the most well-informed men become blinded by their prejudices and are unable to see what lies directly ahead of them.
>
> Arthur C. Clarke, *Profiles of the Future*, 1963

> The discovery of fission has an uncommonly complicated history; many errors beset it. . . . Above all, it seems to me that the human mind sees only what it expects.
>
> Emilio G. Segré
> "The Discovery of Nuclear Fission," December 1988

> The energy produced by the breaking down of the atom is a very poor kind of thing. Anyone who expects a source of power from the transformations of these atoms is talking moonshine.
>
> Physicist Ernest Rutherford, about 1930

SKEPTICS HAVE WRITTEN A HUNDRED OBITUARIES for cold fusion, the unprecedented "miracle or mistake" that burst out of Utah into the public arena on March 23, 1989, but despite many unanswered questions about what "cold fusion" is or is not, evidence for the phenomenon (or phenomena) is now much too compelling to dismiss. Some would call the scientific clues only provocative. I choose to say *compelling.*

With an electric power supply hooked up to palladium and platinum electrodes dipped in a jar of heavy water spiked with a special lithium salt, chemists Martin Fleischmann and B. Stanley Pons were thought to have unleashed one of the wildest goose chases in the history of science. Now there is a significant possibility that they have discovered a quite revolutionary phenomenon that—along with hot fusion—could conceivably turn the world's oceans into bottomless fuel tanks.

Cold fusion is very likely to be real after all, although which aspects of it are valid remains in question. Despite many roadblocks that arose against confirming it as a new physical phenomenon, it is now here to stay. For a time, negative experiments and widespread skepticism seemed to have put cold fusion permanently on ice. Incredulity still runs deep. But cold fusion research is now very much alive in laboratories far and wide. It moves forward through those scientists with intense curiosity and courage to pursue these studies in the face of mountains of ridicule.

It is now reasonably clear that fusion reactions that liberate energy—near but very peculiar relatives of nuclear processes that are the lifeblood of the stars—can occur at room temperature. There is *no chance* whatever that cold fusion is *a mistake. There is the exceedingly remote possibility that "cold fusion" is a collection of many mistakes made in nuclear measurements of many different kinds, in heat measurements of great variety, and in all manner of control experiments. But to believe that hundreds of scientists around the world have made scores of systematic mistakes about the nuclear and nuclear-seeming anomalies that they have reported is to stretch credulity to the breaking point—to distort the meaning of scientific evidence to absurd limits. Cold fusion is not "pathological science" as many have charged, but for critics to continue to describe it as such or to ignore it completely is pathological.*

Current evidence suggests that *nuclear* processes are actually at work in what at first seemed to be merely table-top *chemical* experiments. This is absolutely shocking, and the root of widespread disbelief in cold fusion among scientists. There has been no more iron-clad principle separating chemistry from physics than that chemical behavior *never* leads to nuclear transformations. The tiny atomic nucleus has been inviolate to assault, but now it has been breached by the puffy electron cloud world of chemistry. You see, if the tiny, dense nucleus of an atom were blown up to the size of a golf ball, at that scale its attending fuzzy little electrons would orbit a mile away. Chemistry has only to do with how these distant electrons interact to make connections and disconnections among atoms. Atomic nuclei never become directly involved in chemical reactions and nuclei had not been known to react with one another except in extreme high-energy conditions.

Though the occurrence of cold fusion phenomena at present is erratic, it might some day be tamed and made regular and useful. Many experimenters are finding specific conditions, not reported initially by Drs. Fleischmann and Pons (perhaps not even known to them at the time), that prompt the effects. Furthermore, cold fusion phenomena are now seen in very dissimilar but related physical systems: pressurized gas cells, electrochemical cells with molten metal salts, and metal chips and films alloyed with fusion fuel.

To an extent, the phenomena remain not repeatable *at will*—but repeatable, to be sure, in a *statistical* sense, and sometimes now with very high confidence. (The same has been true in the early development of certain solid-state electronic devices.) There is now convincing evidence for the observation of significant heat *in excess of energy fed in*, bursts of neutrons, radioactive tritium at concentrations elevated above natural background (despite fears of preexisting contamination, there is ample evidence that the tritium is generated by nuclear reactions), possible abundance shifts in some chemical isotopes, and much more. And in a *pièce de résistance* of cold fusion research, in October 1990 scientists in several laboratories confirmed the nuclear *creation* of high-energy nuclei—probably those of tritium atoms—that fly out from titanium chips infused with the well-known fusion fuel, deuterium.

The measurements of power in the form of heat coming from some cold fusion cells is extraordinarily impressive—tens, to over a thousand, times the energy that could emerge from any conceivable chemical reaction. If the numbers from some experiments are to be believed, they add up to tens and even hundreds of kilowatt-hours coming from each cubic centimeter of cold fusion cell electrode material (about the volume of a stack of two pennies)! You know what a kilowatt-hour of electricity is when you pay for ten 100-watt bulbs turned on for one hour. More vividly, a kilowatt-hour is the energy of motion in a 4,000-pound car traveling 140 miles per hour.

Furthermore and most important, there is now a theoretical basis to begin to understand these apparent cold fusion phenomena. The heat-generating nuclear process must be very exotic, indeed, somehow being able to distribute released nuclear energy over a large array of atoms rather than emitting it as discrete high-energy particles.

Soon after the startling announcement at two universities in Utah in March 1989, the idea for this book was born. This might have been a very different work—a chronicle of the birth of a new age of cheap, clean, and limitless power. Though that era may still arrive through some form of controlled fusion—including the very real prospect of *controlled cold fusion*, the story turned out to be far more interesting, in both its scientific aspects as well as in the *process* of science that triumphed in identifying cold fusion as something literally new under the sun.

We have, instead, the saga of the tumultuous birth of a new physical phenomenon—more exactly, a class of scientific phenomena—an origin beset by bouts of optimism, pessimism, and every emotion in between for both proponents of the new wonder and those who vehemently deny its possibility—respected and well-intentioned scientists all. There occurred a veritable scientific roller-coaster ride that has held the scientific world in sway for almost two years. Now that many more facts are available and the furor has quieted down, the story can be told in its delicious and delirious detail. This is an account of the unfolding of a new phenomenon—the scientific process observed.

Through a sometimes tortured, contentious process the truth ultimately triumphs in science. Thus is scientific research done in the real world, not by idealized textbook prescriptions. Science is not conducted by poll nor by appeal to authority, nor always shackled to an imperfect and occasionally obstructive peer review process. Science proceeds through dogged experimental and theoretical effort.

At the beginning of the cold fusion saga, it was my good fortune to be working at the Massachusetts Institute of Technology. I was trained as an engineer, both in aerospace and environmental engineering at MIT and at Harvard, but after having done engineering for some 15 years, writing about science and technology became first an avocation and later a job.

As the chief science writer at the MIT News Office during the period when the cold fusion controversy arose, I found myself at a crossroads of scientific inquiry and intrigue. I heard from all sides in the scientific turmoil that broke loose and had the opportunity to witness firsthand how scientific news was being made. I, too, swang from skepticism to belief, back to skepticism, many times. At the outset, cold fusion seemed both too preposterous to believe and too important to ignore. The urge to chronicle this fascinating chapter in scientific history became irresistible. I have tried to be as faithful as possible in chronicling the complex events in the cold fusion saga and in illuminating difficult experiments and theory. *The opinions and perspective on the cold fusion controversy are entirely my own, however, and are absolutely not intended to represent any official or unofficial university position.*

We will explore the scientific intrigue and infighting that occurred in the cold fusion revolution, which provided much human drama. There were fights to publish and to forestall publication, issues of priority of discovery, funding matters, misinformation and disinformation, rumors that became "fact," questions of academic standing, and even allegations of scientific deceit. The hard lessons in science learned in the quest for cold fusion will depend on the ultimate resolution of the scientific questions, but whatever the outcome, some are already clear:

✳ Spectacular resistance to paradigm shifts in science are alive and well. Plasma fusion physicists were extremely reluctant to consider new fusion mechanisms even though they knew very well that the environments of electrochemical cells and palladium metal atomic lattices were remarkably different from the high-temperature gaseous systems to which they were accustomed.

✳ The majority does not rule in science. It is a gross mistake to draw conclusions about the validity of reported findings by polling the membership of this or the other scientific organization or panel.

✳ It is dangerous and often deceptive to make analogies between one scientific controversy and another. Comparing the cold fusion episode with several notable blind alleys in science—the "polywater" episode of the 1960s–70s, or the early 20th-century "N-rays"—is counterproductive and wrong. I acknowledge, however, that it may also be hazardous to compare the cold fusion debate to heated episodes in science that *did* result in a well-established discovery.

✳ Irving Langmuir's rules for identifying so-called "pathological science" are best retired to the junk heap for prejudice and name calling.

✳ Ockham's Razor is too easily forgotten. In science, the simplest unifying theory or connection is often most appropriate. Better to have a single explanation to bridge a host of apparently related phenomena, than to concoct baroque excuses for why multiple independent experiments may *all* be systematically incorrect. *Any* possible nuclear effect, even a tiny suspected one, such as low levels of neutron particle emissions seemingly unconnected with heat production, should have been a tip-off that other puzzling and erratic effects in similar physical systems might also have something to do with nuclear phenomena.

✳ Use extreme caution in dismissing experimental results just because theory suggests they are "impossible." Theory must guide science, but it should not be allowed to be in the driver's seat—especially when exploring the frontier.

✳ The fear that possible scientific error would be ridiculed, or worse, interpreted as fraud, is stultifying. A witch hunt against cold fusion affected researchers: Some who wanted to work in the field did not get involved for fear of scorn; others hid positive results from colleagues, anticipating career problems; and some laboratory managers refused to allow technical papers to be published on positive results obtained in their organizations. Most incredible, some scientists publicly decried cold fusion, while privately supporting its research.

✳ The peer review process by which articles make their way into journals is not infallible. While peer review is meant to act as a filter against spurious results and sloppy science, mismanaged or unchecked it can be a tyrannical obstacle to progress as well. It is unwise to be persuaded by the editorial position and selection of technical articles that appear in a single well-respected publication.

✳ Vested scientific interests are not easily persuaded to share their resources. Too small a total funding pie, in this case limited federal

> expenditures for energy research, led naturally to rivalry and anti-scientific tendencies that would have moderated with a policy of broader research support. The hot fusion fraternity, like any scientific community with its back to the wall, may find it difficult to draw impartial conclusions about a perceived threat to its dominance.

Above all, I wanted to distinguish between the real, initial scientific shortcomings of Drs. Fleischmann and Pons's work (including their initial incomplete disclosure of relevant experimental protocols) and their fully justified bewilderment in the face of a phenomenon for which they had no satisfactory explanation (other than a firm belief that the evidence pointed to it being nuclear). This required raising numerous questions about the *process* of science and communicating scientific developments to the public.

This may shock the uninitiated or misinformed, but when the science finally works its way to more firm conclusions, it is my view that Fleischmann and Pons, Brigham Young University's Steven E. Jones with his reports of neutrons, and other early cold fusion pioneers may be regarded in the history of science as heroes—very human, imperfect ones. Fleischmann and Pons's most serious failing, which ultimately sandbagged the whole subsequent scientific process, was to suggest initially that their experiment was very easy to reproduce, and that scaling it up to practical, power-producing devices would not be especially difficult. In some sense the Fleischmann-Pons experiment *was* relatively easy to reproduce, but it proved *far* from simple to interpret or to augment. Ironically, Steven Jones is to be faulted for consistently *denying* that electrochemical cells could be producing excess heat from nuclear reactions—an opinion arising from his stubborn disbelief and desire to protect the priority of *his* discovery, not from the results of his own experiments or deep analysis of the thermal measurements made by others.

Yet all three protagonists took their incomplete preliminary findings to the scientific community and kicked it into unprecedented and rapid global action. A U.S. Department of Energy report estimated that initially between $30 and $40 million dollars were spent worldwide on cold fusion research. That estimate is now woefully low, as the pace of research quickens. A recent compilation of reports of only *positive* evidence for cold fusion, which have come from more than 80 research groups in a dozen nations and at five U.S. national laboratories, gives some idea of the scope and seriousness of the activity (see pages 246–248 in Chapter 15).

The cold fusion story cannot be understood without grasping the parallel effort to develop controlled *hot* fusion, one of the most noble and difficult technological quests ever undertaken, now in its fifth decade. Without rehashing the extraordinary history of hot fusion research—a fascinating saga in its own right—included is sufficient background to put cold fusion in proper perspective.

An essential caveat: After reviewing mounting evidence from cold fusion experiments, I am persuaded that it provides a *compelling* indication that a new kind of nuclear process is at work. I would say that the evidence is *overwhelmingly* compelling that cold fusion is a real, new nuclear process capable of significant excess power generation. The evidence for significant power generation, however, cannot be said to be *conclusive.* The word conclusive in science denotes an intimate melding of experimental observation and theoretical explanation. In the case of cold fusion, this cannot be said to have occurred. There is yet no *proved* nuclear explanation for the excess heat. That excess heat *exists* is amply proved.

Teasing a new phenomenon from nature is not easy. Simply review the history of the discovery of fission in the 1930s—the phenomenon was staring physicists in the face, yet fission was slow to be recognized. Or recall superconductivity, which a Dutch physicist stumbled across in 1911, but for which no good theory existed until the 1950s. High-temperature superconductivity, which exploded into the world of physics in 1986–87, is still incompletely understood. Or recall the "cat's whisker" or crystal radio of the 1920s, which wasn't understood until the transistor was invented three decades later. But for ignorance and skepticism, we might have had transistor radios in the 1920s! Or take the totally unexpected phenomenon of lasing, both at optical frequencies (lasers) and at microwave frequencies (masers), and more recently at X-ray wavelengths. Radio waves themselves, predicted in the 1860s and discovered in the 1880s, were another totally unexpected manifestation of matter and energy. Why not "cold fusion"? Nature has marvelous tricks up her sleeves, and it is the delight of the scientist to discover them. Let us see how the power of the stars is coming down to Earth.

Bow, New Hampshire

Acknowledgments

IF AN IDEA HAS A THOUSAND PARENTS, a book may have at least a few hundred. *Fire from Ice* would not have been without the dedicated work of the hundreds of researchers who probed and who continue to investigate cold fusion phenomena, and without the efforts of thousands who strive to tame hot fusion. Proponent and skeptical views alike were the two streams that blended and fused in this work.

My deep appreciation for extremely helpful discussions extends to Professor Martin Fleischmann of the University of Southampton and the University of Utah and to Professor Steven E. Jones of Brigham Young University.

I am particularly grateful to four scientists at MIT with whom I have discussed both hot and cold fusion: Dr. Richard D. Petrasso, Dr. Stanley C. Luckhardt, and Professor Ronald R. Parker, all of the Plasma Fusion Center, and Associate Professor Peter L. Hagelstein of the Department of Electrical Engineering and Computer Science. Other scientists and engineers who have been immensely helpful are: Dr. Bruce Gregory of the Center for Astrophysics in Cambridge, Massachusetts; MIT visiting scientist Dr. Henry H. Kolm, Professor Lawrence M. Lidsky of MIT's Department of Nuclear Engineering; Dr. Vesco C. Noninski, an electrochemist from Sofia, Bulgaria; Associate Professor Donald R. Sadoway of MIT's Department of Materials Science and Engineering; Professor of Physics Emeritus David Frisch of MIT; Dr. Mark Stull of Bedford, New Hampshire; Dr. Frank Sulloway of the MIT Science, Technology, and Society Program; Donna Baranski-Walker of the MIT Technology Licensing Office; Dr. Fritz Will, Director of the National Cold Fusion Institute; and Donald Yansen of Bio-Rad, Cambridge, Massachusetts.

Beyond the East Coast of the United States, I am indebted to Professor John O'M. Bockris of Texas A&M University; Hal Fox of the Fusion Information Center in Salt Lake City; Russ George of LGM Productions; Dr. M. Srinivasan of the Bhabha Atomic Research Center in India; Dr. Howard Menlove of the Los Alamos National Laboratory; Professor Julian Schwinger of UCLA; Dr. David Worledge of the Electric Power Research Institute (EPRI); and my colleague in the writer's art, physicist Dr. Robert L. Forward.

Many science journalists and public information officers have helped directly or indirectly, including: Jerry Bishop of *The Wall Street Journal;* Nancy Enright of the American Chemical Society; Pamela Fogle of the University of Utah; Joel Shurkin of Stanford University; Ed Walraven of Texas A&M University; science reporter Ed Yeates of KSL TV, Salt Lake City; Professor Bruce Lewenstein of Cornell University; Ivan Amato of *Science News;* Robert Cooke of *Newsday;* Irwin Goodwin of *Physics Today;* Ron Dagani of *Chemical and Engineering News;* Robert Pool of *Science;* and graduate students John Travis, formerly of MIT, and Silvia Bianchi, formerly of Boston University, both of whom studied independently the media's coverage of cold fusion.

Throughout the cold fusion episode, Kathy Powers, Paul Rivenberg, and Pat Stewart of the MIT Plasma Fusion Center were helpful in sharing with me the Center's archive of technical information. The crucial support and sage advice of editor David Sobel at John Wiley & Sons and the work of my literary agent Richard Curtis were the foundation of my efforts, as were critical scientific discussions and some initial writing on cold fusion shared with my scientific colleague, Dr. Gregory L. Matloff. I am thankful also for the work of Maria Danzilo, Frank Grazioli, and Judith McCarthy of John Wiley and Laura Van Toll of Impressions. Above all, my family deserves five fusion-powered stars for putting up with this confining passion. I return now to the home planet.

1 Prologue: Desperately Seeking Fusion

Water, water, everywhere,
Nor any drop to drink.

Samuel Taylor Coleridge, *The Rime of the Ancient Mariner*, 1798

Anything that is theoretically possible will be achieved in practice, no matter what the technical difficulties, if it is desired greatly enough.

Arthur C. Clarke, *Profiles of the Future*, 1963

✳ A Genie Shrugs

THE SNOW-COVERED WASATCH MOUNTAINS, so beautiful and unreal in late March, glistened against the intense blue of the skies above Salt Lake City. Spring skiers sported within those hills, unaware of news that was soon to come from the city below and oblivious to an approaching intruder above, in deep space.

For those—superstitious or not—who like to connect life on this world with celestial events, an auspicious or portentous happening: At about 8 hours Universal Time on the 22nd of March, 1989, multimillion ton asteroid 1989FC whizzed by Earth and its Moon, coming within 430,000 miles of our world. It made the closest known pass by a body of such mass since Hermes in 1937—the year before the discovery of nuclear fission.

As the asteroid continued on its path traveling many miles per second, the world turned not even once on its axis. The next day, Thursday, March 23, 1989, brought a glimmer of hope from a city that had grown up near the barren flatlands of the Great Salt Lake in Utah. At 1:00 P.M. in Salt Lake City, chemists Martin Fleischmann and B. Stanley Pons burned their names into the history of the quest for fusion power. Essentially unknown to the hot fusion community, they claimed to have achieved what seemed to be impossible: power-producing fusion reactions at room temperature.

Hours later, a gargantuan tanker left the port of Valdez, Alaska, en route with oil for an energy hungry world. At four minutes past midnight, March 24th, the Exxon Valdez ran aground and spilled 11 million gallons of crude oil into the pristine waters of Prince William Sound. The disaster symbolized the ultimate futility of our dangerous dependence on the planet's subterranean fossil fuels.

The massive oil spill drew deserved national attention and outcry, but it did not eclipse the extraordinary news from Utah about "cold fusion"—a concept that seemed to drop from the sky like an alien intruder straight into the public psyche. At the press conference held at the University of Utah, B. Stanley Pons, professor of chemistry and chairman of the Department of Chemistry at the University of Utah, and colleague Martin Fleischmann, professor of electrochemistry at the University of Southampton, England, proclaimed that they had discovered an amazingly simple method to create power-producing nuclear reactions—possibly fusion—not at hundreds of millions of degrees in imitation of the stars, but at room temperature!

The Genie of fusion shrugged in his ancient vessel that year and amazed the world. The spring of 1989 will long be remembered as a time of unexpected shaking, when extraordinary claims by groups of researchers in Utah and subsequently around the world led scientists to reexamine a decades-long pursuit: the quest to tame nuclear fusion. The struggle has been to bring this power of the stars down to Earth, much as fabled Prometheus snatched fire from the gods. The interest of the scientific community and the public at large was temporarily galvanized by the idea that a new kind of fusion process, immediately dubbed *cold fusion,* might soon lead to a way to get the fusion Genie to stop shrugging and come completely out of his bottle.

Startling events occasionally make us step back to get a better view of our pursuits and to examine cherished assumptions. This often leads to rededication, to unforeseen possibilities, and to new directions. The shaking of complacency now and then in a positive way is healthy, no more so than in the fields of science and technology where intense concentration on an established course sometimes promotes a possibly too narrow focus.

We now know that confirmation or rejection of the remarkable cold fusion claims of 1989 were not to come easily and that unusual doubt and confusion (inevitably termed "fusion confusion") beset a baffled, bemused, and even outraged scientific community. Estimates are that, for a time, more than one million dollars per day—in person-hours and equipment—was expended worldwide to confirm or disprove the claims that nuclear fusion reactions can occur in apparatus no more complex

Dr. Martin Fleischmann holds an electrochemical cell of the kind used in the cold fusion experiments at the University of Utah. (Courtesy University of Utah)

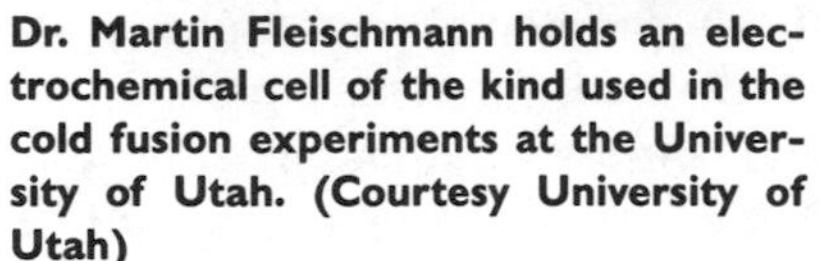

Dr. B. Stanley Pons holds a prototype cell that is larger than that used in his and Dr. Martin Fleischmann's first experiments. (Courtesy University of Utah)

than a laboratory electrochemical cell, or in pieces of metal infused under pressure with a heavy version of hydrogen, the isotope deuterium.

At a bare minimum, it now appears very likely that a wholly unexpected scientific phenomenon has been discovered. If it really is a new mode of fusion, it occurs, quite surprisingly, at room temperature. Moreover, the phenomenon appears to be capable of net power generation, but whether what seems to be an erratic, difficult-to-reproduce process can be tamed for practical applications remains an open and extremely intriguing question.

While the jury is still out on the significance of these developments, there can be little doubt that the larger effort to tame fusion for human needs has received an unexpected and perhaps much needed boost. The public imagination and interest in fusion power has stirred in a way that has never before happened in the relatively unknown quest. The nations of the world have spent billions of dollars to control thermonuclear (hot) fusion in gaslike plasmas whose temperatures sometimes reach several 100 million degrees centigrade, but the average citizen has heard little about the dramatic progress in recent years in this exceedingly difficult scientific and technological effort.

The new developments on the frontiers of fusion research come at a critical juncture in the U.S. and international efforts to control this potentially limitless and extremely benign source of energy. A large and complex laboratory machine, the Joint European Torus (the so-called JET tokamak in England) has just now reached, in effect, the long-sought energy *breakeven* point in "conventional" high-temperature fusion experiments: achieving about as much energy output as input. A few more years and self-sustaining, so-called *ignited*, fusion experiments are destined to produce significant net power, but in a form still not suitable for practical and extended power generation. For hot fusion, the goal of reaching engineering and commercial feasibility lies two or more decades ahead.

To fully understand the implications of cold fusion, it is essential to put fusion power in the widest possible context, and to tell how it may eventually dramatically affect human affairs. The fossil fuel era is nearing an end. No matter what conservation steps are taken, the world's reserves of coal, oil, and natural gas are clearly running down. They will be severely depleted within a single century and will have vanished completely within a few hundred years, if we keep using them intensively. Moreover, the local and global environmental consequences of running full-tilt at power generation with fossil fuels may perhaps be as ominous, if not *more* frightening, than simply running out of power. Whether or not there will be significant global warming as a result of carbon dioxide and other "greenhouse" gas emissions is not the issue. To continue dumping the other noxious end products of combustion into the environment is simply *stupid* given existing and emerging alternatives.

Fusion power offers the prospect of energy abundance over times comparable to geological ages, in contrast to the microscopic blip in human history of reliance on fossil fuel.

If we expect our descendants to live virtually indefinitely on this planet—until perhaps our Sun, our fusion reactor in the sky, "dies" some five billion years hence—we had better plan now to possess a source of inexhaustible power. What will that be? Possibly a source of solar power captured by vast solar cell arrays in space and beamed back to Earth's surface as microwaves, solar power collected by large arrays deployed in desolate areas, or a new kind of nuclear fission power perhaps, a modification of present nuclear reactor technology that may allay even passionately antinuclear fears? This kind of passively safe nuclear reactor, which can be shown to release no radioactivity to the environment even when its coolant is lost, has already been built and is practical.* A new generation of safer fission power plants merely awaits the economic and political wherewithal.

*Professor Lawrence M. Lidsky, MIT: "Safe Nuclear Power," *The New Republic*, December 28, 1987: 20–23; "Nuclear Power: Levels of Safety," *Radiation Research*, Vol. 113, 1988: 217–226.

Despite public fears about present-day fission power reactors, they have *by far* the best track records in safety of virtually all means of generating electricity (remember, even hydroelectric dams break and kill), and with their high-level radioactive wastes safely disposed in subterranean chambers—as must begin to be done in the coming decades—fission reactors are infinitely more benign to the environment than fossil fuel power. But while fission power may take us very far into the future—some hundreds of years or several thousands of years, depending on how fuel sources hold up—even fission has a demonstrably limited future. Fusion is an energy resource that is *virtually infinite.*

✳ Fusion Is Forever

We inhabit a water planet. Though relatively speaking it is less than eggshell-thin, a layer of water covers more than 70 percent of the world's surface. If we could use a tiny fraction of the millions of cubic kilometers of water for fuel to produce power for an energy-hungry globe, it would be infinitely better than achieving the alchemists' goal of turning base metals into gold. One way or another, the vision of harnessing the world's oceans to that end will come true. In researchers the world over, the dream of wrenching fire from ice is alive: fusion power, the fire of stars, taken from icy water.

The clever Prometheus of Greek legend merely stole fire from Zeus, the chief deity, and returned it to humankind. More audacious, fusion scientists have been struggling for four decades—roughly since the birth of the idea of fusion bombs—to steal the fire of stars from ordinary water. Because water is so cold (on a relative scale being but a few hundred degrees above the absolute zero of temperature) taming fusion aims almost literally at teasing fire from ice.

Enough fusion fuel exists on Earth to keep billions of people going effectively forever. It is frozen fire that has existed since the birth of time. When realized, the vision of controlled fusion power will allow us to release energy from deuterium, a special form of hydrogen ("heavy" hydrogen) that exists in a small but potent amount in every drop of water in nature. About one hydrogen atom in every 6,700 on Earth is a hydrogen isotope, deuterium (often written, D). That is, deuterium *is* hydrogen because it has one proton in its tiny, dense nucleus, but deuterium also has a neutron accompanying the usual single proton, making it about twice as heavy as H—ordinary hydrogen (a neutron is only very slightly heavier than a proton). Every water molecule, H_2O, contains one oxygen atom and two hydrogen atoms.

When you look out a window on a rainy day, you are watching fusion fuel falling from the sky. The tiny amount of deuterium in every gallon of ordinary water, about 1/250th of an ounce—not nearly enough to fill a baby's spoon if it were liquid, contains potential fusion energy equivalent to the chemical combustion of 300 gallons of gasoline. A comparison of fusion, fission, and fossil fuel required for a typical power plant is in order: A typical electric power plant of 1,000 megawatt (MW) capacity—meaning one thousand million watts—requires about twenty thousand railcars of coal *per year*—a procession carrying some two million tons and stretching about 400 kilometers! The oil energy equivalent of this is some ten million barrels of crude oil—seven supertankers' worth. The nuclear fission fuel equivalent of this horrendous pile of coal or lake of oil comprises a mere 150 tons of raw uranium oxide—a volume easily carried by about eight tractor trailers. But a single pickup truck could carry the 0.6 ton of heavy water (D_2O) necessary to fuel an equivalent 1,000 MW fusion power plant for one year!

There is obviously more than enough fusion fuel to go around, but before we can use it, we have a lot to learn.

✳ The Fusion Universe

Look up in the sky on a dark night and you will see thousands of bright fusion reactors—the stars. The Sun is the fusion reactor that keeps us alive. If plants were to die for lack of fusion-produced starlight, the animal kingdom would soon follow into oblivion. We can say with confidence that every life-form on Earth—energized as it is by sunlight—is an embodiment of fusion power.

We owe this to the violent collision of the nuclei of hydrogen atoms at the cores of stars where temperatures are reckoned in tens of millions of degrees. These collisions of hydrogen nuclei, simple single protons stripped of their ordinarily attending electrons, promote fusion reactions—the buildup of heavier nuclei from lighter ones. This results in a stupendous release of energy and an "ash" or reaction end product, the nuclei of the next heaviest element, helium—the kind of atom that buzzes within a child's balloon.

A star's fusion reactions produce the necessary temperature and gaseous pressure to counter the tendency of the star to collapse from its own self-gravitation, that is, from under its own weight. But gravity keeps the fusion fuel in a star cooking and contained. For decades, hot fusion researchers on Earth have tried to mimic the Sun by using intense magnetic fields to contain fusion reactions in gaslike *plasmas* at scores of millions of degrees, and more recently by aiming intense laser beams at solid fusion fuel pellets to turn them briefly into glowing plasmas—in effect, miniature stars.

Plasmas are omnipresent in the universe. The visible universe is more than 99 percent plasma: the hot interiors of stars themselves; glowing reaches of material between the stars about to give birth to other stars or luminous from the intense radiation of stars of advanced age; lightning itself; the minute sparks jumping off one's finger after walking on a rug on a cold, dry day; the eerie, glowing auroral displays (Northern Lights); and plasmas within glowing fluorescent light bulbs or neon lights. The word plasma was coined in the 1920s by American physicist Irving Langmuir, who made a metaphoric comparison between the multicomponent blood plasma that carries red blood cells and the species of charged particles in the hot plasmas with which he was working.

Plasmas are gases in which temperatures are so high that negatively charged electrons have been stripped off of atomic nuclei to one degree or another and are swimming within a "soup" of positively charged particles. The overall charge of a plasma is typically zero, but it is a good conductor of electricity, because, like a metal, lots of electrons may roam freely.

Plasmas exhibit some of the most complex, dynamical behavior in nature, because their charged components respond to the forces from electrical and magnetic fields and these motions, in turn, set up their own fields. Not solids, liquids, or gases, high-temperature plasmas constitute a veritable fourth state of matter, the most common one in the cosmos. Rocky planets and moons with their ice, liquid oceans, and gaseous atmospheres, are the exception rather than the rule in the plasma universe.

When the universe was born some 15 billion years ago in the titanic Big Bang explosion at the beginning of space and time, by the end of the first three minutes a high-temperature maelstrom of quarks (the fundamental constituents of protons and neutrons) and other subnuclear particles had cooked up a mixture of about 75 percent hydrogen nuclei (protons) and 25 percent helium nuclei (each with two protons and two neutrons), plus some other trace elements.* Yes, the visible universe consists mostly of fusion fuel and helium ash. Perhaps even more fantastic: All the heavier elements that go into building our planet and our bodies, such atoms as carbon, oxygen, nitrogen, iron, silicon, not to mention more exotic ones such as palladium, platinum, or uranium, were once inside distant stars that exploded billions of years ago. That fusion is central to the scheme of the universe is a striking cosmic fact.

No matter that the kinds of fusion reactions within the Sun and other stars are of a different variety than we might expect to use in a human-engineered reactor. It will probably be much too difficult to fuse protons at high temperature, so hot fusion scientists have sought to fuse together deuterium nuclei and one even heavier hydrogen nucleus *tritium* (containing one proton and *two* neutrons) in various combinations.

---

*Percentages by *mass* not number of atoms.

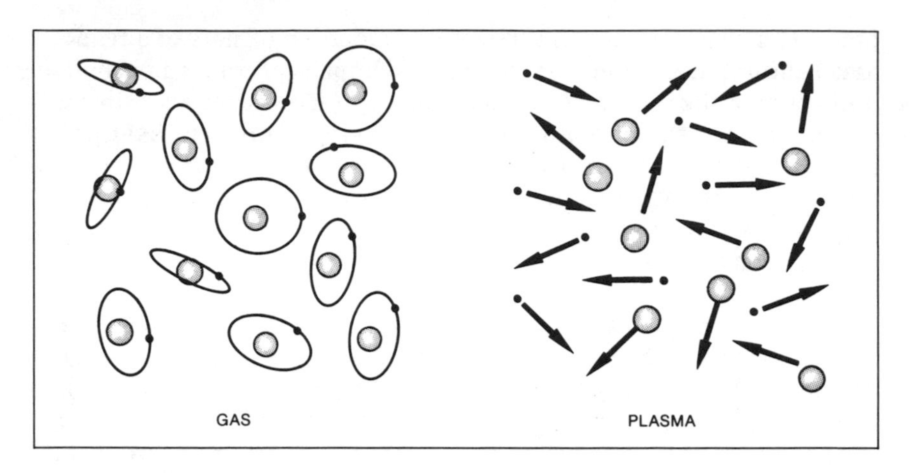

A plasma differs from a gas in which electrons remain physically bound to nuclei and form complete atoms. In a high-temperature plasma, negatively charged free electrons swim in a soup of positively charged ions—nuclei with electrons stripped off. (Courtesy Princeton Plasma Physics Laboratory)

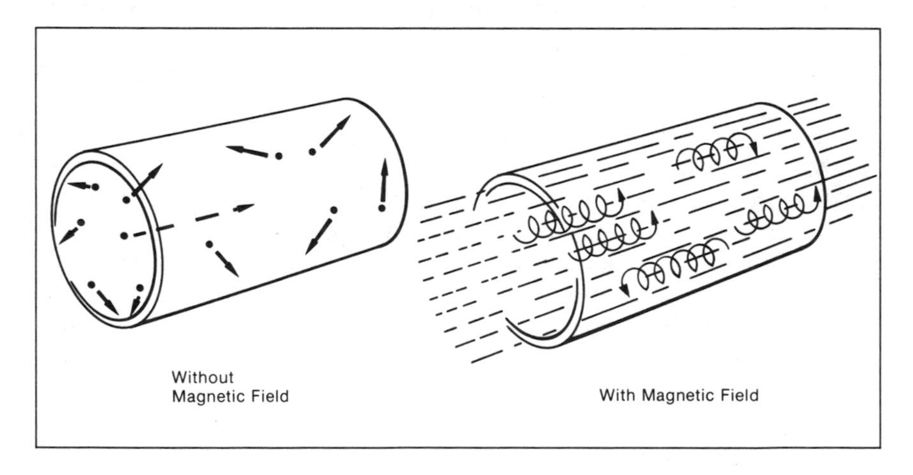

In the absence of a magnetic field, the charged particles of a plasma move in straight lines in random directions. Particles may come in contact with the walls of a containing vessel, thus cooling the plasma and inhibiting fusion reactions. If, however, a magnetic field is imposed on a plasma, the charged particles follow spiral paths about the invisible magnetic field lines and are thus kept from striking the walls of the containment vessel. (Courtesy Princeton Plasma Physics Laboratory)

The absolute zero of temperature is mighty cold: about −460°F (Fahrenheit) or −270°C (Celsius). In most substances, atoms jiggle barely at all near that frigid temperature. At higher temperatures, atoms and molecules move around faster, bumping into one another, their average speed depending on the temperature. Temperature, in fact, is a measure of the average velocity and energy of moving atoms or molecules. Indeed, temperature seems to be central to the occurrence of fusion reactions in nature. This is true because the relative velocity between atoms or their nuclei is one means by which the nuclear ingredients of fusion reactions can be made to overcome the extreme electrical repulsion forces between positive charges that normally keep them apart. That is why it is so difficult to fuse the bare protons of two ordinary hydrogen nuclei.

> It is by far more convenient to use the Kelvin (K) scale of temperature, rather than Fahrenheit (°F) or Celsius (°C). There are no "below zero" temperatures on the Kelvin scale, because temperature is reckoned from 0 K, absolute zero, where minimal atomic or molecular motion is occurring. When we are talking about millions of degrees, as is often done in fusion research, the Kelvin temperature is virtually identical to the Celsius temperature, since a Kelvin and a Celsius degree are of the same size (measure of temperature rise) and the zero temperature for Celsius (0°C, the freezing point of water under normal conditions) begins only 273°C above absolute zero—a small number compared to millions of degrees. (Unlike for °C temperatures, it is customary not to indicate a degree sign "°" before the K.)

✳ Star or Planet?

It is not *strictly* true that without the fusion reactions of the Sun, the temperature of our planet would approach that of deep space—about 3 K. When the rocky Earth and the other planets formed some 4.5 billion years ago from a cloud of primordial debris that was enriched with the heavier elements of exploded stars, radioactive atoms were mixed into the recipe for the planets. The nuclei of these atoms are so unstable that they disintegrate and emit radiation spontaneously, radiation that can slowly but surely heat the body of a planet. The heat flow coming from the interior of Earth is thousands of times less than the power of radiation from the Sun that strikes the planet.*

Now these nuclear processes that contribute to heating Earth's interior are, of course, not fusion reactions. They are simple radioactive decays of one heavy element such as thorium or uranium into lighter

---

*Still, it is interesting that upward through a square of continental surface about 130 feet on edge passes enough heat to power a 100 watt electric light bulb (if the heat were convertible to electricity with 100% efficiency). No one has ever tried to harness this weak flow of energy from radioactivity, except in those rare places where geological formations—hot springs, geysers, and the like—bring greater heat flow to the surface.

elements—ultimately to such stable forms as the element lead. For the most part, these nuclear processes are not even *fissions,* in which atomic nuclei split into two roughly equal fragments, although a small amount of natural fissioning does occur. The recent interest in cold fusion, however, has prompted wild speculation that low levels of natural *fusion* reactions may be occurring deep within the Earth.

So basic a question as, "What is the difference between a star and a planet?", has to do with whether copious fusion reactions either are occurring or ever did happen within an astronomical body. Tiny Earth, Mercury, Venus, Mars, and Pluto are obviously planets. They certainly aren't massive enough to have any abundant "conventional" fusion reactions going on within their cores, nor do they have hydrogen fusion fuel in their central regions. But what about the Solar System's gas giants, Jupiter, Saturn, Uranus, and Neptune? Could these planets more properly be termed *failed* or *borderline* stars?

Certainly Jupiter and its sister giant planets may make at least a remote claim to being stars. Astronomers have measured the electromagnetic radiation coming from Jupiter—both visible light and infrared radiation—and find that more energy is coming out than is going in. Some have speculated that this excess radiation is coming from weak fusion reactions going on within Jupiter. If this were true, we would have hot fusion reactions in stars and cold fusion reactions in planets—from fire to ice, as it were.

However, to be a true star that generates significant energy of its own, astronomers believe that an aggregation of hydrogen and helium, self-contracting from the force of gravity, must have a mass of about 80 times that of Jupiter. This is still much less matter than exists in our own Sun. Jupiter, with 300 times the mass of Earth, has but one-thousandth the mass of the Sun, so to be a star, a body should be no less massive than about 8 percent of the Sun. There has been much interest in the search for these low mass stars that have been dubbed brown dwarfs, because of their presumed very low surface temperatures. In recent years, evidence (albeit not yet conclusive) has accumulated that brown dwarfs with relatively weak fusion reactions in their cores exist, both as companions orbiting other suns and perhaps as independent objects coasting freely through space.

It is important to realize that despite the multimillion degree temperature and high density of the Sun's core, it is still far too cool for the kinds of fusion reactions that scientists have been trying to produce in laboratory hot fusion reactors. (Newspaper articles often say that hot fusion scientists are trying to "tame the power of the stars," unfortunately giving the misleading impression that they are planning to use those very same fusion reactions. They are not.) The temperatures that scientists are seeking are 100 million K and beyond. What is more, energy production in the solar core is actually very weak—only a few watts per ton of "starstuff." The bodily heat output of a resting human being, coming from chemical reactions of course, is by far more impressive! The solar core's great size and mass explain how the total output of the Sun can be so stupendous—4×10^{26} watts. The energy released in *one second* by the Sun could keep our civilization going at its present rate of energy consumption for more than a million years; collecting that power radiated in every direction by the Sun would be another matter.

✳ What Is Fusion?

The idea behind fusion is really very simple. Two light-weight* nuclei come together and stick to one another or *fuse,* forming a nucleus of greater weight than either of the two reactant nuclei. In creating the new nucleus, this fusion process may also include the ejection of one or more subnuclear particles such as a positive electrically charged proton or a chargeless neutron, or other kinds of particles. But the key phenomenon in fusion—its *defining characteristic*—is the formation of a more massive nucleus and the release of energy in a number of forms, whether in the velocity of particles such as neutrons or protons, in penetrating powerful radiations called *gamma rays* (like X rays, only much more energetic), or in other mechanisms that some have hypothesized for cold fusion. The resulting mass of the newly fused nucleus is less than the combined mass of the nuclei that formed it—a tiny amount of mass disappears during fusion and is converted to energy.

The energy release in fusion comes from the conversion of matter to energy by an amount given by Albert Einstein's formula from his 1905 theory of special relativity, $E=mc^2$, that is, the energy release is equal to the mass that is converted multiplied by the speed of light squared. (Light speed must be in units consistent with the mass, such as meters-per-second if mass is in kilograms; then E would come out as *watt-seconds, a unit like kilowatt-hours* that you notice in despair each month on your electric utility bill.)

What form of matter is disappearing in a fusion reaction is far less obvious, but disappearing it surely is. To cite one astonishing example: Every second some four-million metric tons of mass disappear within the Sun's fusion reactor, being converted to energy that eventually emerges at the star's surface! Yet so massive is the Sun that this destruction of mass can occur for billions of years and still less than one-

---

**Mass* is the more general and accepted terminology that physicists use, because, technically, *weight* depends on location (an object weighs less on the Moon than on the Earth), whereas the quantity known as *mass* does not.

ten thousandth of its original mass will have vanished. We too easily forget, but this is what is so remarkable about any kind of nuclear power: The conversion of a minute fraction of the mass of fuel can liberate staggering amounts of energy, all because of $E=mc^2$.

The energy requirement *per proton or neutron* to bind an atomic nucleus together for a long time *generally* becomes *less* in the case of larger nuclei (up to the mass of about iron, which typically has 26 protons and 30 neutrons). This is the so-called *binding energy* of a nucleus. When two light nuclei fuse to form a more massive nucleus, adding up the masses of the resulting nucleus and any particles such as neutrons that may fly off in the process gives a total final mass that is *less than that of the original two nuclei added together.* This mass deficit or loss is what has been converted to the energy of particles and radiations that emerge from the fusion reaction. Fusion reactions, just like fission reactions, *must* involve the loss of mass and its conversion to various forms of energy such as heat and radiation.*

There are many, many kinds of fusion reactions that can occur among light elements, but the following one, for example, is of concern in the engineering of hot fusion reactors because it illustrates how deuterium can be used as fuel (the → means simply goes to or becomes):

$$D + D \rightarrow {}^{3}He \text{ [at 0.82 MeV energy]} + n \text{ [at 2.45 MeV energy]} \qquad (1)$$

Deuterium plus Deuterium (Goes to) Helium-3 plus a neutron

We will have more to say about such reactions in discussing the different technologies that scientists have considered to tame fusion, but it is instructive to understand how to interpret these simple symbolic equations. Don't let them scare you—they are really quite easy and you certainly *don't have to memorize them!* Reaction (1) suggests that two *deuterons* (deuterium nuclei, designated D, just as ordinary hydrogen has its own symbol, H) can combine to form the nucleus of helium-3 (designated ^{3}He) plus a neutron (n). By definition, the element helium has two protons in its nucleus (the number of protons always defines what the element is), and the added neutron gives a total nucleon count (protons plus neutrons) of three, hence the superscript 3. Helium-3, extremely rare in nature (though prevalent on the surface of the Moon, having been transported there by the solar wind), is a variant or an

*For our purposes, it really isn't important to understand exactly why *less* energy per constituent *nucleon*—neutron or proton—should be required to hold this more massive nucleus together by what are called *nuclear forces*. Understand, however, that there is a natural tendency for positively electrically charged protons to repel one another, and it is only the presence of chargeless neutrons along with the attractive nuclear forces that "glue" a nucleus together.

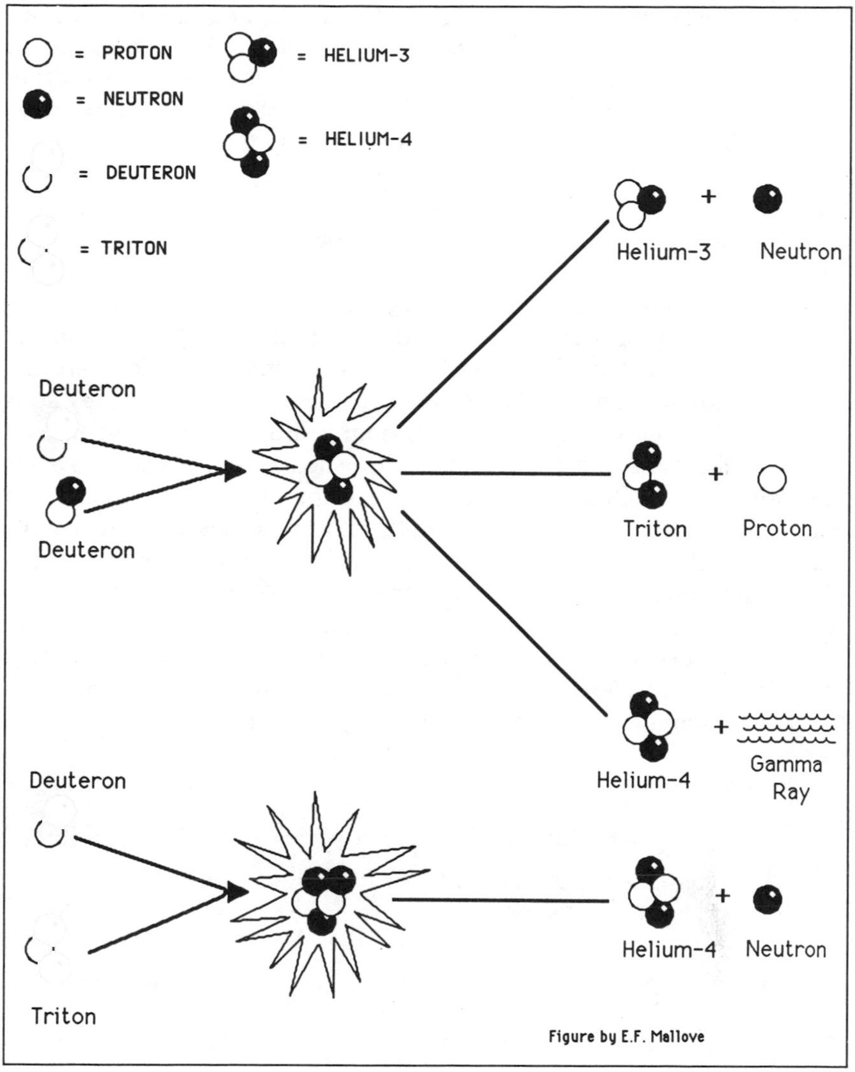

Schematic of four basic fusion reactions among hydrogen isotopes that are of particular interest to fusion researchers.

isotope of the ordinary kind of helium, helium-4 or ^{4}He, which has two protons and two neutrons in its nucleus.

For the nuclear "bookkeeping" in such equations to be correct, the number of individual particles or nucleons (protons or neutrons) on the left side of the equation must equal the number of nucleons on the right side. (Example: Together the two deuterons on the left in reaction (1) comprise four nucleons; on the right, ^3He plus the neutron, n, comprise 3 + 1 or four nucleons. Thus, the equation balances.) The numbers in brackets near each reaction product tell how much energy of motion (*kinetic energy*) is vested in that particle or nucleus after the reaction occurs and energy is liberated. This is the energy that typically may be used in some kind of conversion process toward useful power generation. The numbers represent how many *"MeV"* or *"millions of electron volts"* of energy are in the motion of that particle or nucleus.

An *electron volt* is a very tiny amount of energy. Millions of electron volts are still a small amount of energy (one MeV is about the energy needed to lift up a speck of dust weighing a millionth of a gram a distance of about one-millionth of a meter), but when many reactions are occurring simultaneously among trillions of like particles, the energy adds up! One electron volt is the energy that a tiny electron (with only 1/1836th the mass of a proton) picks up when it is accelerated by one volt—about the voltage difference between the two ends of a flashlight battery. Ordinary *chemical* reactions between individual atoms have energies on the order of a few electron volts (a few eV's), but *millions* of electron volts (MeV's) are characteristic of the energy output of the several nuclear reactants in fusion processes. This explains why fusion reactions involving nuclei are typically millions of times more potent than chemical reactions, which *by definition* only involve the interactions of the tenuous clouds of flitting electrons that surround individual nuclei.

Several other reactions are of major interest to fusion pioneers:

$$D + D \rightarrow T \text{ [at 1.01 MeV energy]} + p \text{ [at 3.02 MeV energy]} \quad (2)$$

Deuterium plus Deuterium (Goes to) Tritium plus a proton

$$D + D \rightarrow {}^4He + \gamma \text{ [at 23.8 MeV energy]} \quad (3)$$

Deuterium plus Deuterium (Goes to) Helium-4 plus a gamma ray

$$D + T \rightarrow {}^4He \text{ [at 3.5 MeV energy]} + n \text{ [at 14.1 MeV energy]} \quad (4)$$

Deuterium plus Tritium (Goes to) Helium-4 plus a neutron

In reaction (2), two deuterium nuclei react and form a *tritium* nucleus (a *triton*), another isotope of hydrogen (two neutrons plus the basic proton that identifies tritium as an isotopic form of hydrogen), plus a surplus proton. In reaction (3), two deuterium nuclei react to form a nucleus of helium-4 plus a high-energy gamma ray. In reaction (4), a deuterium nucleus reacts with the nucleus of the hydrogen isotope tritium. The reaction produces ordinary helium-4 plus a surplus neutron.

The first three reactions occur when pure deuterium fuel is brought to extremely high temperature. The first two of these three reactions, or branches as they are fondly called, are by far the dominant ones that occur with pure deuterium fuel. These two occur with about equal probability. So "burning" deuterium in a fusion reaction gives about an equal number of end products from these two reaction branches: about as much helium-three (^3He) as tritium and about as many protons as neutrons. Much more rarely (with a probability of only about one out of ten-million for every two D's that come together) the third branch occurs, producing ordinary child's balloon helium-4 and a powerful and penetrating gamma ray.

Because in high temperature plasma fusion the so-called branching ratio between reactions (1) and (2) is about one-to-one and because reaction (3) occurs only rarely, this became a major bone of contention in the cold fusion controversy. Hot fusion physicists who were already extremely skeptical of "fusion by chemistry" were loath to abandon so solidly established a finding as the hot fusion branching ratios and took this point as a fundamental article of disbelief. It is perfectly true that in no cold fusion experiment have the traditional branching ratios been found, much less was there any evidence of consistency between these reported reaction end products and the amount of heat being measured.

Even though these are the three reactions with which hot fusion scientists primarily concern themselves in present experiments, for various technical reasons it would be difficult and needlessly expensive to build a hot fusion reactor using pure deuterium fuel, so the practical working reactors that they hope to build would use the more powerful and easy to produce reaction (4) between deuterium and tritium (D + T). The potent neutron coming off the reaction is the key to hot fusion power, because its energy could be absorbed in a surrounding blanket of molten lithium (Li) metal, which would, in turn, heat water to produce steam to run an electricity-producing turbogenerator (Chapter 2).

The fast neutron would also turn some of the lithium atoms into tritium, which could then be extracted and fed back to the reaction chamber and used as fuel. In a sense the tritium part of the fuel would be *self-regenerating* through the conversion of lithium. Tritium is one of the less hazardous radioactive isotopes, in part because it decays so fast—half of it disappearing in only 12.5 years, half of the remaining atoms in 12.5 more years, and so on till it virtually vanishes.* But this

*Tritium doesn't have any extremely powerful penetrating radiation coming from it when it decays, just a single electron—called a beta particle in this kind of decay. The beta particle is easily stopped by a single sheet of paper! To make this electron, one of the neutrons in the tritium nucleus changes to a proton, leaving a helium nucleus behind, specifically helium-3 or ^3He. (An even more evanescent particle with no electric charge, called a neutrino, also comes out of this tritium decay, but the fleeting neutrino is one of the least interacting particles in nature. Neutrinos are a hazard to no one except those wracking their brains trying to find better ways to detect them.)

also means that tritium occurs in almost imperceptibly tiny amounts in nature and must be produced bootstrap-fashion in a working fusion reactor that used D + T. But so be it—this can be accomplished.

You may have heard that the radioactive gas, tritium, is also useful in making part of the fusion fuel for hydrogen bombs. Tritium *is* hydrogen—simply an isotope of hydrogen. (That's why we call thermonuclear weapons hydrogen bombs or H-bombs, though to be accurate we should really call them "T-bombs" or "D-T bombs," because they use deuterium too.) We make tritium now with fast neutrons that emerge from certain fission reactors (in Savannah, Georgia, and elsewhere). Unfortunately, these have lately been in the news because their waste products have been so poorly attended in the weapons program. This has led to serious environmental problems that we must now correct—problems not having to do with tritium itself. Since March 23, 1989, tritium has also made news in the cold fusion controversy, because researchers have claimed to have observed it in numerous cold fusion experiments. If this tritium is really being *generated* and is not the result of contamination, then cold fusion is proved.

Thousands of scientists and engineers around the world have been working for decades to harness the power of these hot fusion reactions. They heat plasmas to hundreds of millions of degrees in elaborate machines designed to produce fusion. To confine the plasma, they typically work with torus-shaped (donut-shaped) vessels called *tokamaks* pervaded by high-intensity magnetic fields, or they assault sometimes frozen pellets of fusion fuel with intense laser beams from many directions at once. The hot fusioneers have reached the threshold of the Genie's inner sanctum and are knocking on his door. Are they seeking fusion desperately enough to break down the final barrier? Will civilization give them the keys—namely, money and time? Or, has a backdoor labeled cold fusion opened far enough to enable them and a new generation of fusion scientists to step in?

2 A Brief History of Hot Fusion

Take Something Like a Star

Robert Frost

With ingenuity, hard work, and a sprinkling of good luck, it even seems reasonable to hope that a full-scale power-producing thermonuclear device may be built within the next decade or two.

Amasa S. Bishop, 1958

It's nice now to be able to go to cocktail parties and tell people you work on fusion and have them know what you're talking about.

Ronald R. Parker, Director of the MIT Plasma Fusion Center, May 1, 1989

IMAGINE THE SHOCK AND DISBELIEF that befell the hot fusion community when a couple of gentlemen from Utah suggested that they could achieve in their basement chemistry lab with cheap, relatively ragtag, cool equipment what billions of dollars and gigantic, complex machines with temperatures of a few hundred million degrees K had not yet accomplished in national laboratories. The struggle to tame hot fusion has lasted for over 40 years and was not easily put aside after hearing claims of "miracles."

Without understanding the gargantuan effort that went before in hot fusion research, trying to grasp the reaction to cold fusion and the course of events that followed is like attempting to comprehend the history of the United States without reference to Europe, Africa, Asia, or Latin America. Moreover, some researchers in the hot fusion community have become involved in cold fusion studies, first as skeptics or debunkers, though some more recently as committed investigators of the new phenomenon. So before recapitulating the events of that epochal month, March 1989, let us return to the 1890s, when some respected people in the house of science could still doubt something as basic as the existence of atoms.

✱ The Prehistory of Hot Fusion

The late Nobel laureate physicist Richard P. Feynman, who died only a year before the Utah announcement, once remarked that if all scientific knowledge were to disappear, except that we could choose to preserve a single fact with which to begin anew, we would best select: "The world is made of atoms." Although since antiquity people had speculated that the universe is made up of atoms, it was only at the beginning of this century that the atomic theory became impossible to doubt. It is strange that as we aspire to tame the power of the stars we are removed less than a century from knowing the very foundation of fusion theory—a world made of atoms. How we came to this realization is a fascinating story that has an eerie connection to the effort to tame fusion power.

In the waning months of 1895, German physicist Wilhelm Conrad Roentgen, in the course of his experiments with electrical apparatus and glass vacuum tubes had stumbled onto a marvelous discovery—X rays. When Roentgen had satisfied himself that X rays were real, he announced his findings within weeks, and soon the world was agog with the incredible news of pictures of bones being made from the living. So startling were X rays, much as the notion of cold fusion still seems to be, that even great men of science—Lord Kelvin in England,* for example, momentarily passed off X rays as rubbish and likely to be a hoax, before seeing for themselves and being satisfied. Roentgen later won the first Nobel prize for physics for his discovery of X rays. Ironically, an X-ray photograph had also been made accidentally in February 1890, but its unlucky maker, A.W. Goodspeed of Philadelphia, had not recognized its significance! Similarly, in the cold fusion controversy some skeptical scientists have overlooked or *ignored completely* equally anomalous experimental observations. These are, of course, only cautionary tales—not meant to suggest that other elusive evidence, such as for cold fusion, must always be believed.

In 1896, Antoine Henri Becquerel at the Ecole Polytechnique in Paris discovered radioactivity, foreshadowing the nuclear age. Becquerel was investigating phosphorescence produced by uranium compounds. He had exposed a compound of uranium and potassium to sunlight for several hours and then placed it on a photographic plate. Developing the plate revealed the outline of the compound's granules—an indication of penetrating radiation.

---

*_Life of Lord Kelvin_, Silvanus P. Thompson, Vol. II, London, Macmillan, 1910.

Becquerel later repeated the experiment without exposing the uranium compound to direct sunlight. He was surprized to find that putting a plate near the chemical sample in the dark and later developing it continued to reveal some kind of radiation emanating from matter. Other scientists, including Marie and Pierre Curie, were quick to investigate the new phenomenon. They soon learned that a number of heavy elements—forms of uranium, polonium, and radium—could decay spontaneously over a period of time and emit various mysterious radiations. Unfortunately, these pioneers lacked knowledge of the long-term effects of these mysterious rays on the human body. Some of them, including Marie Curie, ultimately paid with their lives from the consequences of their exposure.

They had discovered that nature exhibited a process akin to the *reverse* of the alchemists' dream. Rare and precious elements such as radium were decaying to less valuable lead. They could show that in a certain period of years, half of a sample of radioactive element was "decaying" in this manner. What we now know as a radioactive substance's *half-life* is the period during which half of an initial mass of the substance decays. The half remaining decays in the following half-life period, the remaining quarter decays in the next half-life period, and so on, until almost none of the radioactive element remains.

Radioactivity is a statistical process having to do with the atomic nucleus. As you may know, statistical processes—*quantum mechanical* processes—are very much the way nature operates in the microcosm, a fact of great importance in nuclear reactions, both fusion and fission. Radioactivity was explained during the first decades of the century as scientists applied insights from the newly developed theory of quantum mechanics. An individual atomic nucleus has a certain probability of undergoing a radioactive decay, but no magic formula can ever predict exactly when a *particular* nucleus will decay.

Three basic forms of radiation were found to be emitted in radioactive processes—alpha and beta particles and gamma rays. (Evanescent, highly penetrating, and probably massless neutrino particles also are emitted in such processes.) An alpha particle is nothing more than a helium nucleus, which contains two protons and two neutrons. Because of the natural abundance of helium in the cosmos (not on Earth, however), the alpha particle is the most common nucleus in the universe after the ubiquitous hydrogen nucleus—the proton.

Beta particles are simply electrons, which have 1/1836th the mass of the electrically positive proton and the electrically neutral neutron. Most beta particles are electrically negative and are identical to the electrons that orbit atomic nuclei. In some nuclear reactions, positively charged beta particles (positrons) are emitted. These are the antimatter or antiparticles of the negatively charged betas.

A gamma ray, the third fundamental kind of emission in radioactivity, is a high-energy photon—like a photon or "particle" of light but having much higher energy. Gamma rays, part of the electromagnetic spectrum that includes light, radio, and television waves, can penetrate deeply into materials. They have a much higher frequency and therefore much shorter wavelength than photons of light. Gamma rays share with visible light and other electromagnetic radiation the same "Jekyll and Hyde" wave-particle duality—in some experiments they act like waves, but in others they behave more like particles.

During the early part of the 20th century, physicists continued to make remarkable progress in understanding the inner workings of atoms. In the period 1906–1908, by bombarding materials with alpha particles from radioactive substances, British physicist Ernest Rutherford came to the startling realization that atoms must indeed have a very tiny, dense, and positively charged nucleus. Despite the misleading impression that cartoon diagrams in elementary chemistry and physics texts give, a nucleus is so incredibly small that if it were expanded to the size of a pea, the outer limits of the atom's cloud of electrons would be a few hundred meters away.

In another feat of enormous significance, Rutherford accomplished the first artificial transmutation of elements in 1919. He changed nitrogen to oxygen by bombarding nitrogen with alpha particles from a radioactive source—a fusion reaction in its own right. Then in 1932, through the experiments of British physicist James Chadwick, science learned there was a second nuclear particle, the neutron—slightly heavier than the proton but with no electric charge. So these early experiments in nuclear physics painted the basic picture of the atom that survives to this day: Negatively charged electrons flit about in orbital zones surrounding a much more compact and massive nucleus, which contains positively charged protons and chargeless neutrons.

The first attempts to probe into the nucleus with alpha particles were not generally very successful, because the electrostatic repulsion, or *coulomb repulsion* as it is also known, caused most collisions to be elastic. The positively charged alpha particle and target nucleus simply bounced apart without ever touching—charges of the same sign repel. The discovery of the neutron changed all this. Physicists now had a powerful, electrically neutral "bullet" with which to probe the secrets of the tightly compressed atomic nucleus. They soon learned about binding energy—the energy invested in nuclear forces that overcomes the coulomb repulsion and keeps the positive protons from flying apart. The coulomb repulsion is also the root of the problem in getting nuclei to come together in an artificial fusion reactor. Fortunately for us, the coulomb barrier prevents fusion from occurring willy-nilly all the time, and allows elements to pretty much preserve their identities—even in the face of very high-temperature natural processes, such as lightning, that create bare or nearly bare positively charged nuclei. This electrical repulsion has a strong tendency to keep nuclei with their positive charges from getting close enough to allow the *attractive* nuclear forces to overwhelm the coulomb force.

✳ The Fission Prelude

As we all know, putting fission into large-scale application came first, even though fusion reactions on a microscopic level were known for decades before there was a hint about fission. In 1938, Otto Hahn and Fritz Strassman in Germany bombarded uranium with neutrons and created the first *recognized* artificial fission of an element. It was Lise Meitner, who had been working with Otto Hahn before fleeing Germany, who helped interpret the results of the seminal Hahn-Strassman experiments as evidence for nuclear fission. As World War II dawned, physicists had arrived at the basic recipe for both a fission power reactor and a fission bomb: Begin with a quantity of uranium metal, ^{235}U (a uranium isotope with 92 protons and 143 neutrons). Make the uranium metal object sufficiently large so that the few neutrons released in the natural radioactive decay of a uranium nucleus cannot escape the volume within this *critical mass*, and so interact with other uranium nuclei. This will cause fission of the first target nucleus into such elements as xenon or strontium, and as a by-product will give rise to additional energetic neutrons. The neutrons from the splitting nucleus, or fission, fly off and interact with other uranium nuclei—causing more fissions and surplus neutrons, ad infinitum. This is the well-known *chain reaction* mechanism, the basis of fission power.

In a fission bomb, a significant fraction of the uranium nuclei split in less than a millisecond after the critical mass rapidly assembles, as it is compressed by chemical explosives from initially separated parts. (This is one definite case in which chemical reactions *cause* nuclear reactions!) What happens is the conversion to energy of perhaps 1/1000th of the mass of the uranium atoms that split. The energy manifests itself as visible and high-energy electromagnetic radiation (gamma rays and X rays), the kinetic energy of charged particles, and energetic neutrons. In a power reactor, the uranium (or other fissionable material, such as plutonium) is extended over a much larger volume and an element such as carbon or heavy water is used as a *neutron moderator* to slow the speed of the neutrons to make them interact better with uranium nuclei.

President Franklin D. Roosevelt, impelled by the weight of scientific opinion, initiated the secret Manhattan Project on December 6, 1941—auspiciously, the day before Pearl Harbor. Less than a year later on December 2, 1942, at 3:45 P.M. at the University of Chicago, Enrico

Fermi and his colleagues observed the first self-sustaining fission reaction in a uranium-graphite pile. The course was inexorably set for the explosion of the first fission bomb at Alamogordo, New Mexico, on the morning of July 16, 1945, at 5:30 A.M. The following month the war ended with the nuclear explosions that incinerated Hiroshima and Nagasaki.

Physicist Emilio Segré, who died at age 84 less than one month after the March 1989 announcement in Utah, the previous December had recounted the discovery of nuclear fission—on its 50th anniversary—before a meeting of the American Physical Society.* Segré had worked with Enrico Fermi in Rome in 1934 on experiments that bombarded uranium with neutrons to attempt to produce what they thought would be the first artificial element beyond uranium, element 93—one that by prediction would be similar chemically to rhenium. But this strong *expectation* of a result prevented the discovery of fission for five years before Hahn and Strassman in Germany ultimately recognized it. They too would make the same mistake by expecting what the Fermi group had likewise anticipated. Segré also recalled other lost opportunities: how another scientist had suggested the possibility of fission happening in their work, but whose writing was ignored; and Swiss researchers who may have seen the fission fragment evidence but who instead thought something was wrong with their detector. But the biggest problem was the *expectation* of seeing an element heavier than uranium, and not paying attention to the possibility of lower mass atoms that turned out to be the telltale fission fragments.

Segré said of writings by Hahn and Meitner on the road to the discovery of fission, "Their early papers are a mixture of error and truth as complicated as the mixture of fission products resulting from the bombardments. Such confusion was to remain for a long time a characteristic of much of the work on uranium." Segré recalled, "My own feeling at the time was that there was a mystery in uranium." In a remarkable statement printed in the historic December 22, 1938, paper in *Naturwissenschaften* announcing the fission discovery, Hahn and Strassman wrote, "As 'nuclear chemists' working very close to the field of physics, we cannot yet bring ourselves to such a drastic step, which goes against all previous experiences in nuclear physics." When the great physicist Niels Bohr heard of the new insights on fission, he was reported to have exclaimed, "Oh what idiots we have all been! Oh but this is wonderful! This is just as it must be!" As Segré concluded in his talk, "Above all, it seems to me that the human mind sees only what it expects." Scientists who go far afield to explore puzzles and anomalies

---

*Emilio Segré, "The Discovery of Nuclear Fission," *Physics Today*, July 1989: 38–43.

often bump into obstacles, but every once in a while they run into a remarkable phenomenon waiting to be discovered. It was just so with fission, might it be true also with cold fusion?

✳ Fusion Comes to Earth

A mere four years after Hiroshima, August 1949, the Soviet Union became the second world power to explode a fission bomb. The race to develop the much more powerful fusion or hydrogen bomb had begun. On November 1, 1952, the United States detonated the first fusion bomb, punching a gaping crater in the coral Pacific atoll, Elugelab.

If the fireball that rose over the Pacific was not exactly a man-made sun—ordinary hydrogen was not being fused to helium—how had fusion been brought to Earth? In hydrogen bombs, as well as in experimental first-generation hot-fusion reactors, the reactants are mixtures of the hydrogen isotopes, deuterium and tritium. These combine with far greater ease than the bare protons that are the nuclei of ordinary hydrogen. But how was it possible to heat mixtures of deuterium and tritium atoms to stellar temperatures? In a star, gravity alone does the trick as the enormous pressure of overlying material contains and sustains the fusion reactions that inevitably begin at the core. But the energy production in a stellar core is ordinarily very dilute and weak. This would not do for a controlled fusion reactor, much less for a fusion weapon. To make such a super-stellar bomb, a different mechanism was obviously needed.

Physicists had found that the temperatures and pressures inside the fireball of a fission bomb explosion approximated, for a fraction of a second, temperatures akin to those in the core of a star. Stars can sustain high core temperatures and pressures for billions of years, bombs for perhaps 0.001 second. Human technology could not yet aspire to fuse hydrogen directly, but it could fuse the more reactive deuterium and tritium if they could be heated in a fission explosion.

Though details remain highly classified, it is possible to piece together a plausible understanding of a modern hydrogen bomb from unclassified writings: Chemical explosives detonate and drive pieces of uranium-235 or plutonium metal together, creating a high temperature fission explosion. Then neutrons from the fission reaction bombard lithium-6, producing tritium and helium-4. The tritium thus produced reacts with deuterium, originally packed into the bomb, and results in the rapid release of fusion energy—a thermonuclear explosion.* Certain

*The reason tritium had to be partially produced at the time of detonation, rather than being prestored, is that tritium, the heaviest isotope of hydrogen, is strongly

specially shaped bomb components, made of materials such as beryllium or uranium-238, can act as reflectors and absorbers of X rays and neutrons to yield an amplified fusion explosion. Optimizing the process—making fusion bombs smaller, more powerful, and reliable—is what weapons research has been all about in the past decades.

Although it is not known how powerful a thermonuclear device is possible, in the early 1960s before the treaty banning atmospheric explosions, the Soviet Union tested an H-bomb that approached the equivalent of 100 megatons (million tons) of the chemical explosive TNT. Most "garden variety" fusion bombs are in the energy range of several hundred thousand tons to a few megatons of TNT equivalent. The 20 thousand-ton equivalent fission bombs that destroyed Hiroshima and Nagasaki were thousands of times less powerful, yet between them they killed more than 200,000 people.

Though most people are aware of the destructive potential of thermonuclear bombs, very few realize that bomb explosions were briefly considered as a means to produce electricity—a kind of brute force approach to controlled fusion. In the PACER study during the 1970s conducted at Los Alamos National Laboratory, researchers proposed to detonate bombs with yields in the range of 10 to 100 kilotons of TNT equivalent in a cavity at least a mile underground. Water placed in the subterranean chamber would then be superheated to steam by the explosion and channeled to a turbine to generate electricity. Neutrons from the blast might also have been employed to "breed" fission reactor fuel from the abundant form of uranium, ^{238}U.

Surprisingly, PACER appeared to be an economical kind of power generation, at least on paper. What killed the project, of course, were environmental concerns of the type that continue to plague the fission power industry, only much more enhanced. Some feared that the underground cavity might collapse and release radioactivity. And if water from the cavity percolated closer to the surface, drinking water would have been contaminated. It is clear that if fusion is to ever satisfy a substantial portion of our energy needs, it must do so in a manner infinitely more benign than underground blasts.

✳ Magnetic Confinement Fusion

Hot fusion requires temperatures of many millions of degrees to be successful, that is, to liberate more power than is used to create those

---

radioactive—unlike very stable deuterium. Tritium decays into helium-3 plus an electron, and its half-life is a mere 12.5 years. For this reason, tritium is essentially nonexistent in nature. Ordinarily, tritium for "seeding" the initial fusion explosion in a bomb is produced in special fission reactors such as the one at Savannah River, Georgia.

high temperatures in the first place. Temperature is nothing more than a measure of the average speed with which atoms or nuclei move in all directions as they bounce off one another in frenzied three-dimensional billiards. Atoms stripped of their electrons—naked nuclei—have a much greater chance at high speed, and therefore high energy, to overcome the mutually repulsive forces of their positive charges, the coulomb barrier. At high temperature, the random collisions of pairs of nuclei bring them close enough together so that they can stick permanently due to the extremely short-range nuclear forces. Thus, new nuclei are formed in the fusions; neutrons, protons, and so on are transformed or cast off, and much energy is liberated.

High temperature alone, however, won't buy us hot fusion. The brew of naked nuclei and stray electrons—the plasma—must stay together long enough to make enough fusions occur per second to add up to more than breakeven power, at least as much power being liberated as that required to heat the plasma. For hot fusion to work, plasma must be confined and kept extremely hot, so hot that if the plasma were very dense (reckoned in particles per cubic centimeter), its multimillion degree temperature would sear to smithereens a vessel made of any conceivable material. If a high-temperature plasma of *low* density were to expand and cool, or contact physical parts of whatever was holding it in place, the game would also be over. Its energy would be sapped, its life gone.

There are many ways, however, for the hot fusion Genie to work his magic. And one of the most promising is to have him reveal his magnetic personality. In fact, the very first way that scientists considered to control fusion was through magnetic fields, to confine plasma quite literally in a "magnetic bottle" and keep it from touching the surrounding physical vessel. Later, years after the laser made its debut, other scientists got the idea to create dense, high-temperature plasmas by bombarding tiny solid or gas-containing pellets of fusion fuel to create momentarily little fusion blasts—thermonuclear microexplosions. This genre of hot fusion is for various reasons called *inertial confinement fusion* or ICF. Many researchers think ICF is a promising route to controlled fusion—usually if they happen to be working on it already. But magnetic confinement fusion (magnetic fusion, for short) is the more tried and soon-to-be true approach. It is highly probable that magnetic fusion will be the kind of hot fusion likely to succeed first in being practical. Certainly, the most time and money have been spent on it.

Magnetic confinement fusion had a very strange beginning. Argentine dictator Juan Perón announced on March 24, 1951, that his country had mastered controlled fusion, bypassing completely the development of fission power that was then in vogue in various nations. Perón, a Germanophile, had set up an island laboratory for a certain obscure

German scientist, Ronald Richter, who supposedly had brought the secret fusion work to fruition, or so headlines in the United States had allowed. The press was much less circumspect in those days about amazing scientific claims.

Few details were immediately available, but Princeton University astronomer Professor Lyman Spitzer, Jr., apparently read of the work in *The New York Times*, whose reporter was properly skeptical. The front page article was headlined, "Perón Announces New Way to Make Atom Yield Power." The claim was false, indeed, and Perón's German scientist apparently paid for his scientific error by being jailed for misleading the Argentine dictator. Spitzer was already studying interstellar plasmas, and in an incidental way was involved in developing the U.S. hydrogen bomb. Until then, no one had really thought much about controlling fusion reactions. But within days, the word from Argentina caused Spitzer,* during his vacation on the ski slopes of Colorado, to invent single-handedly the idea of using magnetically confined plasmas to control high-temperature fusion reactions. One might say that a crackpot fusion claim at a press conference in Argentina, which was wildly touted in the U.S. press, gave rise to the hot fusion program that is still working toward its goals.†

Magnetic fusion has, indeed, come a long way since the early 1950s and is now verging on the magic breakeven point with recent successful experiments on a European machine called JET, the Joint European Torus. Even though a demonstration power-producing hot fusion reactor may still be several decades away, the light at the end of the tunnel seems finally in sight, and this time it seems to be a real light. This is especially dramatic considering the long struggle that the worldwide program has gone through.‡

The nuclear physics of hot fusion—how the nuclei and particles act when they come together—is relatively simple and straightforward. Instead, it is the complex and elusive physics of high-temperature plasmas

---

*Incidentally, Lyman Spitzer, Jr., also originated the idea of sending a large telescope into space, which culminated decades later in the 1990 launch of the Hubble Space Telescope by space shuttle Discovery.

†Skeptics were even in those day quoted prominently. The *Times* report quoted physicist Wernher Heisenberg: "I do not believe that at the present time something new in atomic research has been developed in Argentina which United States scientists did not know long ago." Otto Hahn, a codiscoverer of fission agreed with him. Highly skeptical Dr. Walter Whitman, head of the Chemical Enginering Department at MIT said, "You can't have an atomic explosion without uranium. I believe Perón's claim is very unlikely."

‡Anyone interested in the dramatic events in the quest for magnetic fusion are advised to explore three books, one by Joan Lisa Bromberg, *Fusion: Science, Politics, and the Invention of a New Energy Source*, Cambridge, MA, MIT Press, 1982, another by T.A. Heppenheimer, *The Man-Made Sun: The Quest for Fusion Power*, Boston, Little, Brown, and Company, 1984, and a more recent book by Robin Herman, *Fusion: The Search for Endless Energy*, Cambridge, England, Cambridge University Press, 1990.

in magnetic fields that has blocked the road to breakeven. In the United States, responsibility for the magnetic fusion outgrowth of fusion weapons research was vested in the Atomic Energy Commission (AEC) from 1951 through 1974. Thereafter, AEC became ERDA (Energy Research and Development Administration) for a short time, and in 1977 the present DOE (Department of Energy) was born. Initially, because of its association with weapons work, controlled fusion research was highly classified and code-named Project Sherwood. After all, there were and are possible applications of controlled fusion in generating nuclear materials for weapons. Most of this classification ended in 1958, however, after it became clear that controlled fusion did not promise easy, cheap breeding of fission material for bombs.

The money spent annually on the program was ramping up slowly, from under $1 million in 1951 when less than 10 people were in the effort. National laboratories such as Los Alamos, university-affiliated laboratories, and even a few private companies became involved in the U.S. hot fusion effort. In the 1980s, wealthy publisher of *Penthouse* and *Omni* magazines, Robert Guccione, even chipped in $16 million of his own money to an ill-fated venture by fusion pioneer Robert Bussard, who wanted to develop small, disposable reactors (of the tokamak design). The fusion program all along experienced major swings from optimism to pessimism as technical barriers presented themselves and the federal funding spigot functioned as erratically as an untamed plasma. The 1958 declassification of the U.S. program at the Second Geneva Conference on the Peaceful Uses of Atomic Energy showed that the U.S., Soviet, and British programs were at comparable levels. Since that time, international cooperation in controlled hot fusion has been the rule, Cold War or no. The superpowers realized that to conquer the exotic enemies of plasma stability, they had to band together.

The scientific outlook had its swings that mirrored the numerous political battles. Research was aimed at achieving in any of a number of competing configurations of magnetic bottles, a combination of temperature, density, and duration that would make hot fusion work. A plasma with a density of 3×10^{14} particles per cubic centimeter would have to be held for one second at about 100 million degrees K to reach energy breakeven.*

---

*The so-called "Lawson parameter" became a well-known figure of merit for hot fusion. To ignite a deuterium-tritium plasma (the fuel combination of choice) and make it burn with fusion fire continuously, researchers calculated that the Lawson parameter would have to reach 3×10^{14} second-particles per cubic centimeter at a temperature of a few hundred million degrees K. This meant that any combination of plasma density (nuclei per cubic centimeter) multiplied by confinement time at high enough temperature would make low-level fusion change from a sputtering match to a burning flame.

In the heady 1950s, a working reactor seemed to be not more than half a decade away, but experiments began to show that researchers had been far too optimistic. Plasmas were much trickier than was thought at the outset, whatever the configuration or intensity of the confining magnetic fields. Researchers had tried cylindrical magnetic bottles with magnetic end-caps—so-called mirror machines. There were twisted pretzel-shaped devices, all surrounded with complex electrical wire windings to provide high magnetic fields. There were even some torus-shaped devices, called stellarators—forerunners of future more successful machines called tokamaks. But the fusion conditions reached in the 1950s in pure deuterium plasmas in any of these devices were some 10,000 times lower than necessary.

Then in 1968, the year American astronauts first orbited the Moon, the Soviets announced a spectacular achievement of their own: a tenfold increase in plasma conditions through a magnetic fusion device they called a tokamak—an acronym from the Russian phrase meaning "toroidal chamber with magnetic coil." The Soviet claims were met with great skepticism, until a British team's firsthand measurements in Moscow during the summer of 1969. Hot fusion scientists, it seems, have been traditionally and properly skeptical of their own results, which explains their skepticism toward any "miracles" claimed by outsiders.

But the miracle of the tokamak was soon believed, and the donut-shaped machine that came out of the Kurchatov Institute of Atomic Energy in Moscow became the rage in magnetic fusion research. Most magnetic fusion machines today are, in fact, of the tokamak design, and it is with a tokamak reactor that the hot fusioneers aim to produce practical, controlled fusion power.*

One of the biggest problems with any kind of magnetic fusion device is to project how a relatively small prototype might operate if it were scaled up to a larger power-producing size. There is no firm agreement on how to predict plasma performance under different conditions. Due to the complexity of plasmas, many of these predictions are empirical—

---

*Tokamaks typically have two components that make up their final magnetic field. The invisible magnetic field lines that run through the inside of the donut (in the direction a tire tread runs on a tire) are established by electromagnets wound around the body of the toruslike bandages. High current running through these magnets, which in working reactors would ultimately be superconducting magnets, create this so-called *toroidal* field (because it goes in the direction of the torus). But a tokamak also has a poloidal magnetic field that runs through the "hole" in the donut. This comes from electric current that travels through the plasma—again, in the direction of the "tire tread." The current gets established in the plasma through a number of different possible means, such as an external electrical transformer clamped around the ring or through high-intensity electromagnetic waves (rf energy) that can be shot into the tube. The combination of the toroidal and poloidal fields twists around the ring and allows charged particles to gyrate among the invisible lines of force so as not to unduly come into contact with the tokamak walls.

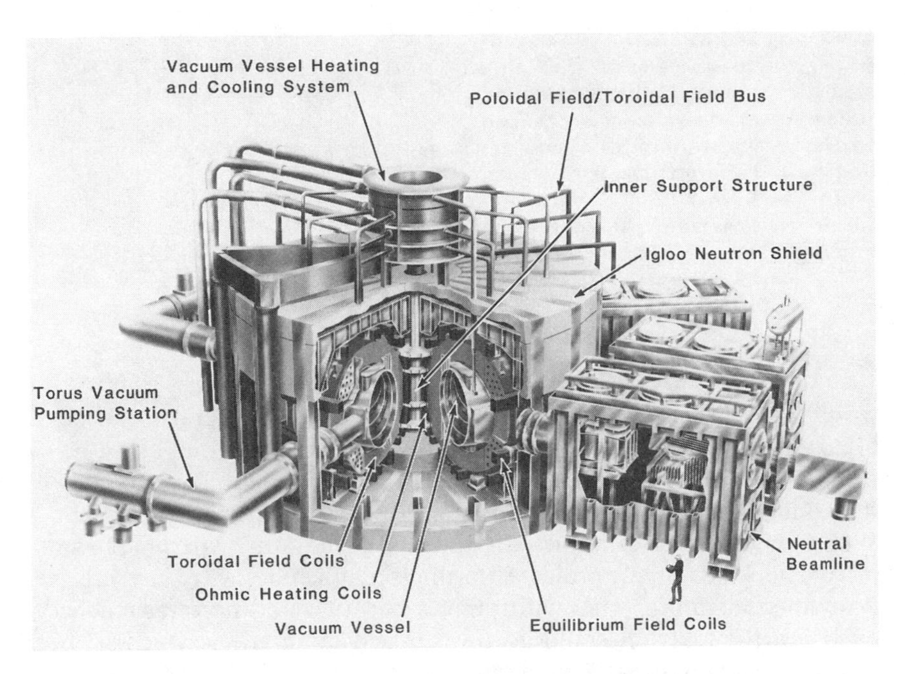

The Tokamak Fusion Test Reactor (TFTR) at the Princeton Plasma Physics Laboratory—a cutaway view. (Courtesy Princeton Plasma Physics Laboratory)

derived from experiments rather than from theory alone. Furthermore, the way in which plasmas will "burn" when they finally reach ignition conditions is a region of grave uncertainty. No one knows exactly what will or will not happen when "the match is lit." Though tokamak researchers are optimistic that ignition will bring out no unusual show stoppers and that ignition might even enhance plasma stability, there is room for doubt.

With a new burst of optimism, the hot fusion program accelerated dramatically in the 1970s, boosted not only by technical progress, but by multiple oil crises, and environmental concerns. A veritable alphabet soup of powerful magnetic machines emerged: TFTR, MFTF, TMX, TMX-U, JET (Europe), JT-60 (Japan), T-15 (USSR), and so on. Even as plasma physicists were bogged down in the continuing fight against plasma instabilities, elaborate planning began for designs to build full-scale power plants. This resembled somewhat the Wright brothers drafting plans for a commercial airliner before their 1903 flight! But there is no denying that these exercises in working power plant design have helped to illuminate some of the very difficult problems that need to be worked on if magnetic fusion is ever to be practical.

The general outline of how a hot fusion reactor based on the tokamak design would work is as follows. The toroidal plasma chamber

A tokamak and its electric power generating system—a schematic. The fusion plasma emits neutrons that heat molten lithium metal (shown as *coolant*). The hot coolant passes through a heat exchanger and converts water to steam, which in turn runs a turbine attached to an electric generator. (Adapted courtesy Argonne National Laboratory)

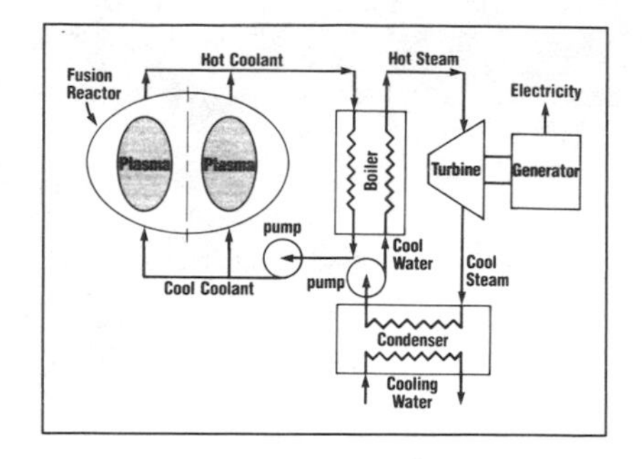

would have a high purity "first wall" around which there would be a blanket of circulating molten lithium metal to capture the fusion neutrons and convert their motion to heat. The magnetic coils to set up the confinement fields would surround this wall. Various attachments, such as high-energy electromagnetic wave generators or particle beams of various kinds might project into the torus the energy required to get it up to temperature. After ignition or a burning condition was reached, however, the reaction could go on by itself for many seconds—if not indefinitely—as long as deuterium and tritium fuel was continuously injected into the plasma. Deuterium-tritium fusion produces a helium-4 nucleus (alpha particle), which serves to heat the plasma further. This benign helium ash would have to be extracted from the plasma to keep the reaction running.

Tritium produced in the lithium blanket from neutron bombardment by the fusion reactions would be extracted and fed back to add to the deuterium-tritium fuel in the inner fusion reaction chamber.* To provide a continuous supply of fuel, it is expected that frozen pellets of DT would be injected at high speed into the magnetically confined plasma. A double-loop fluid heat exchanger would connect to the molten lithium blanket to draw out energy for the production of steam and finally electricity in a steam turbine.

Tokamaks that are used for experimental purposes are large and very expensive, the TFTR (Tokamak Fusion Test Reactor) at Princeton having cost over $300 million. It has the form of a 7-foot diameter tube bent to form a donut with a 24-foot outside diameter. In 1988, TFTR reached a plasma temperature of 300 million degrees K, more than twice what its predecessor tokamak at Princeton had only a decade before. The Joint European Torus at Culham Laboratory near Oxford in En-

*The tritium breeding reactions from neutron bombardment of lithium are:

$$^{6}Li + n \rightarrow {}^{4}He + T + 4.78\ MeV$$
$$^{7}Li + n \rightarrow {}^{4}He + T + n + 2.47\ MeV$$

gland cost the European community some $500 million to build. At the Plasma Fusion Center, the largest laboratory on the MIT campus, researchers have been developing tokamaks for some time. There in 1983, Alcator C was the first tokamak to exceed the minimum condition (Lawson parameter) for break even, but the plasma temperature of only 17 million degrees K was much lower than the 100 million K that would have been required to actually breakeven (with D-T fuel). To give an idea of the widespread acceptance of tokamaks, about a dozen major experimental tokamaks are operating today.

Apart from technical difficulties, the main problem hot fusion has faced from the word go, particularly as the 1970s merged with the 1980s, is that human beings have very short planning horizons. On a long time scale (centuries) the need for fusion power is absolutely compelling, but not so over a span of mere decades. There has always been the feeling about magnetic fusion that we could afford to take our time. As a result, the stop-and-go tension within the program has produced alternately leisurely academic studies of plasma behavior, and crash programs to attain specific reactor engineering goals. The hot fusion effort has been nothing like the Manhattan Project when it ran full tilt.* The fusion budget struggles under the load of reduced funding, even as the magnetic fusion community has its eyes on building a series of very successful tokamaks that could culminate in working fusion reactors long before 2050.

✳ Small Stars Are Born: Inertial Confinement Fusion

Inertial confinement fusion returns to the stars for inspiration, much more so than does magnetic fusion. (An excellent treatment of the subject is "How to Make a Star: The Promise of Laser Fusion," by Erik Storm, May 1986, Lawrence Livermore Laboratory.) Imagine a scene deep inside the vacuum of an extraordinary test chamber—one that has not yet been fully realized, but which has every prospect of coming true: A pea-sized hollow plastic capsule filled with a few milligrams of deuterium and tritium is suspended by invisible fields. Suddenly a gigantic bank of lasers fire dozens of beams at the little sphere, and several

*The Magnetic Fusion Energy Engineering Act of 1980 recommended a doubling of the U.S. magnetic fusion budget within seven years. At the time of the legislation, Congressman Michael McCormack (D-Washington) said, "The practical development of nuclear fusion power will be the most important energy-related event in human history since the first controlled use of fire." The impressive funding was not to be. In fact, the financial fortunes of fusion declined during the 1980s. As is common in large Federal research programs, official review followed upon review, followed upon reevaluation. Even as these words are composed, yet another review and projection of magnetic fusion plans is under way (Chapter 18).

hundred trillion watts of light energy blast the fuel pellet from all sides in perfect synchrony. Momentarily this is tens of times the combined power of all the electric power plants on Earth.

In a flash (several billionths of a second), the fuel capsule reaches a density over 20 times that of lead and attains a temperature over 100 million degrees K. The density of the compressed pellet produces a plasma some 10 billion times the density of the tenuous magnetically confined plasmas. Pressures rise to a trillion times atmospheric pressure at sea level. A burst of fusion occurs as a tiny star is born, lives, and dies in a fraction of a billionth of a second. Hundreds of times more energy comes out of the micro-star in the form of fast neutrons than went into making the pulse of laser light used to compress the capsule. The implosion of the tiny pea releases in a fraction of a second as much energy as would come from burning several gallons of fuel oil. If this could be done over and over again in rapid fire succession (almost as fast as gasoline explosions occur in an automobile engine—20 times/second), the energy of the escaping neutrons could be converted to heat, and we would have another means to control fusion.

The fusion fuel in the pellet was compressed and confined by the fleeting outward blast of plasma that the laser beams created, hence the designation *inertial confinement fusion* or ICF. Inertia keeps the tiny star together, but only if the distribution of laser light is exceedingly uniform on the little sphere. When the laser light blasts the pellet surface, heated plasma rushes outward at hundreds of kilometers per second, the mechanical reaction giving the effect of hundreds of ultra-advanced rocket engines all pushing inward.

ICF was a relative latecomer to the controlled fusion program, emerging into public view only in the early 1970s from its secret birth in the nuclear weapons research program. Soviet researchers performed the first demonstration of laser fusion in 1968. Studying the plasma fusion behavior in laser-compressed fuel pellets has provided insight into the physics of more powerful thermonuclear explosions. Just as magnetic fusion has already produced fusion reactions in its plasmas, so has ICF, but ICF has also not crossed the magic energy breakeven line. No one believes that an operational ICF power plant could be built any sooner than an only slightly less daunting commercial tokamak reactor. Both would be multibillion dollar installations and would require a few tens of billions of dollars to develop. Many would say that a working ICF plant would be even more formidable to create than a tokamak power reactor. Imagine trying to ignite and burn five to ten such micro-stars per second to achieve perhaps a 1,000 megawatt power level! During an energy maelstrom, how could the pellets be positioned accurately enough to make the laser beams strike them evenly each time?

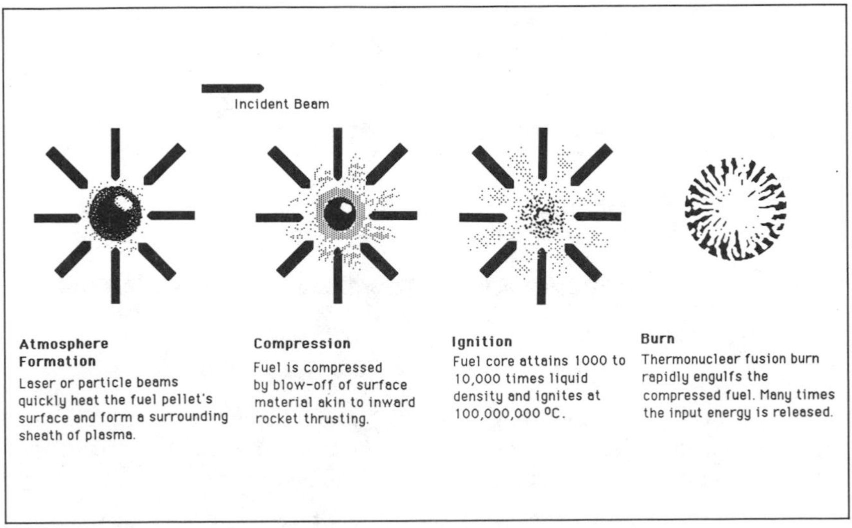

Incident Beam

Atmosphere Formation
Laser or particle beams quickly heat the fuel pellet's surface and form a surrounding sheath of plasma.

Compression
Fuel is compressed by blow-off of surface material akin to inward rocket thrusting.

Ignition
Fuel core attains 1000 to 10,000 times liquid density and ignites at 100,000,000 °C.

Burn
Thermonuclear fusion burn rapidly engulfs the compressed fuel. Many times the input energy is released.

The basic mechanisms of inertial confinement fusion (ICF).

A working ICF plant would be similar to a magnetic fusion reactor in its general approach to capturing and using the fusion neutrons: lithium blanket, tritium production from lithium via neutron bombardment, heat extraction from molten lithium, and so on. In principle, inertial confinement fusion would achieve breakeven by having a much higher particle density than magnetic fusion plasmas, though during a confinement interval billions of times less than a magnetic fusion plasma. Less money is being spent on ICF fusion than on magnetic fusion, because there is a somewhat shaky consensus (Chapter 18) that magnetic fusion is the way to go in hot fusion development and commercialization.

The hot fusion community is rightly proud of its goals. Who can argue with them? The allure is limitless, cheap fuel, with no production of greenhouse gases; high power density reactions (unlike solar power); and potential "worst-case" environmental impact demonstrably low compared with conventional fission reactors. Despite the slings and arrows that have assaulted its budget, it is run by a supremely confident crew. Hot fusion seemed to be able to survive, phoenixlike, for decades.

3 Claiming the "Impossible"

Stan and I often talk of doing impossible experiments. We each have a good track record of getting them to work. The stakes were so high with this one, we decided we had to try it.

Martin Fleischmann, quoted in University of Utah press release, March 23, 1989

To answer the world's material needs, technology has to be not only beautiful but also cheap.

Freeman Dyson, *Disturbing the Universe*, 1979

✳ March 23, 1989

WAS MARTIN FLEISCHMANN, soon to celebrate his 62nd birthday, ready for the challenge of standing the world of science and technology on its head? It was more than a little unsettling. Could he imagine the shocking impact of what he was about to say? Did he or Stanley Pons think that the world would comprehend, much less believe their message about possible eternal, trouble-free energy? If so, Martin Fleischmann didn't show unease, either in person or in print. "What we have done is to open the door of a new research area," Fleischmann confidently stated in the University of Utah press release that was handed out at the afternoon press conference on March 23. "Our indications are that the discovery will be relatively easy to make into a useable technology for generating heat and power, but continued work is needed, first, to further understand the science and secondly, to determine its value to energy economics."

Were it not for their solid scientific credentials and reputations as excellent electrochemists, Fleischmann and Pons would have been roundly laughed out of court. Then the scientific community would have yawned and forgotten them, so preposterous seemed their contention. Imagine, fusion reactions in a jar of water (albeit 99.5% pure heavy

Dr. Martin Fleischmann, at the podium during the press conference at the University of Utah on March 23, 1989, holds a larger version of his original smaller cold fusion cell. (Courtesy University of Utah)

water), with wires and electrodes hooked up to the equivalent of an automobile battery! It was the kind of experiment a bright high school student might be able to perform, or so Pons and Fleischmann had said. Whatever the merits of their experiments, the announcement provoked a sustained media blitz the likes of which has attended few other scientific events. The closest recent parallel, many would recall, was the hoopla surrounding the announcement in late 1986 of the discovery of high-temperature superconductivity; and that didn't even come close.

The fusion community was primed to be skeptical, regardless of the credentials of the claimants. But the Utah duo were clearly not crackpots or blatant frauds. Dr. Fleischmann was a widely respected and extensively published electrochemist and a professor at the University of Southampton in England. Dr. Pons, then 46, also widely published, headed the Chemistry Department at the University of Utah. He had earned his doctorate in chemistry at the University of Southampton, grew to know and respect Professor Fleischmann there, and through the years the two had remained close friends and colleagues.

But fusion researchers remembered all too well the occasional bogus claims of dramatic breakthroughs that span the history of fusion research. At least a half-dozen other claims have been made about "fusion just around the corner."

Professor Ronald R. Parker of MIT's Plasma Fusion Center said in the spring of 1989, "You have to realize that in fusion research, we fairly frequently get communications from people who claimed to have produced fusion in their 'X-Y-Z machine,' and they have a lot of ideas that seem to come out like a stream of consciousness. Finally, it ends up that a green man from Mars actually told them about it!" Though very skeptical of the Utah claims, Ron Parker was, of course, not attributing this kind of lunacy to Fleischmann and Pons, but he was explaining the context in which the scientific duo's claims had arrived.

✳ Fusion by Electrochemistry?

In March 1989, Fleischmann and Pons said that they had pulled off something truly mind-boggling, certainly in light of the multibillion dollar international hot fusion effort that had been going on for so long. Their experiments, they said, seemed explainable only as the result of some kind of nuclear fusion reactions that worked, quite miraculously, near room temperature. They had begun by immersing an electrode of precious palladium metal into a glass jar containing heavy water—chemically identical to ordinary water, H_2O, except that in place of the ordinary hydrogen atoms are atoms of the isotope deuterium, so the molecule is denoted, D_2O. Between this (negative) palladium electrode and the other (positive) electrode of platinum, they had attached a low-voltage battery. They previously had added some trace amounts of another chemical to make electric current—electrons—flow in the heavy water between the electrodes. This was lithium-deuteroxide (LiOD—a single lithium, oxygen, and deuterium atom linked together), analogous to the more commonly known compound, lithium hydroxide (LiOH), which has ordinary hydrogen, not deuterium.

Fleischmann and Pons claimed that deuterium atoms chemically broken from the water molecules (by electrical forces in the cell between the positive and negative electrodes—anode and cathode) were being driven by the voltage deep into the crystal structure of the palladium. Some of the deuterium was merely bubbling up as gas (the double atom molecule, D_2) and escaped the cell near the negatively charged palladium rod. Similarly, in an ordinary light water electrochemical cell, simple hydrogen gas, H_2, bubbles up and out. The oxygen part of the D_2O would likewise bubble out as oxygen gas, O_2, and leave the liquid near the positively charged platinum wire, the anode.

This much no one could dispute, because the basic electrochemistry of such a simple cell was obvious. Palladium metal was well known for its ability to absorb huge quantities of hydrogen—or what is almost the same chemically, deuterium. Open any standard chemical text and you will find that palladium absorbs 960 times its original volume of room-temperature hydrogen gas.

But here is where trouble begins. Fleischmann and Pons said that the apparatus could produce significant amounts of excess power continuously—much more power than was being fed in from the electric battery. In one instance they claimed four times as much power out as

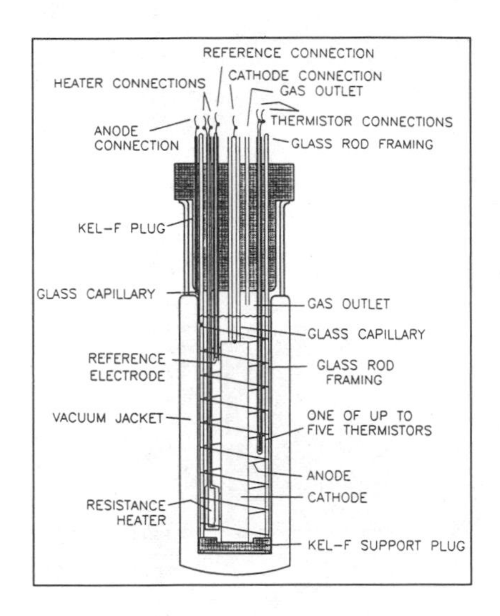

Cross-sectional view of a Fleischmann-Pons-type cold fusion cell set up for heat measurement as a dewar calorimeter with a glass vacuum jacket that is integral to the cell structure. The thermistors measure temperature electronically; the electrical resistance heater is used to calibrate the cell; and the platinum anode spirals on a glass rod frame around the central palladium cathode. (Courtesy Drs. Martin Fleischmann and B. Stanley Pons)

went in.* This assertion was very explicit in their announcement of March 23: "This generation of heat continues over long periods, is proportional to the volume of the electrode, and is so large that it can only be attributed to a nuclear process." Moreover, they said they had detected neutrons emanating from the glass cell that could only mean that some kind of nuclear process was occurring within. These were coming out, they said, at many thousands per second, still a level only three times the natural radiation background. As if that were not enough, Fleischmann and Pons claimed to have detected within a cell the rare hydrogen isotope, tritium, in amounts above natural background. Tritium, we have seen, is another hallmark of certain fusion reactions, and if truly being *produced* in their cell, would strongly suggest that some kind of fusion was taking place. Their press release carefully noted, however, that the tritium and neutrons were from "side reactions" to the main heat-producing nuclear process.

In fact, they *explicitly* said that had achieved fusion.† Stanley Pons, who often seems to be nervous when he speaks in public, had said this

*In their paper, Fleischmann and Pons presented excess power results with respect to three explicitly defined kinds of "breakeven." The "four times input" claim was for but one of their experiments in which the D_2 gas produced in electrolysis would be used in a **hypothetical** reaction to produce extra power. Without this additional reaction, the actual excess power obtained in their experiments as a percentage of input electric power ranged from 5 to 50%. As a percentage excess of the pure **resistance** heating of the cell contents, their experiments yielded 5 to 111%.

†The title of the press release was: 'Simple Experiment' Results in Sustained N-Fusion at Room Temperature for First Time: Breakthrough Process Has Potential to Provide Inexhaustible Source of Energy

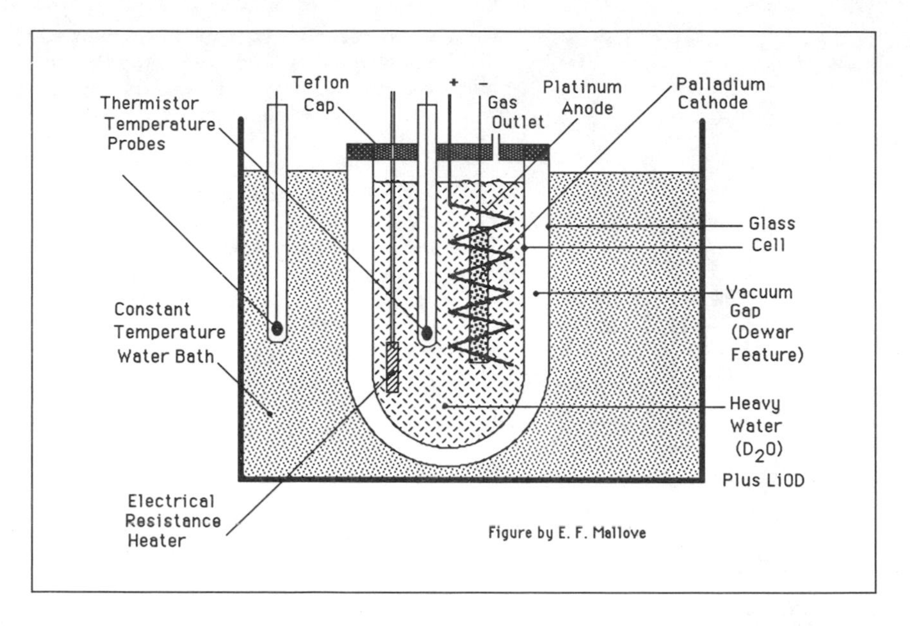

The essential features of a Fleischmann-Pons-type cell, shown in schematic view.

in his opening remarks at the press conference: "Basically, we have established a sustained nuclear fusion reaction by means which are considerably simpler than conventional techniques. Deuterium, which is a component of heavy water, is driven into a metal rod—exactly like the one that I have in my hand here—to such an extent that fusion between these components, these deuterons in heavy water, are fused to form a single new atom. And with this process there is a considerable release of energy; and we've demonstrated that this can be sustained on its own. In other words, much more energy is coming out than we are putting in."

Fleischmann and Pons emphasized that it was the enormous amount of liberated heat energy that made them believe it must be coming from a nuclear process. Interviewed on the MacNeil/Lehrer Hour, a seemingly ill-at-ease Pons said, "It certainly is a breakthrough in the field of nuclear fusion." Asked how he knew that it was nuclear fusion, he replied, "First by the enormous amount of heat that was generated. There is no known chemical process or other process that we are aware of that could explain such huge amounts of energy. And subsequent to that we have detected particles that are associated with nuclear fusion reactions over and above normal background. . . . We've sustained cells for several hundred hours over the last few years."

Fleischmann said on the same program, "The main indication which we had that we had nuclear fusion was the extremely large release—and *continued* release—of heat energy from the electrode." What led them to suspect a nuclear process? "The conditions in this electrochemical cell are completely different from the conditions which are now investigated in the conventional approaches to nuclear fusion. I think one can best explain it in simple quantitative terms, by saying that if you pass an electric current into the cathode under the conditions which we have used, then, if you try to achieve the same conditions in the cathode by compression of the gas, you would need a billion, billion, billion atmospheres—a billion, billion, billion times the pressure at the surface of the Earth. It is this enormous compression of the species in the lattice which made us think that it might be feasible to create conditions for fusion in such a simple reactor.

"In any investigation there are always problems of the science, and then there are the technical problems," he said. "We do not see such massive technical problems in this instance as there might be in some of the other [fusion] approaches which have been tried so far." Both chemists said it would take one or two decades to develop practical applications. "The reaction would be clean, the fuel supply would be plentiful, and it could in this embodiment, we think, be carried out in a very simple manner," said Fleischmann.

Both researchers made extravagant implications and claims about scaling up the poorly understood process. At the press conference, Fleischmann had said of a cell and electrode that he held in his hand—bigger than what they had used in their research: "This experiment has to be approached with some caution, because if this device worked as rapidly as the small electrodes, which we've run, this would be generating about 800 watts of heat." Pons had said, "I would think it would be reasonable within a short number of years to build a fully operational device that could produce electric power or drive a steam turbine. . . ." Immediately after the formal press conference, Pons told Ed Yeates of KSL TV in Salt Lake City, "Depending on the kind of cell you use where we are working now on a few watts, we could go to many orders of magnitude [powers of ten] higher than that in a properly designed cell—a couple of orders of magnitude higher than that in a very simple cell."

The scientific community now had to take up the challenge of cold fusion, said Fleischmann on MacNeil/Lehrer. "We have been skeptical about it now for five years. As I explained earlier this afternoon, you can never prove a scientific discovery right, you can only prove it wrong. I think that others will have to examine our results, extend them, and try and see whether our interpretation is in fact correct—partly correct or wholly correct. So only time will show whether this is so and whether we can take the next step towards developing the technology."

But don't be challenged *too* fast! "I would strongly urge everybody to wait until the work is described in the scientific literature," warned Fleischmann, "before they attempt any such experiments. In any experiments on nuclear energy one has to be cautious, and I think I would commend people to wait until the publications appear in May." He was alluding to the possibility of melting, vaporizing, or exploding electrodes—a phenomenon that they had encountered on at least one occasion.

The claimed event had occurred possibly in 1985 in one of their first attempts to perform a cold fusion experiment. A cell that had been operating quiescently for a few months and whose electric current had been reduced before being left unattended, exploded violently one night, destroyed part of a laboratory protective enclosure, and gouged a hole four inches deep into the lab's concrete floor. About half of the tiny cube-shaped electrode (a cubic centimeter of palladium) was found to have vaporized! (This would have required the metal to have approached a temperature of 3,000 °C.) The dramatic event, which Pons and Fleischmann did not attribute to a mere chemical explosion from the buildup of deuterium gas and some exotic chemical reaction in the electrode, motivated them to redouble their efforts. They were quite serious in thinking that they had literally achieved fusion *ignition*. Had the explosion not occurred, they might even have abandoned the project.*

Fleischmann was unequivocal in describing the event at the press conference: "At a fairly early stage, we devised an experiment in which part of the electrode fused. Now the melting point of palladium is 1569 degrees centigrade—1554, I stand corrected. My memory is not what it used to be. And there are two terminologies in fusion research. One is breakeven and the other one is ignition. Now we didn't want to believe that we had ignition—because we wanted to continue our work." The audience laughed at that. "But what we then did, of course, is to turn everything [the current] down as hard as we could. And we've been slowly coming up towards the point where we might be closer again to the condition where there might be ignition. But the thing which really triggered the whole thing off fairly early on was that we realized that you could generate a lot of heat—a lot."

Needless to say, the scientific community was astounded by all these extraordinary claims from Fleischmann and Pons—forgetting even the claim about presumed ignition, which others have tried to explain as a

*Martin Fleischmann told me that by the summer of 1988, when he was certain that their experiments had demonstrated excess heat, he was "not convinced that it didn't have defense implications." Because of this concern, Fleischmann and Pons would later write to Dr. Ryszard Gajewski of DOE: "We don't think this work should be published."

chemical explosion.* In view of the Utah scientists' credentials, the reaction of that community was initially respectful, but highly skeptical. The comment of nuclear engineer Professor Ian H. Hutchinson of the Plasma Fusion Center at MIT on CBS Evening News was typical: "What I have heard is that they are saying reactions are taking place at room temperature, which in itself is not that remarkable. What would be remarkable—indeed *very* remarkable—is if the reactions are taking place at a rate sufficiently large for generating net energy. So while I am open to being convinced, I am highly skeptical." Describing how he felt about the announcement, he said, "Suppose you were designing jet airplanes and then you suddenly heard on CBS News that somebody had invented an antigravity machine." Hutchinson's colleague at the Plasma Fusion Center, Dr. Kim Molvig, was more emphatically negative and told *Science* magazine that week, "I'm willing to be open minded, but it's really inconceivable that there's anything there."

A fusion researcher at the University of Rochester, John Soures, commented skeptically that if the Fleischmann-Pons cell were putting out a few watts of power, trillions of neutrons per second would be emanating from it. "It would require some shielding to prevent people in the vicinity from being irradiated. . . . You wouldn't put it out in the open. It would certainly be a safety hazard. . . . It certainly raises some questions in my mind." Robert G. Sachs, former director of the Argonne National Laboratory and now professor emeritus of chemistry at the University of Chicago was upset that the findings were being released at a news conference. "That makes you suspicious," he told the press.

More open to the possibility that Fleischmann and Pons actually had stumbled onto something of importance was physicist Philip Morrison, Institute Professor Emeritus at MIT. Morrison, who had worked on the Manhattan Project, later had turned his legendary energies to work on nuclear arms control activities and had long been a highly respected writer and thinker about science and science education. Professor Morrison said, "There's a chance that it's the most important development in a century, and there's a chance that it's nothing at all." That week he told David Chandler of the *Boston Globe*, "Based on the information I have, I feel it's a very good case . . . but not conclusive." In an optimistic assessment that was to prove remarkably shy of the truth, Morrison added, "We certainly will know the answer in a few weeks."

---

*See ideas in Peter A. Rock et al., "Energy balance in the electrolysis of water with a palladium cathode," *J. Electroanalytical Chemistry*, Vol. 293, October 25, 1990: 261–267. But Fleischmann told me: "You could possibly get this sort of phenomenon chemically, but it is unlikely. We have taken electrodes out of the cell and stuck them on the bench and this thing does not happen. Wires may do so—very thin wires may do so—but massive palladium does not do that."

Dr. Edward L. Teller, director emeritus of the Lawrence Livermore National Laboratory (LLNL) and one of the major figures in developing the American hydrogen bomb, was surprisingly upbeat too (possibly because LLNL as a Department of Energy lab had gotten advance word of the Utah work, through the proposal Fleischmann and Pons had submitted to DOE.) On the CBS Evening News on the day of the announcement he said, "It seems that controlled fusion works, indeed not in free-space, but in the confinement of a metal in which the hydrogen nuclei can move freely. It appears that the reaction does go forward and there may be a real breakthrough." A day after the announcement Teller described the Utah work as sounding "extremely promising." He said in his thick Hungarian accent, "Initially, my opinion was that it could never happen. I'm extremely happy now because I see a good chance that I was completely wrong." (Quoted in the *Deseret News*, Saturday, March 25, 1989.)

If respected chemists had claimed to have changed water into gold by employing batteries and simple electrodes, physicists and chemists would not have been more startled, bemused, and even outraged than they were following that seminal news conference in Salt Lake City. Within hours, the announcement had set off a global conflagration of scientific debate and both experimental and theoretical activity. The University of Utah, its public information office, and a number of science departments were inundated with hundreds of telephone calls from scientists and journalists attempting to get more information. Initially, a few were able to speak with the principals, but as literally hundreds of calls mounted and exhaustion set in, that became impossible.

Given advance notice of the announcement through a scientist colleague of Fleischmann, whose son, Clive Cookson was technology editor, the *Financial Times of London* ran the first of what were to be thousands of news accounts of the Utah work in the coming weeks and months. The front page story by Cookson blared, "Test Tube Nuclear Fusion Claimed." A sidebar to the story actually showed a detailed schematic of the the Utah cold fusion cell.

Nervously and excitedly fingering newly faxed copies of the *Financial Times* article, prospective theorists and experimentalists at laboratories and universities gathered to speculate. At the MIT Plasma Fusion Center, physicists and engineers consulted handbooks detailing the properties of palladium and wondered—even amidst their disbelief—about the miraculous process by which a lattice of palladium atoms might promote deuterium fusion. It was an academic exercise, for there were few if any believers. Something just *had* to be wrong in the reports coming from Utah.

Network television newscasts, including CNN, covered the cold fusion story that night. Dan Rather on the CBS Evening News led his half-hour program with the story, saying, "Some of the world's leading researchers have spent decades and millions of dollars trying to achieve this [fusion] in the laboratory. Scientists in Utah tonight believe they have taken a big step forward in a little test tube." The first words from Utah heard on CBS were shocking, but paradoxically added some credibility to the claims: "Stan and I thought that this experiment was so stupid that we financed it ourselves." Later the MacNeil/Lehrer Hour ran its longer segment on the story. Some scientists, including Professor John O'M. Bockris of Texas A&M University who would later play a major role in confirming some of the Fleischmann and Pons results, got their first introduction to cold fusion from that MacNeil/Lehrer report. So did Julian Schwinger, Nobel laureate in physics, who was immediately convinced that the matter had to be taken seriously. He went on to develop a detailed theory of how "impossible" cold fusion could work.

Fleischmann and Pons claimed to have operated one of their cells for over a hundred hours, producing 4.5 watts of total power for every watt of electrical power input. This performance was far beyond the power breakeven point that plasma physicists and engineers had been trying to achieve for decades with their magnetic confinement and laser fusion techniques. Fusion scientists were further incredulous at the Fleischmann-Pons claim because the level of power reported (26 watts per cubic centimeter of electrode in one case) would seem to translate into trillions of individual fusions per second, each producing an energetic neutron and consequent potentially lethal radiation exposure to experimenters.

Fleischmann and Pons had not mentioned anything about massive radiation shielding, though they hinted ambiguously that precautions should be taken in trying to duplicate their work. Pons acknowledged the shortfall of neutrons, but had no explanation as to why this should occur. And the two researchers did implicitly encourage others to duplicate their results. Fleischmann said, "We don't know what the implications are. The science has to be established *as widely as possible to challenge our findings.* [author's italics] But it does seem there is a possibility of realizing sustained fusion in a relatively simple device." At the press conference Pons had described their motivation for getting into cold fusion work in the first place, "The experiment was so simple that at first it was done for the fun of it and to satisfy intellectual curiosity."

The press, egged on by skeptical members of the scientific community, would soon make much of the allegation that the scientific process was being violated by disclosure of such a potentially momentous discovery at a mere press conference. "Science by press conference" became the derisive term. Moreover, many in the scientific community made the supposedly damning allegation that Fleischmann and Pons' work had not even been reviewed by peers, and it had certainly not yet appeared in any technical publication.

In truth, Fleischmann and Pons had submitted a paper on their work to the *Journal of Electroanalytical Chemistry* on March 11, a full 12 days before their press conference, and it was accepted in final form for the journal's April 10 issue days before the press conference. Fleischmann and Pons certainly erred, however, by not providing at the press conference copies of their technical preprint, along with a clear statement that their work was to be published in either a specific technical journal, or was under review at a journal or journals whose names were being withheld pending final review. This would have done much to defuse the mounting sea of criticism that would overwhelm though not quite drown them. Scientific "etiquette" also dictated that they not mention this journal because they hoped an alternate version of their paper, which they were about to send to the more prestigious journal, *Nature,* would be accepted. So the two had to make evasive statements to the press that were misinterpreted. Pons told Malcolm W. Browne of *The New York Times* that the article would be published in a technical journal, but he would not disclose its name. Even several supportive colleagues of Fleischmann and Pons considered the overall method of disclosure "outrageous" and "appalling."

At the March 23 press conference and in the University of Utah's press release, Pons and Fleischmann acknowledged that graduate student Marvin Hawkins from LaJara, Colorado, had worked with them on the research, but they did not include him as a coauthor of the paper submitted to the *Journal of Electroanalytical Chemistry.* However, Hawkins had actually been working with them only since the previous October, making the matter of coauthorship somewhat delicate. In a subsequent issue of the *Journal* (*Journal of Electroanalytical Chemistry,* Vol. 263, May 10, 1989: 187–188), the researchers devoted two pages to correct known technical errors in the April 10 version of their paper. Included was the statement, "M. Fleischmann and S. Pons regret the inadvertent omission of the name of their coauthor, Marvin Hawkins, Department of Chemistry, University of Utah, from the list of authors." (Since Fleischmann and Pons were the driving powers in the research as well as the central figures in the emerging controversy, subsequent references to the University of Utah work (prior to March 23, 1989) are to Fleischmann and Pons.)

Reports circulated that the work would be published within a few months in the respected scientific journal, *Nature.* To this day, it is not certain how this information on the *Nature* submission came into the hands of the press. Fleischmann and Pons adamantly say that they did not divulge it. A day after the press conference, they, in fact, submitted the draft of a shorter version of the paper to *Nature* but had not received approval for its publication. The article never did appear in *Nature* owing to criticisms by the magazine's scientific reviewers, requests for additional data that Fleischmann and Pons could not immediately provide, and *Nature's* desire to have the paper expanded by a few thousand words. In any event, within the week of the press conference, copies of their draft paper to the *Journal of Electroanalytical Chemistry* began appearing at laboratories around the world via fax machine.

✳ A "Preposterous" Experiment Is Born

Before announcing table-top cold fusion to the world, Fleischmann and Pons had spent more than five and a half years on their cold fusion research and had invested about $100,000 of their own money, mostly that of Dr. Pons. They had conducted experiments in Pons's basement laboratory in the Henry Eyring Chemistry Building on the University of Utah campus. How had they conceived this seemingly wild approach to fusion, that by their own recollection seemed to have a "one in a billion chance of working" at the outset?

In the late 1960s, Fleischmann had become aware of certain anomalies in separating hydrogen and deuterium isotopes in palladium. A decade later, Pons had come across other anomalous phenomena in isotopic separation in electrodes. The two were also strongly influenced by a 1977 RAND Corporation report, which discussed the exotic metallic phase of hydrogen that should exist at ultra-high pressures—possibly being an elevated temperature superconductor, a very efficient rocket fuel, or a powerful explosive. The confluence of their thinking apparently occurred in 1984, so the two have said, during a drive across Texas and later on a walk up Millcreek Canyon on the outskirts of Salt Lake City, not far to the north of a town called Mount Olympus! Conversations in the kitchen of Pons's suburban Salt Lake City home pushed them more and more to the unthinkable: to actually put their ideas to the test.

Their vision was wild, but it originated from mundane matters—new kinds of chemical synthesis and storing large amounts of hydrogen in a metal lattice. The latter is connected with advanced hydrogen-powered automobile propulsion. Fleischmann and Pons knew that the "effective pressures" of hydrogen within metal lattices were well-nigh astronomical, if the high "chemical potentials" in the lattice were appropriately interpreted. Think of astronomical pressures, and immediately fusion comes to mind—particularly when hydrogen isotopes are the atoms being pressured. It was remarkable that Fleischmann and Pons were bold enough to try out the crazy idea only a few months later. A claimed initial hint of success drew them ever deeper into uncharted territory. They were "hooked," as many others would be in 1989 and beyond.

Even though they knew the resources required to do their experiment would be minimal, Fleischmann and Pons wondered where to obtain additional funding should they uncover anything interesting. They imagined it would be impossible to raise any money, since the experiment seemed on its face so far-fetched. Fleischmann would testify to Congress in April 1989, ". . . it was almost incorrect to ask for financial support for a project which had a low probability of success . . . there is a sort of limit beyond which we did not even want to drive our friends to the limits of credibility." And he has suggested wryly that their kitchen roundtable thinking might have been catalyzed by a liquid substance of another sort—a bottle of Jack Daniels whiskey. Whatever their "mad" idea's origin, they kept to themselves and carried on their work secretly.

Who were these scientists who had so suddenly burst forth into the public eye and who began to be called the "thermodynamic duo" by those who knew what the word meant? (Thermodynamics is the science of thermal properties—the flow of heat and energy as well as entropy.) The pair seemed then and still seem an unlikely combination, but somehow they click. Fleischmann is by far more at ease in public and personally outgoing, ever the brilliant jokester and man of good humor. He exudes confidence as he speaks in a well-modulated British style suffused with an undefinable continental accent. By contrast, Pons appears shy, retiring, and on his guard. He speaks in dry tones with his slight North Carolina drawl. Then there was the odd geographic tie. An England-Cambridge, Massachusetts, connection would have been more instantly recognizable. But England-Utah was unheard of!

Martin Fleischmann, now a naturalized British subject, was born March 29, 1927, in Karlsbad, Czechoslovakia, to Jewish parents. The family came to England to avoid inevitable persecution by the Nazis. Martin went to high school in Sussex, England, during the war, attended Imperial College in London after the war (1947–1950), and later distinguished himself by achieving at age forty the professorial Chair in Electrochemistry at the University of Southampton. Fleischmann has been called a genuine Renaissance man with a reputation for brilliant and creative ideas—not all of which pan out, but such is the nature of creativity. Surely, when one listens to or is in the presence of Martin Fleischmann, one feels that the image of an exceptional polymath fits him like a glove.

Since 1986, Fleischmann has been a Fellow of the Royal Society, an honor given only to the most distinguished of scientists. The author of over 200 scientific papers—a number of them with Pons as collaborator—and a number of portions of textbooks, Fleischmann won the Royal Society of Chemistry's medal for Electrochemistry and Thermodynamics in 1979. He was president of the International Society of Electrochemistry (1970–1972). In 1985 he was awarded the Palladium Medal (how appropriate!) by the U.S. Electrochemical Society. Fleischmann, married since 1950, is a father of three (a son and two daughters), and a grandfather of four. His leisure interests run the gamut from skiing, walking, and music to an appropriate avocation for a chemist—cooking. (Those few readers who still may think of Martin Fleischmann as a quack will be happy to know that he lives on Duck Street in a nice English town.)

B. (Bobby) Stanley Pons is about young enough to be a son of Martin Fleischmann. It was mildly ironic that Pons was born in 1943 in the small town of Valdese in the North Carolina foothills, because on the day of the cold fusion announcement, the huge oil tanker Exxon *Valdez* (same pronunciation as Valdese) was coming to grief on the rocky Alaskan coast. There soon appeared a MacNelly (*Chicago Tribune*) cartoon connecting cold fusion with the oil spill. An oil-soaked bird adrift on a buoy was remarking to a similarly blackened seal or sea lion, "Any more word on how those fusion experiments are going?"

Pons's Italian Protestant ancestors had fled religious persecution in the old world. Now, less lethally but in some fashion, Stan Pons was about to be assaulted by many members of the scientific community. He would need a lot of stamina to fight back. In his youth, as at present, Pons was very athletically oriented, engaging in track and football. The cold fusion brouhaha immediately took away from his love for skiing, which in calmer times he had pursued in the Wasatch Mountains, sometimes with Fleischmann. Pons was also drawn to the world of chemistry as a child, as many youngsters had also been, encouraged by parent-bestowed chemistry sets and the like.

Pons attended Wake Forest University in Winston-Salem, North Carolina, graduating in 1965, and began advanced studies at the University of Michigan at Ann Arbor. But with his doctorate almost in hand in 1967, he, the eldest of three brothers, left school to work in his father's prosperous textile mills and to manage a family restaurant in North Palm Beach, Florida. Eventually, his love for chemistry drew him back to active science. With the encouragement of faculty at University of Southampton in England, he entered its graduate program in chemistry and received his Ph.D. there in 1978. Martin Fleischmann was one of his professors. After being on the faculty at Oakland University in Rochester, Michigan, and the University of Alberta in Edmonton, Pons came to the University of Utah in 1983 as an associate professor, becoming a full professor in 1986, and Chairman of the Department in 1988. He has authored or coauthored over 150 scientific publications. Pons has been married to his wife Sheila since 1973. An interesting curiosity, both the chemists' wives are Sheilas. This is Pons's third marriage; he has six children.

Prior to the cold fusion announcement, when Fleischmann came once or twice a year to Utah to enjoy the mountains and collaborate with Pons, he would stay at the Pons residence for months at a time. Besides cooking up experiments into the wee hours of the morning, both scientists foster reputations as gourmet cooks. In December 1989 at the Pons home, the duo gave a lavish holiday bash for well-wishers, for which they spent the previous week baking and cooking. Fleischmann is a connoisseur not only of heavy water, with or without measurable tritium content, but he is also an accomplished wine-taster and lover of good beer.

✱ A Tale of Two Universities

Lurking behind the circumstances of the March 23 press conference was a much deeper story, a tale of scientific competition, rivalry, and intrigue between researchers at two neighboring universities. Within a week of the announcement would come another dramatic cold fusion disclosure from physicist Steven Earl Jones at the private Brigham Young University in Provo, Utah, less than 45 miles to the south of the first announcement. Jones would not claim to have created power-producing cold fusion reactions, but he did say that he had detected neutrons emerging from an electrochemical cell that was vaguely similar to Fleischmann and Pons's.

Though he did not hold a press conference, Steven Jones and colleagues at the Department of Physics and Astronomy at Brigham Young University and at the University of Arizona did submit a paper to *Nature* in which they claimed to have detected neutrons from fusion reactions within metal electrodes immersed in heavy water.* They mentioned nothing about generating excess heat or tritium. But theirs was in no way the same kind of experiment carried out by Fleischmann and Pons. They employed a strange broth of different chemicals in heavy water and worked with thin foils of palladium rather than bulky electrodes.

The fusions they reported were said to happen in only tiny numbers (a few hundred per hour). This was far below the quantities that would need to have occurred in the Fleischmann-Pons experiment to explain its energy production by conventional fusion reactions. The Jones group was convinced, however, that it had created fusion. In their technical paper they attributed various geological occurrences of an isotope of helium (helium-3) and tritium to deuterium fusion reactions occurring

*S.E. Jones, E.P. Palmer, J.B. Czirr, D.L. Decker, G.L. Jensen, J.M. Thorne, S.F. Taylor, and J. Rafelski, "Observation of cold nuclear fusion in condensed matter," *Nature*, Vol. 338, 27 April 1989: 737–740.

within the Earth. Like Fleischmann and Pons, they attributed their alleged room temperature fusion to enhanced "quantum tunnelling" that supposedly allowed deuterium nuclei to overcome the normal energy barriers that separate them.

Unlike Fleischmann and Pons, however, Steven Jones was well known to physicists and the hot fusion community, which gave him a credibility that Fleischmann and Pons could not match. That Jones came out with a dissimilar but closely related item of cold fusion news at about the same time, ironically, may have boosted the credibility of Fleischmann and Pons in their claims. But there was initial confusion about what Jones was asserting, because of his well-known earlier work on cold fusion of a different sort—the concept called *muon-catalyzed fusion* (Chapter 6).

Much of the difficulty that ensued between Fleischmann and Pons on one side and Jones on the other—a friction that has now lessened considerably—can be understood in part from a chasm of personality differences. Jones is the youngest of the threesome, having been born in 1949 and raised a Mormon, with all that his religion's outlook and demanding codes of conduct implies. Jones was a missionary in Europe for the Church of Latter-Day Saints and abides by the faith in not drinking alcoholic beverages, coffee, or tea. He is the father of seven children. His frameless glasses give him an upstanding, almost Boy Scoutish bearing; he speaks in a soft voice and with hesitation at times, grinning and laughing frequently. Jones pursues his science with religious fervor, almost literally. His University stationery bears witness, inscribed as it is with the Brigham Young University motto, "The Glory of God Is Intelligence."

For about a decade, Steven Jones and his colleagues had been pursuing muon-catalyzed fusion, a technique that they already had shown experimentally to produce low-intensity fusion reactions at room temperature within a sample of deuterium—certainly a kind of cold fusion in its own right. It was an idea that had come from researcher F. C. Frank at the University of Bristol, England, and Andrei Sakharov in 1948. A muon is a subatomic particle with a negative charge that is produced in a nuclear particle accelerator when protons are slammed at high energy into other atomic nuclei. In the hierarchy of the subnuclear zoo of particles, it is closely related to the electron, though it is some 207 times more massive. The hot fusion community had regarded Jones's muon-catalyzed work with respect, though they believed it was not nearly as promising for energy breakeven as high-temperature magnetic or inertial confinement fusion.

Differences in the Brigham Young University and University of Utah experiments aside (and these were not so clear in the first few days after the Utah announcement, because BYU was then making no disclosures), it was hard to believe that such an astounding coincidence of technique and geography could occur, but it most surely did. Two groups had come up independently with the same basic concept: the electrochemical loading of palladium with deuterium to attempt to induce fusion. Yet the concurrence in the timing of the announcements was anything but coincidental. Fleischmann and Pons have since said that they had originally intended to wait until about September 1990—a full 18 more months—to announce their discovery, but not so Jones. He was all set to go public early in May 1989 with a talk at the American Physical Society Meeting in Baltimore. There was more than met the eye behind the seemingly precipitous and incompletely prepared disclosure by Fleischmann and Pons.

The two had gone about their research secretly for five and a half years since 1984, blissfully unaware of Jones's work, though perhaps they might have seen his name and BYU affiliation in connection with his muon catalyzed fusion work. Jones and his colleague Johann Rafelski had written a popularized account of the subject in *Scientific American* in July 1987 (Chapter 6). Likewise, Jones was unaware of Fleischmann and Pons. But in September 1988 he was asked by Dr. Ryszard Gajewski (Ri-shard Guy-eff-ski) of the Department of Energy (DOE) to review a grant proposal for funding a research project that Fleischmann and Pons had submitted to DOE after their personal funds dried up. Jones and his colleagues had already received close to $2 million dollars since 1985 for various fusion studies, so it was natural for DOE to choose him as a reviewer. The section of DOE responsible for funding this kind of frontier work was the Division of Advanced Energy Projects, directed by Gajewski, a man who would soon be embroiled in the controversy about federal funding of cold fusion.

Jones's eyes must have opened wide and his jaw probably dropped when he first handled their proposal and saw what the University of Utah duo were up to. It was unmistakably the same generic kind of experiment that he had already performed and from which he had obtained his low-level neutron data.

Jones gave the $322,000 proposal his honest and favorable review, meaning that he thought it was worthwhile to pursue. Shortly after the Fleischmann and Pons announcement, in fact, a DOE spokesman would say that the $322,000 had been approved (March 2 was the actual approval date). Many times, scientists who receive proposals to review disqualify themselves as reviewers if the proposed work is very similar to their own, but Jones did not do so. A week after the storm broke on March 23, Jones expressed regret that he had reviewed the proposal—not because of any ethical violation that he considered he had made, but because of the ambiguous situation his review had put him in.

A perception fostered by the press was that after Jones became aware of Fleischmann and Pons's work in 1988, he contacted them and suggested that the two groups should collaborate, specifically to use Jones's spectrometer to detect neutrons that might be coming from the Fleischmann and Pons experiment. Jones denies that he made any such direct call to the University of Utah. He claims, instead, that he phoned Ryszard Gajewski and suggested that Gajewski might want to tell Pons about the BYU work. Jones thought this was an appropriate way to open a Utah/BYU collaboration. Furthermore, says Jones, Gajewski was the one who called Pons and then Pons called Jones in December 1988. Jones maintains that Pons asked him to send information on the BYU neutron spectrometer, a request with which he complied. Jones claims that he did not ask for more information in return from Pons.

In late April, Pons had a much different perception that he related to G. Christopher Anderson of *The Scientist*: "In all my scientific life, I have never seen a situation where a proposal was sent to a certain person, who calls up and says, 'Tell me more,' and who then immediately reveals himself as the reviewer and suggests collaboration. I had no idea when he was going to go public." (*The Scientist*, May 1, 1989.)*

For his part, Pons appeared not to want the collaboration that Jones was edging toward; he obviously felt that it would take away some of his and Fleischmann's claim on original discovery. Jones still maintains that Pons was interested enough in hearing about details of the Jones neutron spectrometer to want to learn more. Jones says that Fleischmann and Pons did not at that time reject outright a collaboration.

Soon followed a meeting at BYU on February 23 involving Jones, Pons, and Fleischmann.† Jones had previously committed himself to speak at the American Physical Society's meeting in Baltimore on May 4, at which time he planned to report on his electrochemical fusion experiments. He had already submitted an abstract of his paper to the APS, and at the February 23 meeting told Fleischmann and Pons all about this intention. But he was tempted a bit perhaps to do otherwise—not to tell them about his scheduled meeting. An electronic mail message from Jones to another researcher says, "I pondered whether to be open about this, but thought of the golden rule, in truth, and therefore was open with them. I suggested back-to-back publications. I still feel this was the Christian thing to do."

Though Fleischmann and Pons wanted to pursue their own work until their data and theory were more definitive—some 18 more months

---

*Jones has angrily accused Pons and other University of Utah officials of spreading the story that he, Jones, pirated ideas from his review of the University of Utah proposal. Jones says that Pons, in fact, apologized to him for this on or about February 21, 1989.

†Jones showed them his neutron spectrometer, actual data, and geological evidence for fusion.

down the road—Jones wanted to publish sooner and did not intend to call off his speaking arrangement for May. It is noteworthy that Jones appears to have been the driving force in pushing for rapid public disclosure. He submitted his talk abstract to the American Physical Society on February 2 concerning "evidence for a new form of cold nuclear fusion which occurs when hydrogen isotopes are loaded into various materials. . . ." He did, in fact, ultimately give the May 4 talk in Baltimore, but it occurred against the background of the Fleischmann and Pons public announcement.*

There was another meeting at BYU on March 6, this time with the two university presidents (Chase Peterson of the University of Utah and Jeffrey Holland of Brigham Young University). Jones made it clear that he doubted the Fleischmann-Pons heat results were from a nuclear process. He exhibited notarized log-books dating back to April 7, 1986, to validate his discoveries. Jones was "shocked," he now says, that President Peterson then asked him to put off his invited talk. He agreed to cancel a scheduled colloquium that he was to give at BYU two days after this meeting. And one of his graduate students canceled a talk that was scheduled for some other research conference. Since the two groups had come up with similar ideas, though they could not agree to a substantial delay in the time for public disclosure, they agreed to submit their papers jointly to *Nature* magazine—literally via the same Federal Express packet—on March 24, 1989. But people at the University of Utah were becoming nervous about protecting their priority of discovery and about any financial benefits that might accrue. Since they could claim to have a process that was producing excess power and Jones could not, the Utah people undoubtedly felt compelled to go forward at least with patent applications. The BYU people apparently felt no such pressure. Two University of Utah patent applications were filed only days before the March 23 announcement. And about March 21, they made the momentous decision to make a public disclosure. This was after Fleischmann and Pons's technical paper was submitted to the *Journal of Electroanalytical Chemistry* (*JEAC*) on March 11.

We may never know precisely what considerations drove the decision to hold a press conference on March 23, rather than go forward with the original plan of simultaneous submission to *Nature*. President Peterson was undoubtedly the major driving force behind making the disclosure. It was he who earlier had phoned Cornell University emeritus professor of physics, Nobel laureate Hans Bethe, seeking his opinion about the work of the Utah scientists. Bethe reportedly recommended

*Martin Fleischmann told me, "Stan and I were in favor of publishing (our) paper in the *Annals of Utah Science*, which is not read by anybody. We did not want this hoo-hah!"

delay and caution in announcing anything. University spokespeople and Fleischmann and Pons themselves would later say that the fear of leaks to the press and patent concerns were factors. In retrospect, no one could point to any *specific* press leak that was about to occur. Certainly, since Fleischmann and Pons had no agreement with Jones specifically not to submit to another journal too (a fact that Fleischmann and Pons now assert, but which Jones disputes), they were within the "letter of the law" to submit to *JEAC*. When a few days after they had submitted it they had an indication it would be published (and why wouldn't it, given their previous publication history in that journal and their familiarity with its editors?), they must have felt all bets were off. They were no doubt becoming suspicious of Jones's intent and wanted to safeguard their priority of discovery.

There was already a background of rivalry between the two universities. For his part, Jones now blames University of Utah President Chase Peterson for pressuring Fleischmann and Pons to make the March 23 announcement. As University of Utah physicist and Vice President for Research James Brophy would remark later about this decision, "We decided to stop at the point where they could demonstrate fusion without explaining it." Indeed, Fleischmann and Pons at the press conference could not and did not explain the nuclear process that was causing their excess heat. Nothing is wrong with that, of course, but they seriously erred by not making it very, very clear that the nuclear products that they claimed to have detected did not directly add up to explain the heat. This open admission would have cut the level of skepticism at least in half. They may have reasoned this would weaken their case. It would have! But it also would have established a more scientific, less charged framework in which others could try to verify or dismiss their results.

With no direct advance warning to Jones, Fleischmann and Pons held their press conference on the 23d. Not surprising, neither Jones's name nor his work were mentioned at the press conference by Fleischmann and Pons. Yet Stanley Pons *did* mention the other kind of cold fusion—muon catalyzed—but failed to mention Jones's well-known work in that field. Jones actually knew at least the day before of the impending press conference, because University of Utah public information officer Pamela Fogle made calls to roughly a dozen major news organizations telling them what was to be reported the next day. BYU Professor Grant Mason, Dean of the College of Physical and Mathematical Sciences, called University of Utah's James Brophy on March 22 and said that if the press conference was held "BYU would interpret this as a violation of the agreements between the two universities."

At least one reporter, Jerry Bishop of *The Wall Street Journal*, contacted Jones, because of his geographic proximity and muon-catalyzed fusion background, to ask what was going on at the University of Utah. This led Bishop to report the impending news conference in the *Journal* on March 23 under the headline, "Development in Atom Fusion to be Unveiled." Bishop's historic message to the world: "The University of Utah told reporters yesterday that it will hold a press conference this afternoon to announce that its scientists have achieved a 'sustained thermonuclear reaction at room temperature.' A University spokeswoman adamantly refused to give any further information." Even though "sustained thermonuclear reaction at room temperature" was a contradiction-in-terms—Bishop should have removed the "thermo" prefix—it was clear that big news was about to be sprung.

Left with that uncertainty, Bishop speculated in his article that the University of Utah work would have something to do with an advance in muon-catalyzed fusion or cold fusion, the very term that Jones had been using for his work. This marked the first entry of the phrase *cold fusion* into the wide public arena. One line of Bishop's background report on the 23d was prophetic: "Any claims of a major breakthrough would stir considerable controversy and send physicists rushing to their labs to try to duplicate and confirm the Utah experiments." The *Financial Times* of London published nearly as complete an account the same day with an added twist: a diagram of a cold fusion cell—crude, but nonetheless the world's first.

Hearing that the press conference had occurred, Jones no longer felt constrained and submitted his paper to *Nature* the very day of the announcement. Besides being none-too-pleased that the University of Utah press conference had occurred at all, Jones was very upset—and rightly so—that Fleischmann and Pons did not mention his work when they went public. Jones considers this omission, in particular, to be a breach of the agreement reached between the scientists and university presidents at their March 6 meeting. The University of Utah's Vice President for Research James Brophy has said, "It is true that the first BYU heard of it [the news conference] was from press calls, which was not our intent. We fully intended to inform them in advance." (*Deseret News*, March 28, 1989.)

Fleischmann and Pons did not know about Jones's submission of his paper on the 23d, and as previously had been agreed, they sent colleague Marvin Hawkins to the Federal Express departure point at the SLC airport on the 24th to wait for Jones's paper to arrive and to be inserted in the packet. With Jones not showing, off went the paper to *Nature*—a version shorter than the one submitted to *JEAC*.

✳ Immediate Aftermath

The public reaction by Jones was initially muted as BYU attempted to take the "high road" in the unfolding drama. On the day of the University of Utah press conference, spokesman for BYU Paul Richards told JoAnn Jacobsen-Wells, a reporter for the Salt Lake City *Deseret News*, "Both groups made simultaneous discoveries, but BYU is not planning any public announcement until reports appear in scientific journals." The next day, BYU did tell the world that Jones's paper had been submitted to *Nature*, and the university did release the paper's abstract, which said in part, "We have also accumulated considerable evidence for a new form of cold nuclear fusion which occurs when hydrogen isotopes are loaded into crystalline solids without muons." A BYU spokesperson also said that the university was now planning to file a patent application, even though one had not been submitted earlier, only to protect its researchers' integrity, it is said.

Somehow word got out at the same time that Fleischmann and Pons hoped to have their paper published in *Nature* in May. One could assume that it would be of such interest that *Nature* would feel compelled to publish it, but how wrong they were. There was some positive reaction from *Nature* as evidenced by the comment of the magazine's Washington editor, Dr. David Lindley, ". . . whether it turns out to be something tremendous or a novelty, I feel lucky they chose us. *Nature* is a commercial magazine. It is our bread and butter that exciting papers are sent to us."

For his part, Jones knew where he stood experimentally in relation to what Fleischmann and Pons were claiming. For a long time he had been thinking about alternate ways to bring about fusion, beginning with his muon-catalyzed fusion work. But in 1986, he and colleague Clint Van Siclen published a paper titled "Piezonuclear fusion in isotopic hydrogen molecules" in which they speculated that by squeezing together mixtures of hydrogen, deuterium, or tritium molecules at millions of atmospheres pressure with a mechanical press (with diamond "anvils"), it might be possible to investigate very low fusion reaction rates. The"piezo" in piezonuclear simply means to squeeze together.

As early as April 7, 1986, Jones and another colleague E. Paul Palmer had entered the plan for a possible electrochemical fusion experiment in one of their research group's laboratory notebooks. They set up a cell and by May 27, 1986, the group thought it saw the first indication of gamma rays from fusion reactions, but because they were at such a statistically insignificant level above background, the team lacked enough evidence to publish. By the fall of 1988, after they had built and used their neutron spectrometer, confidence was high enough to consider publishing, even though there was no satisfactory theory to explain why neutrons should be emerging from their cell. This is important, because Jones would get a rather smooth reception by the press and other scientists, while Fleischmann and Pons were roundly criticized for not having a satisfactory theory to explain *their* results.

Local government entered the fray too. Utah Governor Norm Bangerter said the day after the University of Utah news conference that he would call a special session of the Utah legislature, probably in April, to request $5 million to support and buttress Utah's new fusion direction. In the background was the knowledge that if the patent applications were successful and Fleischmann and Pons really had come up with something revolutionary, Utah could become a very wealthy state. By university policy, Fleischmann and Pons stood to receive a third of the patent royalties, the Chemistry Department would get another third, and the University the remainder.

The supportive statements coming out of Utah were quite unlike the circumspect or negative remarks being heard outside the state. President Chase Peterson said, "There is a slight chance we are wrong. But if this is fusion, it will rank up there with such things as the invention of fire." (*Salt Lake Tribune*, March 25, 1989.) And one of the state Regents, Charles Bullen, said, "This is one of the greatest scientific breakthroughs in the history of mankind." Peterson turned out to be prophetic when he said at the press conference, "The full story of the research Professors Pons and Fleischmann will announce today will not be known for months or years, as others confirm and challenge and enlarge their ideas and their data."

Outside the protective cocoon of Utah, the story was different. The first week following the Fleischmann-Pons announcement was a blur of activity for scientists the world over who were drawn into the vortex of the effort to reproduce electrochemical fusion. Much ongoing work was temporarily put on the back-burner as eager researchers lunged into crash programs to attempt to confirm or prove incorrect the Utah work.

Newspapers, electronic bulletin boards, telephone calls, and fax machines became the means to receive the latest information about the experiment. Science writers and media people also scurried about, trying to satisfy the thirst for information. Scientists around the country studied videotapes of the MacNeil/Lehrer Hour, CNN coverage, and the CBS Evening News, attempting to glean details that they did not yet possess. It would be Friday, March 31, before the first fax copies of the Fleischmann-Pons paper began to appear and multiply wildly on the phone network. Even so, many experimental details were still missing from the paper, which resulted in a continuous guessing game to determine the "best" way of doing the experiment. How long did one have to wait for the reaction to start, for example? Exactly what kind of preprocessing, if any, needed to be done to the palladium electrode before the experiment would work? Many scientists were angered by the dearth of details. Said physicist Robert L. Park of the Washington Office of the American Physical Society, "These guys called a press conference and didn't even release a scientific report. That's outrageous." (*The Chronicle of Higher Education*, April 7, 1989.)

The distinction between the popular media and the scientific press seemed for a while to disappear. Researchers and media people alike hung on the words of Jerry Bishop in *The Wall Street Journal* each morning. *The New York Times* had much more sporadic coverage, though most of its articles seemed to present a relatively balanced account of the scientific controversy. After its initial article on March 24, the *Times* waited for its renowned Tuesday "Science Times" section to report anything else on the hot controversy (Chapter 16).

At MIT, the frenzy began on Thursday evening, March 23, when a group of students decided to try to duplicate the experiment and actually made one of the first attempts. Other scientists at MIT's Plasma Fusion Center and in the Department of Chemistry got together to work intensely but informally to investigate the extraordinary claims made in Utah. The group included about ten faculty and student researchers led by Ronald R. Parker, director of the Plasma Fusion Center, and Professor Mark S. Wrighton, head of the Department of Chemistry. At Caltech the same general pattern emerged: students starting the experiments, followed by an interdisciplinary team of professors and students working together. At Lawrence Livermore National Laboratory, a group of researchers including Edward Teller gathered to plan their own replication effort. Knowing the solid reputations of Fleischmann and Pons, a group of electrochemists at Texas A&M University immediately began their attempt. The group would soon be one of the first to come in with supporting evidence. Leading the team, an open-minded Dr. Charles Martin said, "A lot of people say this doesn't make a lot of sense, but a lot of science doesn't make sense initially." (*Deseret News*, March 25, 1989, reprinted from the *Los Angeles Times*.)

The political kettle began to boil in Utah as the Governor's recommended figure of $5 million to support cold fusion began to sink in. There were many advocates, but some began to wonder whether enough was known about the Fleischmann and Pons work to rush in with so much state funding. Yes, the stakes were potentially high, but Utah's total state-supported education budget for nine universities was only $350 million. Questions arose about who would control the funding. There was the naïve belief in Utah that much federal money would soon start to flow into the program. But could the state hold onto its position as the world center for cold fusion research? What about commercial interests—how could they be served? By Easter Sunday, March 28, James Brophy had received some 200 inquiries from companies expressing an interest in commercializing the breakthrough. Brophy was predicting that within three to five years there would be small cold fusion power plants, and it would only be several decades before widespread use. He was talking about 100-horsepower units that could power cars and up to 50-megawatt electric power plants. Remarkably, Brophy had been extremely skeptical when first told of the cold fusion research, but he had personally investigated it and became convinced.

President Chase Peterson was simultaneously injecting a small element of caution, saying to the press that though the experiments looked good there could still be some kind of "glitch" or "hang-up" that the researchers were unaware of. Cold fusion reactors, he said, might not be safe and could turn out to be economically nonviable, in which case the state of Utah's support should not continue.

Reports were circulating that the retired two-time administrator of NASA, Dr. James C. Fletcher, would be coming back to Utah to direct the cold fusion program. Fletcher had resigned (effective April 7) as NASA head after the recent first successful space shuttle flight following the Challenger disaster in 1986. Despite Fletcher's optimistic statements to the press about cold fusion, he did not choose to head the new cold fusion organization. Instead, he became an advisor to the project.

A state politician, Eldon Money—Senator Money—went so far as to propose that the Utah public make voluntary contributions to cold fusion in exchange for future tax breaks based on its financial success. Some stalwart Utah Republicans supported this idea by a Democrat, feeling that it was not proper for the state to risk taxpayer money on such speculative ventures. One condition to participate in Senator Money's plan: An interested citizen would have to certify that he or she knew the research might not bring a return on investment, and that they could afford to lose their money.

The *Deseret News* editorialized supportively, "University of Utah chemist B. Stanley Pons and his British colleague, Martin Fleischmann, have fired the starting gun in what could be a scientific and technological race of unprecedented size. It would be ironic and very sad if Utah were left in the dust while others seized the work of these pioneers and ran away with it." Likewise did the *Salt Lake Tribune*, "After certain ground rules are laid, unanimous approval of the [$5 million] appropriation should be expected. After all, there is substantial evidence that the University of Utah Chemistry Department Chairman B. Stanley Pons and University of Southampton Professor Martin Fleischmann have found the key to a cheaper, cleaner, limitless energy supply. . . . Besides reflecting well on the caliber of local education, the discovery counterbalances a series of Utah embarrassments over alcohol regulation and notorious crimes the past few years."

In Utah there was still insularity and euphoria. *The New York Times* may have largely passed them by and physicists elsewhere may have had their serious doubts, but Stanley Pons, for one, was optimistic. In that first week, he could honestly say to Tim Fitzpatrick of the *Salt*

Lake Tribune, "The physicists that I have spoken to see our point and have been quite intrigued and complimentary." (*Salt Lake Tribune*, March 28, 1989.) While many in the outside world wondered about Fleischmann and Pons's sanity, Pons was already contemplating the safety of cold fusion applications. Though he talked of scaling the process up to practical power reactors, he believed for safety reasons alone more had to be known about the cold fusion mechanism.

People were waiting with bated breath for the first signs of confirmation. Pons was the source of a rumor that first week that Los Alamos National Laboratory might already have gotten some positive results from their initial attempt at replication. Where his information came from, no one knew and certainly Los Alamos was not making any announcements then. Physicists on the outside may have been led astray by the experimental simplicity and casual remarks about how easy it was to do. But those that had access to Utah newspapers would have a much different impression. Pons said quite clearly, "We have maintained that the deuterium-deuterium reaction is not the main heat producer. . . . There are other components in the system. . . . Lithium is a fine candidate right now as far as I'm concerned." (*Salt Lake Tribune*, March 28, 1989.)

By then, *Nature* had received both the Fleischmann and Pons paper and the one by Jones and others. David Lindley, the assistant physics editor for *Nature*, who was based in Washington, appeared ebullient in that first week, saying of the Fleischmann and Pons paper, "We're thrilled to have it. Even if it turns out to be wrong, we like to have the first look at it. Assuming all goes well and it's published in *Nature*, that's very good for us." (*Salt Lake Tribune*, March 29, 1989.) He pointed to *Nature's* alleged quick turnaround time, touting the fact that *Nature* could publish within four or five weeks, while specialized journals took months. Strangely enough, *JEAC* was able to publish the Fleischmann and Pons paper on April 10. *Nature* took until April 27 to publish Jones, was never to publish the Fleischmann and Pons paper, and began to take a decidedly negative view toward cold fusion. Lindley's words in that first week seem so improbable in the light of his anti-cold fusion writings a year later (Chapter 13).

The first week of public awareness of cold fusion was coming to an end and Steven Jones felt it was time to speak. On March 30, despite his reservations about what was then becoming known as "science by press conference," he was driven to prove that he too had been working on electrochemical fusion. Jones said that since May 1986, DOE had funded his work. He related that in his experiment he had only detected about a dozen neutrons per hour, the inefficiency of his detector implying that his experimental cell was actually producing up to 1,200 neutrons per hour—far removed from the 40,000 per second that Fleisch-

Dr. Steven E. Jones, Professor of Physics at Brigham Young University, Provo, Utah, who announced in late March 1989 that his group had detected low-level neutron emissions coming from electrochemical cells with heavy water. (Courtesy Brigham Young University)

mann and Pons were claiming. Despite their differences and hard feelings, Jones expressed hope that the two universities could work together on cold fusion. Beyond the superficial commonality, the experiments at the two universities were quite dissimilar. Jones had not even thought to try to measure heat, so certain was he that low levels of neutrons were the most he could expect, with no chance of finding significant heat-producing reactions.

Jones flew to New York City and on Friday, March 31, on the safe turf of a "Plasma Physics Colloquium" at Columbia University, he staked out his claim to cold fusion. Schermerhorn Hall was packed with all manner of technical and media people—a sure sign that a news conference, in effect, was in progress even though it was ostensibly a science colloquium. He began his talk with an oblique attack on the two Utah chemists who had stolen some of his fire, "It's rewarding to see that there's still intellectual interest in physics." The overflow crowd loved it. He spoke first on muon-catalyzed fusion, however, not what everyone really wanted to hear about. Many people in the audience, however, were fingering copies of the Jones paper that had been submitted to *Nature*, whose abstract was so upbeat about the evidence for cold fusion in crystalline solids. It was reported that fax copies of the Jones paper were inadvertently distributed. An honest-to-goodness press conference was held *after* the colloquium in which Jones tried to pour cold water on hopes that practical cold fusion was just around the corner. When asked how long it would be before some kind of cold fusion became practical, he replied, "It's hard to say, but a reasonable answer might be *twenty years to never* [author's italics]." Since then, he has repeated that message countless times.

Jones disclosed that he, unlike Fleischmann and Pons, had set up a clever neutron counter that detected neutrons *directly*. Fleischmann and Pons had measured neutrons "second hand" or indirectly, relying on the counting of gamma rays that were the presumed reaction by-products of the cold fusion neutrons. These slammed into hydrogen nuclei in a container of ordinary water that surrounded their energy cell. Jones could count neutrons for a period of eight hours or so, after which his titanium or palladium electrodes were evidently clogged with coatings of other elements from the cell such as iron.

On the same day that Jones spoke at Columbia, Pons was giving a seminar at the University of Utah—one that almost had to be called off because his technical slides had apparently been stolen during the March 23 news conference. (One week later they mysteriously showed up in the Chemistry Department's library, the "borrowing" probably being the handiwork work of a collector of historic memorabilia.) The crowd overflowed two seminar rooms in the Chemistry Building. Dr. Gerald Byrne, chairman of the University of Utah Metallurgical Engineering Department recalled that it was a "mob scene" with people arguing in the entrance way. Such was the first week of the cold fusion era.

Pons could not explain from what specific fusion reactions his considerable excess heat was coming; he continued to say that he could conceive of no process other than some kind of fusion to explain the heat. But he acknowledged that products from the deuterium-deuterium reaction that they had monitored could account for only about a billionth of the measured heat. Recalling the "vaporization" of the cube-shaped palladium electrode in one of their tests (the paper had said "WARNING! IGNITION?"), he went so far as to warn would-be experimenters not to risk potential explosion hazards by using sharp-pointed electrodes, rapid temperature changes, or substitution of tritium for the deuterium. He reiterated that conventional fusion reactions must only be a small part of the source of the heat, and that other unknown nuclear mechanisms were probably occurring. In the audience were skeptical University of Utah physicists Michael Salamon and Haven Bergeson, who a year later would publish in *Nature* with other colleagues a paper very critical of the Fleischmann and Pons work. They were polite that day, but basically unmoved.

George L. Cassiday, a professor in the University of Utah Physics Department wrote in a letter to the *Salt Lake Tribune*, "At room temperatures, the nuclei would almost have to achieve nuclear densities in order to fuse at the rates necessary to generate the quoted power yields. The only place where I've seen such densities achieved are in rooms full of media persons and publicity moguls anticipating the pronouncements of great scientific breakthroughs purporting the salvation of mankind." At one point, he joked, "... someone over in the Park Building probably flushed the toilet and the reduced pressure throughout the poorly maintained U. of U. campus affected their apparatus readings in some perverse way."

4 *A Frenzy of Replicators*

Don't sell your oil stocks yet!

Steven E. Jones, March 1989

It's not appropriate that, just because an observation doesn't fit into a theory, to say that the observation is incorrect.

B. Stanley Pons, *Deseret News*, April 11, 1989

The discovery of cold nuclear fusion in condensed matter opens the new possibility at least of a new path to fusion energy.

Steven E. Jones, *Science*, April 7, 1989

✳ The Days After

THE AIR WAS ELECTRIC WITH ANTICIPATION in the weeks following the initial announcement, particularly so in the first week. Scientists were agog in efforts to cope with what was either one of science's greatest surprise discoveries, or one of its most bizarre dead ends. They were egged on by one or more statements made by Fleischmann and Pons. On a television program, "The Wall Street Journal Report" (Sunday, March 26), Pons had said, "If someone really wanted to do [the experiment], I expect they could in a couple of weeks after publication of the data." And he had joked off camera that he was going to let his son try it. Since people had been led to believe that the experiment was so easy to do, tension mounted as scientists and laypeople awaited reports of the first confirmation of the Fleischmann-Pons effect. Swiftly enough one came, not from America but from a university in distant Hungary.

The Hungarian news agency, MTI, wired the Reuters News Service on Saturday, April 1. MTI reported that Gyula Csikai and Tibor Sztaricskai at Kossuth Lajos University at Debrecen had set up the Fleischmann-Pons experiment on Friday and had detected neutrons coming from it, but it was unclear whether they had measured BYU levels of neutrons or the much higher ones reported by the University of Utah. The researchers had videotaped television broadcasts, studied the tapes, and then set up their experiment. Obscure to Americans as was the source of the report, palpable relief was felt in Utah. James Brophy called it the "second confirmation" of the experiment; the first, in his estimation, being Jones's neutron evidence. But BYU's spokesman, Paul Richards, immediately took issue with the latter assertion, signaling what would continue for some time to be a war of words between the two universities.

And it was reported that Steven Jones himself was fostering an impression that Fleischmann and Pons may have tried to pre-empt him after seeing and being impressed with his work. It was rumor against rumor. Brigham Young University was now going to file patent applications of its own—not because they felt it was an inherently wise course to follow or that there were likely to be commercial applications from Jones's work, but simply to "protect its own position."

For most of the world, the unsubstantiated report from Hungary was too questionable and preliminary to be believed, so the highly charged air of the waiting period persisted. An erroneous report surfaced that at Brookhaven National Laboratory in Upton, New York, a place well equipped to measure neutrons, physicists had detected some evidence of cold fusion. The scientists had noticed something unusual, but not with sufficient certainty to make a report. Meanwhile, other laboratories plunged into a crash program to test the Fleischmann and Pons contention and the much more modest, though still scientifically revolutionary claims of Jones. The Princeton Plasma Physics Laboratory began their experiment the day after the Utah announcement with no effect immediately apparent. For them (as of this writing) neither neutrons, tritium, or heat ever came—at least no sign of these in which the researchers had any confidence. The same would be true of many other respected laboratories, Caltech, the MIT Plasma Fusion Center, and on and on.

✳ Taking the Plunge

The MIT experience could not have been atypical of many of the world's now round-the-clock palladium and heavy water brigades. But owing to MIT's reputation, its work did become much more visible. There were even hints from the University of Utah that it hoped MIT would be among the first to confirm cold fusion. On the announcement day, Ronald R. Parker, director of the MIT Plasma Fusion Center, happened to be visiting his colleagues at Princeton and first heard about the Utah

Dr. Robert Huggins (l), Professor of Materials Science at Stanford University, examines cold fusion experiments in his laboratory with Turgut M. Gur, Senior Research Associate. Huggins led one of the first groups that obtained apparent excess power in electrochemical cells with heavy water, and at the same time found no excess power in light water cells. (Courtesy Stanford University)

claims. Parker considered it strange that his group "had learned about the 'most significant development in the last century in physics' " from the *Financial Times of London* and *The Wall Street Journal.* Initially, he was inclined to consign the report to what his vivacious secretary Pat Stewart euphemistically calls her "squirrel file" (i.e., 'nut' file). The Utah report was incredible, and it fit the pattern of bizarre fusion claims over the years.

But like Jerry Bishop of *The Wall Street Journal,* Parker knew that Jones had been working for a long time on muon-catalyzed fusion. He viewed Jones's earlier work as having "good standing" and began to believe that there might be some unknown and perhaps promising relationship between the University of Utah and BYU work. There was naturally much confusion at MIT and elsewhere about this connection. But when Jones went public on March 30, Parker was quite willing to believe that Jones at least had something real. He recalled, "It was clearly quite different. Jones was talking about a neutron per hour, a very nice interesting physical effect. Fleischmann and Pons were talking about heat and watts of power—a very different matter. So initially there was also some confusion over who was saying what and what the relative credibility of the two groups was." But Parker's next reaction—the inevitable one for any scientist—was, "If it's that simple, then why not do it and see what they were seeing?"

On the evening of the announcement day, some MIT students had tried the experiment on their own, but were ill-equipped to wring out a firm conclusion. A large gathering had assembled on Friday morning in the PFC seminar room to view the videotape of the previous evening's news programs, CBS Evening News and the MacNeil/Lehrer Hour. The group watched in silent disbelief, intense curiosity mixed with profound skepticism. An unspoken fear of being displaced hung in the air. Professor Ian Hutchinson closed the session with the jocular warning,

"Don't send out your resumés yet!" People gathered in small groups, excitedly exchanging gossip and theories.

Parker's team was getting ready to try the Fleischmann and Pons experiment on Friday and found out about the efforts of the MIT chemistry team, which by then included Professor Mark S. Wrighton, head of the Chemistry Department, who the following year would become Provost of MIT. Also aboard were more than a dozen other students, faculty, and researchers. One of them, electrochemistry postdoctoral researcher Dick Crooks, knew Stanley Pons through Crooks's earlier misplaced application for employment at the University of Utah. MIT doctoral candidate in chemistry Martin Schloh had independently begun his own experiment Friday evening. On Saturday the 25th, the Plasma Fusion Center and the chemistry team combined efforts, put together a cell Saturday night, and started hunting for neutrons and heat. They thought that all one had to do was the electrolysis—"flip the switch, stand back, and count neutrons."

Parker recalls that "the biggest issue was that the radiation safety people really didn't want us to do this experiment, because a quick calculation shows that when you have a watt of neutrons coming out, that's a pretty serious hazard to life and health—in that order! Fatal doses. (With a little academic string-pulling, the radiation protection rules were bent to permit the MIT effort to go forward.) The question was how did Fleischmann and Pons live to tell about it?" But there was every indication that Fleischmann and Pons were, indeed, alive and well and had taken no elaborate shielding precautions. No 10-foot thick walls of concrete for them. So why the inconsistency?

The unified MIT team forged ahead on March 27 and 28, beginning to set up a number of cells with different treatments of the palladium, different electrolytes, and so forth. But there were so many combinations of possible conditions it was impossible to be sure that the correct experiment was being attempted. The group played and replayed video tapes of network coverage of the Utah apparatus to gain understanding. Even without the Fleischmann and Pons paper in hand, the MIT group thought it knew their "recipe" for a cold fusion cell. One PFC researcher, Marcel Gaudreau, had caught a flight to Salt Lake City on Easter Sunday afternoon and began looking for more clues. Meeting James Brophy, he was able to glean much needed information. He almost had a chance to talk with Pons, but Pons was too busy attending to the non-stop phone calls.

The MIT group had initially gone down the wrong avenue in aiming for big palladium electrodes first, thinking "bigger electrodes, bigger effects." They began to use many different sizes, including ones that were a quarter of an inch in diameter. The initial thought was that it

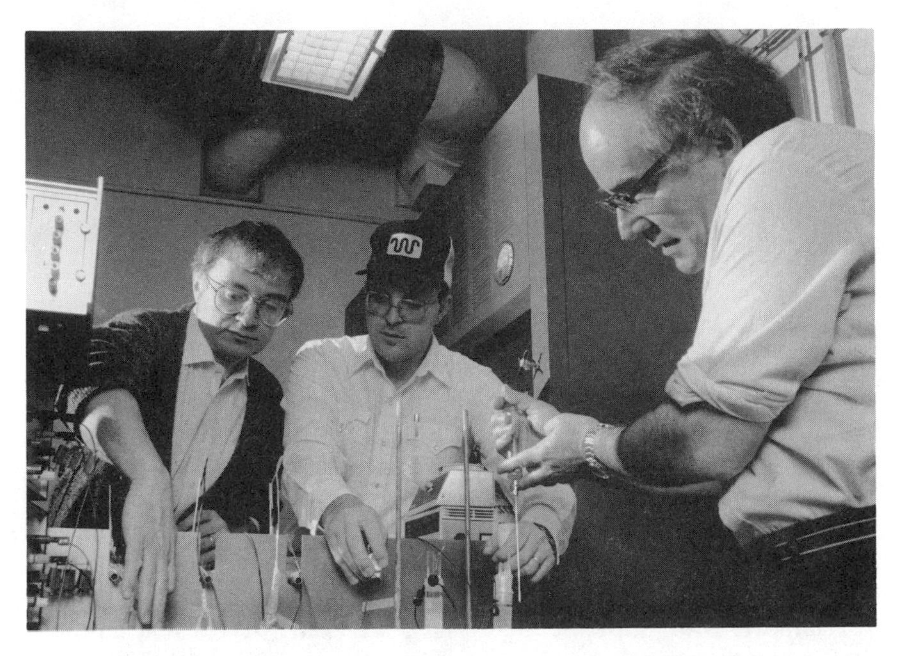

Dr. B. Stanley Pons (l), graduate assistant Marvin Hawkins (ctr), and Dr. Martin Fleischmann (r) in their University of Utah laboratory with equipment that activates and monitors experimental cells. (Courtesy University of Utah)

was a "volume effect"—the result would be proportional to the volume of the electrode.

Ron Parker recalled, "The first night was an all-nighter. Everybody was sitting around waiting for the effect to take off—a lot of sleep was lost. I had to go to Oak Ridge [National Laboratory] that morning and left around 6:00 A.M., but I kept calling back." Three teams of scientists at Oak Ridge were already working on cold fusion experiments. An intensity of spirit prevailed that was best captured in the later words of an Oak Ridge experimenter, "We don't know whether we are doing the right experiment, but we're willing to come in at midnight and do it."

With time, the MIT group's interest began to wane as it kept running the cells, monitoring their temperature, and checking for fusion by-products—neutrons and gamma rays. By the end of the week many believed (as most probably had from day one) that it was all "hogwash." When I visited the lab on one occasion to see how things were progressing, one researcher turned with a smile accompanied by evident venom toward the whole business and said—"Don't quote me, but it's crap!" Another experimenter glanced at the absolutely flat blue ink temperature curve rolling off the chart recorder, departures from which could indicate the onset of excess power. Going about her work with care and precision, she confidently predicted nothing would emerge. She was absolutely correct as far as the MIT group's interpretation of its own experiments would go in the next few months.

Frustration was setting in. But by mid-week—Wednesday the 29th, the PFC received via fax an unsolicited copy of the Fleischmann and Pons paper that was to be published in the *Journal of Electroanalytical Chemistry*. The group could not see much of anything new that they had not already incorporated in their experiments, other than a slightly different electrode size perhaps. The sentiment in the group was that it was a "bad paper"—very skimpy on details.

But the challenge of the Fleischmann and Pons paper was staring them straight in the face. Fleischmann and Pons were saying of the energy production in their experiment, "It is inconceivable that this could be due to anything but nuclear processes." Day and night the watch was kept. The experiments had been going on for ten days and still nothing. Professor Wrighton told the press all that he could, "If nuclear fusion occurs, it is at a very low level and our detectors aren't sensitive enough, or it takes longer than ten days, or it doesn't work."* If reminded at the time, he might have added the caveat that all three assessments were based on two key assumptions: (1) The MIT group had faithfully duplicated every peculiar condition of the Fleischmann and Pons experiment and had interpreted its own experimental results properly and (2) If there existed a new phenomenon, it occurred predictably enough to manifest itself "on demand."

Clouding the issue was the word then being put out by Utah spokespeople that the Fleischmann and Pons electrolysis experiment could require a number of weeks of electrode charging with deuterium to begin to show an effect. Frustration was growing at MIT. Professor Hutchinson at the PFC said, "We've heard a number of different numbers concerning this [waiting period] and they seem to get longer as time goes on." Then MIT Provost, physical chemist John Deutch, told *Energy Daily*, "I see no understood physics or chemistry which will explain [cold fusion]. There is every reason to doubt its authenticity." Professor Wrighton told the Associated Press, "We are very skeptical. . . . We've done the experiment now for ten days and we don't see anything exceptional. . . . We see no physical basis at the moment for thinking that nuclear fusion is going to occur. . . . And we certainly have no evidence that there is a big effect, one that would have technological consequences in the near term."† He noted, however, that one could not yet conclude that the approach was worthless.

---

*To Boston-based Associated Press science reporter, Daniel Q. Haney, April 6, 1989.
†AP wire reports, April 7, 1989.

Other major research groups still had not claimed any positive results, in particular Los Alamos National Laboratory and Lawrence Livermore National Laboratory. In fact, Livermore's first attempt to replicate the experiment failed spectacularly; the fusion flask exploded inexplicably and littered the laboratory with glass. Yet Stanley Pons was at the same time telling the press that he had spoken to a Los Alamos Laboratory researcher who had gotten positive results. Who could be believed? Though a handful of physicists and chemists were already at work on theories to explain the possible new phenomenon, most investigators could think of no convincing way to explain how fusion reactions could occur in an electrochemical cell. Professor Hutchinson drew what seemed to be the inescapable conclusion, "Whatever it is, it is not fusion in the normal sense."

✳ "Confirmations" Roll In

The publicity maelstrom, hundreds of telephone calls, and the brewing storm of criticism was driving Stanley Pons out of the public light. Like a lottery winner, he had to request an unlisted telephone number–the first of three in the year that followed. It was difficult or impossible to pursue cold fusion research with so much distraction, much less direct his department effectively. So in the second week after the announcement Pons stepped down as chairman of the University of Utah Chemistry Department and an interim head took over.

He didn't step far out of the limelight, however. On April 5, Pons offered a new piece of evidence to buttress his claims. He had tried the experiment, he said, with ordinary water instead of heavy water (H_2O rather than D_2O) and found "no significant [excess] heat." Skeptics like University of Utah physicist Michael Salamon were beginning to say that if using light water produced heat, that would be proof that the effect was chemical rather than nuclear. Salamon made a statement to Tim Fitzpatrick of the *Salt Lake Tribune* that no self-respecting cold fusion skeptic would ever again utter. Reacting to Pons's null result with light water he said, "It's indicative of fusion, but it's not proof of fusion." Today, still a disbeliever, he would probably like to retract those words.

The physical basis for Salamon's comment is that two bare protons are even less likely to fuse at room temperature than deuterium nuclei. The Sun, for example, fuses four protons together and makes helium, but it must employ one of Nature's masterful tricks, using other nuclei such as carbon as nuclear "catalysts" (reaction accelerants that are not consumed). Other than the hydrogen atom nucleus with its single proton, there is no such thing as a nucleus made exclusively of protons. Neutrons act as a kind of nuclear "glue" to prevent the positive charges of nuclei from flying apart.

Dr. Stanley C. Luckhardt, Principal Research Scientist at the MIT Plasma Fusion Center adjusts apparatus for testing the Fleischmann and Pons claims. (MIT Photo by Donna Coveney)

True enough, light water does have small amounts of deuterium in it, but light water used as solvent in an electrochemical cell with a palladium electrode gives no evidence of any of the claimed cold fusion effects. This assertion that Pons made in April 1989 would continue to hold true in almost all known experiments that anyone would subsequently carry out, yet skeptics then and now continue to dismiss this important piece of evidence: Heavy water works, light water doesn't.* They say, as Salamon did in April 1989, that heat with light water would be evidence *against* fusion, but the converse—no heat with light water and heat with heavy water—would *not* be evidence *for* fusion.

Monday, April 10, brought startling good news from the Lone Star state. It seemed that researchers at Texas A&M University at College Station, Texas, had "confirmed" the Fleischmann-Pons cold fusion work. Without waiting to publish their findings in a journal, 10 researchers in the Chemistry Department went public at a Monday morning news conference with their initial findings of excess heat—as much as 60 to 80% greater than the input power, they said. Sunday evening the researchers completed work on a technical paper that they were said to have mailed that evening to the now favored *Journal of Electroanalytical Chemistry*. Michael Hall, head of the department, said on Monday: "We have confirmed the most important part of their [Fleischmann

*There have been a few isolated reports of exceptions to this rule, e.g., Takaaki Matsumoto, "Cold Fusion Observed with Ordinary Water," *Fusion Technology*, Vol. 17, May 1990: 490–492.

and Pons's] observations, which is excess heat generation. But we're not sure what is going on yet. Our measurements do not prove it is fusion yet." That was more than enough confirmation for Utah, however. The *Deseret News* headline beamed, "Texas A&M confirms U. breakthrough." Eastern papers were far more reserved in their treatment, for example, "Fusion Test Matched, but Mystery Persists" (*Wall Street Journal*, April 11).

The Texas A&M group making the announcement was headed by Professor Charles Martin, an electrochemist, and thermodynamicists Mr. Bruce Gammon and Dr. Kenneth Marsh. They said that first signs of excess heat had appeared Friday evening and had continued for over 40 hours by Sunday evening. And the supposed excess heat was continuing that very moment! They said that their reaction started almost immediately, perhaps because they had liberated hydrogen from the palladium initially by baking it at 600°C in a vacuum oven, thus making room for deuterium to enter the lattice of palladium atoms. The announcement had a significant bolstering impact. Charles Martin said, "I can't dismiss entirely that it is some chemical reaction rather than a nuclear event. The likelihood is probably very small, but one has to investigate that thoroughly." An upbeat Marsh cautioned, "This isn't good science—yet. You don't do good science in a week and a half."

The sorry story behind this announcement, however, was that a mistake with an important grounding wire connection had been made, and the researchers realized very soon (possibly even on the day of the press conference) that their announcement was grossly premature. Yet they did not immediately retract their claims! In the ensuing media feeding frenzy on cold fusion, this embarrassment went largely unnoticed. The paper from this group led by Charles Martin was, of course, never published. Charles Martin continued some 40 more experiments through the end of 1989 in an effort to find excess heat, but his group at Texas A&M—one of five at the University that worked on cold fusion—never met with success. Martin subsequently left Texas A&M and is now at Colorado State University. After his departure, he would figure in another part of the cold fusion controversy, the question of tritium (see Chapter 14).

Even further to the east, another "confirmation" was unfolding at yet another news conference, held within hours of the one at Texas A&M. Associate professor of nuclear engineering James A. Mahaffey at Georgia Institute of Technology said, "We think we have confirmed the Utah experiment to prove cold fusion." The five-person Georgia Tech research team, like Fleischmann and Pons, had detected neutrons—at least they thought they had. Their instrument picked up 600 neutrons per hour, while the background count was found to be only 40 per hour. It seemed indisputably the mark of fusion. Team member Dr. Bill Live-

Dr. Kenneth Marsh (l), Professor Charles Martin (ctr), and Mr. Bruce Gammon (r), one group from Texas A&M University that on April 10, 1989, reported the first U.S. confirmation of excess power in a Fleischmann-Pons-type cell. The team quickly realized that an overlooked electronic problem completely invalidated its results; the group quietly withdrew its widely heralded claims. (Courtesy Texas A&M University)

say, a metallurgist, beamed, "There's no question it's fusion. . . . I still don't believe it, even though I see it." Said Mahaffey, "It happened so soon, we thought it was an equipment malfunction." That was ominously portentous, because within days the GIT team would realize to its chagrin that its neutron counter had registered false readings due to a well-known temperature sensitivity that was overlooked.

Getting happier by the hour, Pons reacted to the news from the two other universities with uncharacteristic excitement. He told the *Salt Lake Tribune*, "I always knew this would occur in time, without doubt." His and Fleischmann's own skepticism about their results, closely held for five and a half years, seemed to be coming to an end. Fleischmann and Pons were particularly happy to hear about the Georgia Tech neutron measurements, because they well knew the difficulty and finickiness of the neutron counter's art. Like the physicists, they were puzzled that their relatively high number of neutrons–40,000 per second–did not jibe with the rate of heat production under the assumption of conventional fusion reactions. To say that they felt relieved would have been an understatement.

Pons told the press, "This means there will be an increased effort to determine what the source or nature of this new reaction is. The reaction is most likely nuclear. It can't be explained in any other way, but the scientific community will have to try to understand the process involved." Between answering all manner of inquiries, Pons himself was squeezing in every opportunity to work on a new set of his own experiments. As others raced to replicate the Utah work, he had begun running a new series of 18 experimental cells.

Several days after the announcement, Fleischmann had flown to Harwell Laboratory in England to give its researchers advice on setting up an experiment. Before leaving he told *Science* magazine, "It has become apparent gradually that although we do in fact see the accumulation of tritium and we do see neutrons, they are only a small part of the overall picture. So there must be other processes going on. The unsatisfactory part about this whole business is that we do not yet really know what those processes can be." Contrary to a view that gradually crept into the press—that Fleischmann and Pons were claiming to have duplicated at room temperature "the reactions that powered hydrogen bombs and the stars"—both were very open in admitting their ignorance. The only point they did not concede was that some kind of nuclear process (perhaps one heretofore unknown) had to be causing the effect. Too much energy was coming out for it to be otherwise, so they said.

✳ Utah Money

A financial story paralleling the scientific one was emerging. Even before the good news from Texas A&M and Georgia Tech, Governor Bangerter's hastily called special legislative session met on Friday, April 7, to act on the financial future of Utah-grown cold fusion. Sentiment was building to appropriate funds, and schemes were even discussed to work around the budgetary cap that affected the balance of the fiscal year (ending June 30). Not every legislator favored quick funding, however, and the citizen's group calling itself the Utah Tax Limitation Coalition opposed it. In a state with about three-quarters of the population belonging to the Mormon Church, there was also the delicate issue of Brigham Young University, the Church-operated school where Jones did his work. Some lawmakers wanted to have all Utah institutions, both public and private, share in state research funding, even though a Utah law passed several years earlier prevented that. More to the point, how was anyone to know whether the magic number $5 million was too little or too much? To get that number the governor had simply "off the top of his head" doubled a figure that University officials had bandied about.

Despite the governor's willingness to drop money on the University of Utah, most people realized that a mechanism had to be found to certify that the research was sound enough to merit state aid. Some wanted to wait until the phenomenon was proven by independent confirmation. But the governor told the lawmakers, "There are always people who can suggest a hundred reasons not to do something. I ask you to find one reason why it can be done." Unknowingly, he had put his finger on a key difference in motivation between cold fusion proponents and skeptics in the scientific community. Those who believed that the phenomenon might be real were looking for that single reason, and those who found it fundamentally unbelievable sought to find every explanation for why it could not occur. Both attitudes are part of the essential tension within science when it confronts a difficult mystery.

At the end of a single day of deliberation the vote was nearly unanimous (24 to 1 in the Senate and 72 to 2 in the House). The Utah legislature empowered the governor to appoint a nine-member Fusion Energy Advisory Council to allocate $5 million to cold fusion researchers at the University of Utah, but funds could not be released until the Council voted that adequate scientific confirmation was in hand. The legislation provided that the state attorney general's office would receive $500,000 for legal services connected with the patent process, yet even this money would be withheld pending Council approval.

The patent process was already moving along at a steady clip. Two patents had been applied for even before the announcement and another was being planned. In its rush to establish its position, the University of Utah would within a year send about 10 patent applications to Washington and numerous others to foreign countries. The urgency was understandable, because U.S. patent law gives inventors or institutions only one year from the first origination of an invention to file a patent application. The official approval process may then take over a year and often much longer. In the coming year many other institutions and individuals would file for dozens of cold fusion patents.

The University of Utah was planning to market whatever came out of the research through nonexclusive licenses to use the patented technology and there was more than enough commercial interest. Several hundred companies—including General Electric and Westinghouse—and individuals would eventually sign nondisclosure agreements with the University of Utah. They were coming in droves already—both large and small firms, domestic and foreign. A group from Brazil flew in unannounced seeking information. Electrochemist John Bockris at Texas A&M, a longtime colleague of Fleischmann's, told *Science* magazine that if the discovery panned out, ". . . the University of Utah will be the richest university in the country in five years."

Elsewhere, prospects seemed not so sanguine. Physicist Richard Muller at the Lawrence Berkeley National Laboratory was reported to have bet 25-to-1 on at least four, $1 propositions that the University of Utah work was plain wrong. At MIT, one scientist casually offered million-to-one odds and found an immediate taker, but wisely backed off the gamble. But Utah had placed its bet on cold fusion. Certainly there was fear that the whole episode might turn out to embarrass the state, that it would become "just another Utah mistake"—like the problematic artificial heart and prematurely announced nonworking Utah X-ray laser. In the end, the stakes were simply too high to ignore. "A risk worth taking," the *Deseret News* editorialized, the day after the affirmative vote in the legislature.

The quest to explain cold fusion was becoming a thorn or an obsession for skeptics and believers. Scarcely a day went by without some new scientific report (or rumor) from the cold fusion front, as hundreds of laboratories around the world attempted to discover what was or was not going on in "fusion jars."

Seven thousand miles from Salt Lake City, the island of Sicily in the Mediterranean became one of the more exotic venues in the early days of cold fusion. The Ettore Majorana Center for Scientific Culture arranged a conference attended by about 100 scientists that met on April 12 to consider cold fusion. Steve Jones came through the invitation of Nobel laureate in physics T. D. Lee. Members of CERN (the European Center for Nuclear Research), including Nobel laureate Carlo Rubbia were there too.

Despite this open discussion, physicists by and large remained unconvinced. Many scientists had the impression that the cold fusion claims were almost literally like modern-day alchemy. Whether it was Fleischmann or Jones, both groups were suggesting that they had transmuted deuterium into some other atom—some form of helium or tritium, or whatever. They claimed to have done it not with high-energy accelerators or a powerful fusion machine, but with simple chemical apparatus. It simply didn't ring true to the ears of the scientific establishment. As astronomer Carl Sagan had said in a quite different context—the field of UFO reports and investigations—"Extraordinary claims demand extraordinary proof," and that is precisely what researchers were hoping to see before they believed the remarkable reports of fusion. For their part, cold fusion proponents might have used (and in fact later *did* use) a countering argument that is commonly quoted by radio astronomers searching for messages from extraterrestrials: "Absence of evidence is not evidence of absence."

5 Dallas and Beyond

We've looked at this work for five years. We cannot find any fault with it. At that time you have to say, "Well, if it's correct, it is important, therefore we have to publish it—publish and be damned."

Martin Fleischmann
Interview on Radio Netherlands, April 1989

They don't have to believe me. I will just go back to the lab, do my experiments, and build my power plant.

B. Stanley Pons, *Deseret News,* April 21, 1989

I think that a strange piece of research will strike people as being strange. You have to get used to it. You have to live with it. It's like an old bicycle. You have to grow old with it.

Martin Fleischmann
Congressional testimony, April 26, 1989

AS APRIL PROGRESSED, the honeymoon was ending and an angry sea of skeptics was rising. Few scientists were willing to believe that the heat allegedly coming from "cold fusion" cells was anything but a mistake, a misinterpreted measurement, or a quirky new chemical phenomenon. Pons and Fleischmann were down a billionfold in the number of neutrons necessary to satisfy physicists that the heat was of nuclear origin. Of course, if they had gotten that blizzard of neutrons, they and many of their coworkers in the Chemistry Building would have shared the fate of a scientist killed by a massive flux of neutrons in 1945 during the development of the atomic bomb—an agonizing death.

But not all physicists were doubters. Professor Runar Kuzmin, a physicist at Moscow University, upon hearing of the Utah experiments, set up some tests himself. In a measure of how genuine *glasnost* had become, the Soviet news agency, Tass, reported on April 12 that physicist Kuzmin had successfully reproduced the Utah work. He claimed not only to have detected neutrons, but to have observed heating of the heavy water to boiling. Convinced that he was correct, despite doubts in the West, Kuzmin dryly remarked in the Tass report, "In theory, nuclear fusion at room temperature can be used as a source of energy, but a whole series of experiments on the physics and chemical mechanisms would have to be conducted."

Among the doubters in the West and East were not only physicists, but chemists. One from neighboring BYU, Professor Lee Hansen, dispatched a letter to the *Journal of Electroanlytical Chemistry*, explaining why he believed the Fleischmann-Pons effect was either a misinterpreted chemical reaction, heat coming through the electrical wiring, or heat actually being produced by the so-called "Peltier" effect—something that happens when a voltage is put across two dissimilar metals. This was the common litany at the time. (Throughout the cold fusion saga, however, chemists were as a rule more disposed to believe Fleischmann and Pons than were physicists.) Though the media may have touted too prominently the age-old chemist versus physicist rivalry as a source of the growing "fusion confusion," there was some truth to the notion. In many areas, chemists do think, work, and react in very different ways than physicists. They are accustomed to considering systems whose complexity and inscrutability is closer to that of living organisms. Physicist are disposed to seek simplicity and elegant, fundamental laws to explain their work and aren't happy with the "messiness" of chemistry. Furthermore, in their heart of hearts physicists may look down on chemistry as only a "subset" of their more fundamental discipline, which lays down the rules that govern the chemists' complexes of atoms.

✳ Texas Chemistry

This chemist-physicist difference was nowhere more evident than in the reception accorded Stanley Pons at the 197th Annual Meeting of the American Chemical Society (ACS) in Dallas. It occurred, April 12, 1989, the same day that Moscow was telling the world of Kuzmin's success. Fleischmann, though invited to Dallas, was on the island of Sicily, telling that part of the world about cold fusion. Compared to the Baltimore gathering of physicists coming up in May, which would all but roast Fleischmann and Pons alive, Dallas was a love-fest and heroic celebration of the duo's alleged achievement.

If the 1987 all-night high-temperature superconductivity meeting of the American Physical Society (APS) in New York was the "Woodstock of Physics," Dallas was a "Woodstock of Cold Fusion" for chemists. The part of the meeting featuring Fleischmann and Pons was hastily organized by scientist Valerie J. Kuck of AT&T Bell Laboratories, who by coincidence had arranged a special high-temperature superconductivity session at the ACS meeting almost exactly two years earlier. Many people were already beginning to make comparisons between the two highly publicized scientific "breakthroughs."

Between 7,000 and 8,000 chemists packed the Dallas Convention Center Theater—a 10,000 seat circular arena. Many came to the meeting specifically for this well-advertised session, "Nuclear Fusion in a Test Tube?" Reporters were not allowed to photograph or videotape during the three-hour scientific session, but a press conference did follow the main event. (Later, the ACS would offer for sale a videotape of the meeting—$450!) Pons was asked by one reporter, "Are you Prometheus, Pandora, or Piltdown man?" "No comment," was his reply.

There were five main speakers, including someone to keep the fusioneers "honest," Dr. Harold P. Furth, then director of the Princeton Plasma Physics Laboratory, the first technical speaker on the program. President of the ACS, Clayton F. Callis, introduced the session, calling it a "precedent-setting event," both in attendance and general interest. Remarking about the billions of dollars that had been spent on hot fusion research, Callis said, "While much has been learned about plasma physics, and while much progress has been made, the goal has remained elusive and the large, complicated machines that are involved appear to be too expensive, and too inefficient to lead to practical power. Now it appears that chemists may have come to the rescue." His remarks, clearly not calculated to endear him to physicists, were met with great applause and cheering by the audience of proud chemists. Any physicist present might have laughed inside, however, because Callis uttered the classic mispronunciation of nuclear—"nuc-u-lar." Dr. Kuck received a hearty laugh and enthusiastic applause when she requested that "members of the press refrain from asking questions" during the question and answer period for the thousands of assembled technical people.

Furth enlightened the chemists with the basics of hot fusion theory and technology. He admitted that it was "conceivable that the branching of reactions would be different in cold fusion from warm fusion, but it is a little surprising because the only experience previously with cold fusion is muon-catalyzed fusion (Chapter 6), and for that the behavior is exactly as for warm fusion, and there is no evidence that the helium-branch goes." He said, "Nuclear physicists at the moment are puzzled by the aspects of the cold fusion experiments, which is that the neutrons seem to be down by a very large factor—variously reported as a billion or more relative to what one would expect from the energy release. And that is troublesome, because that would call for some unusual prominence of straight helium production, perhaps with the lattice carrying away the excess momentum, rather than the gamma ray. And that's

Dr. Harold P. Furth, Director of the Princeton Plasma Physics Laboratory at the time of the cold fusion announcement, an outspoken skeptic who demanded—and eventually got from the experiments of cold fusion proponents—null results from light water cells and positive results from heavy water cells. (Courtesy Princeton Plasma Physics Laboratory)

puzzling first of all. Secondly, what's more puzzling is [why] the usual reactions [are] suppressed if the deuterons are so cozy that they are doing [making] the helium.

"Instead of everybody sticking palladium into heavy water, some fraction of the population should be sticking palladium into 'half-and-half'—light water and heavy water. The point is, of course, to look for the 5.5 MeV gamma ray of the p-D [proton-deuteron] reaction alongside this 2.45 MeV neutron of the D-D reaction.

"I think really the most fundamental thing that needs to be done at the moment isn't meditations by nuclear theorists on whether this is crazy or plausible, or whatever, but a little work by the American Chemical Society, where you repeat the experiments producing excess heat in heavy water—reproduce that exactly using ordinary light water, and compare the excess heat. I think that's really the key to it all and we are waiting anxiously to see it." That evidence later came in spades, but going against his pronouncement in Dallas, Furth's disbelief in cold fusion did not change, probably because he did not believe the evidence.

Dr. Alan J. Bard, an electrochemist from the University of Texas at Austin, addressed the "Fundamentals of Electrochemistry"—information for the uninitiated, a short course in electrochemistry in 20 minutes. "I'd like to finish by thanking Stan Pons and Martin Fleischmann," said Bard. "They've given the electrochemists in this world a very interesting few weeks, sleepless nights, forced us to learn more about nuclear reactions and radiation detection than we ever thought we would be interested in. I think that physicists have suddenly discovered electrochemistry, which is all to the good. I think the most important thing I've noticed from myself and my colleagues is it's forced more older professors and senior scientists back to the laboratory and behaving like graduate students than has occurred in the past 20 years." Professor of Chemistry and Chemical Engineering Ernest B. Yeager of Case Western Reserve University particularized electrochemistry to that of the endlessly mysterious hydrogen-palladium system, which includes the deuterium-hydrogen system, because deuterium *is* hydrogen.

Pons's appearance in Dallas was buoyed, of course, by the recent good news from Georgia Tech and Texas A&M, and from Hungary. As Dr. Kuck introduced him, she provided high drama: "We were just informed by a Dallas radio station that the University of Moscow has just announced that it has successfully repeated the Pons-Fleischmann experiment." Pons referred jokingly to the Utah apparatus as the U1 Utah Tokamak. He flashed a slide showing a cell within a plastic basin and he quipped, "We chose Rubbermaid™ early on because we didn't have much money." Not everyone was having success, however. The word was "nothing yet" from MIT and many other laboratories. On the day of the Dallas meeting, a report came from the *San Jose Mercury News* that the initial attempt at replication by researchers at Lawrence Livermore National Laboratory ended with the experiment "blowing up," littering the laboratory with shards of glass. Perhaps it was only a hydrogen (deuterium) gas explosion coming from the test cell, but readers of this explosive news may have also recalled the warnings by Fleischmann and Pons about the hazards of runaway cold fusion.

Harold Furth said that "World-class physicists will not take the Utah findings seriously until the process is explained and demonstrated with control experiments." Whether he realized it or not, he implied that no scientist should "take seriously" the appearance of any remarkable new phenomenon without first being able to "explain it." Theory was beginning to take precedence over experiment.

In what was to be the first public airing of a theory to account for electrochemical fusion, chemist Katherine Birgitta Whaley of the University of California at Berkeley spoke before the thousands of chemists. She suggested that deuterons could act like a class of nuclear particles called *bosons* to provide electrical "screening" that would facilitate cold fusion. About her boldness in advancing a theory to explain an as yet unproved phenomenon, not a peep of protest was heard, in contrast to the way later cold fusion theorists were challenged.

Meanwhile back in Washington, the Utah congressional delegation was trying to orchestrate a visit to the University of Utah by some of their colleagues to witness a cold fusion demonstration. Senator Jake Garn (R-UT) held a news conference in his office and proposed an April

29 demo at the University of Utah. Representatives Wayne Owens (D-UT) and Howard C. Nielson (R-UT) were there too. Representative Owens had earlier proposed, but had not yet submitted, a bill to establish a National Fusion Research Center at the University of Utah. Along with the Secretary of Energy James D. Watkins and then presidential science advisor Dr. William H. Graham, the plan was to have President Bush's Chief of Staff, John Sununu, come also.

Sununu, the former governor of New Hampshire, holds a Ph.D. from MIT in mechanical engineering, had been an entrepreneur and an engineering professor at Tufts University, and was known to be following cold fusion developments with great interest—the informed interest of an engineer with experience in the nuclear field. Governor Sununu, a great friend of the nuclear industry, had been a driving force to get New Hampshire's Seabrook Station nuclear power plant operating, even as it was besieged for so many years by antinuclear forces. There is little doubt that he was discussing the cold fusion storm with President Bush at this time.

✳ Beyond Dallas

By tax-paying time, April 15, there were a host of new developments. On April 14, a very embarrassed and disappointed Professor James Mahaffey held a press conference at Georgia Tech to retract his widely trumpeted neutron claims of only a few days earlier. "What can I say?" he asked. The sensitivity of the group's neutron counter to rising temperature was to blame. When Mahaffey announced his neutron "confirmation," he had said, "It's like being there when fire was discovered." The embarrassing error in the work by the Charles Martin group at Texas A&M remained submerged.

The fruits of weeks of theorizing about mechanisms for cold fusion began to appear. University of Utah chemistry professors John T. Simons and Cheves T. Walling announced that they had submitted a theory to the *Journal of Physical Chemistry.* They released only the barest essentials to the public for fear of being preempted before the paper's acceptance. They believed that the end product of deuterium fusion in palladium was good old toy-balloon helium-4 (^4He). This main reaction would be neutronless, thus finally explaining why Pons and Fleischmann were still alive. Many others had been thinking about this explanation too, including Nobel laureate Julian Schwinger, who because he could not yet contact Pons, resorted to making his suggestion in a letter to the *Los Angeles Times.*

Ordinarily D-D fusion leading to ^4He produces a nasty, highly penetrating gamma ray, but the Walling-Simons theory purported to find a way to get the resulting energy into the palladium lattice. Walling and

Simons invoked the well-established physical process of *internal conversion,* in which an electron from the ^{4}He atom is struck by the gamma ray from the nucleus and is then boosted to high energy. Energetic electrons expected from the Walling-Simons mechanism would slow down—"thermalize"—within the palladium, but would also penetrate inches into any surrounding water, giving off a bluish glow from light called Cerenkov radiation. That prompted Pons to try a simple impromptu experiment: turning off the lights in his laboratory. Alas, no blue glow, but was it dark enough? The Walling-Simons theory made another prediction: Deuterium might react with the trace amount of ordinary hydrogen in heavy water and produce helium-3 (^{3}He), thus explaining why a tiny amount of heat-producing fusion might occur even in ordinary water with its trace amount of deuterium. Did this motivate the evasiveness of Fleischmann and Pons—their unwillingness to agree that a light water cell would constitute an appropriate control experiment?

Put in a defensive posture, Pons was certainly premature in announcing some supporting evidence for the Walling-Simons theory, but on April 17 he said at a press conference that he and his associates had in recent days detected helium-4 in a cold fusion cell. Simons told the press, "At first it seemed too good to be true, but I don't believe you can produce that much helium without something nuclear going on." This data would remain quite ephemeral for many more months, owing to the great difficulty in making definitive measurements of ^{4}He. (Because of similar masses and charges, trace amounts of ^{4}He are readily confused in some kinds of instruments with deuterium gas, D_2.) Pons put himself needlessly further out on a limb by suggesting that the ^{4}He being detected was about the right amount—a trillion helium atoms per second, according to the Walling-Simons mechanism—to explain the levels of heat being measured. In the same breath, Pons alluded to talks with some 60 laboratories that he claimed were getting positive results.

That week, the world also had learned that Associate Professor Peter L. Hagelstein in MIT's Department of Electrical Engineering and Computer Science had burned the midnight oil over many days to generate four theoretical papers to explain cold fusion. He submitted all four to *Physical Review Letters*, where they were never published, no matter how numerous his revisions and improvements.

Hagelstein, an expert on laser physics, whose renowned work on the X-ray laser of "Star Wars" fame occurred under unusual circumstances, had come up with something so new and remarkable, that it was almost as startling as the claim of cold fusion itself! He proposed that fusion was occurring "coherently" within the palladium lattice, whatever that meant to a world still reeling from an overdose of heavy water and palladium. Initially, Hagelstein was positing that one reaction product in cold fusion could be ^{4}He, but there would be tritium, and neutrons too as side-reaction products. His theory was still in its formative stages and evolving rapidly at the time, but news of it came to light that week through an MIT news release with official approval. (More on this new theoretical direction later.) MIT also announced that it had filed several patent applications for technology based on Hagelstein's ideas.

This caused no end of confusion, because no matter how specific the news release had been about patents being *on technology based on theories*, many wondered how MIT could have the chutzpah to attempt to patent theories. Pons himself, hearing of Hagelstein's work, told the *Deseret News*, "I didn't know you could patent theories." In its periodically appearing "Fusion Scorecard" chart, the *Deseret News* truthfully but somewhat mischievously listed MIT under "Not confirmed (but filed patents)." East-West rivalry—perhaps better described as both coasts against the interior—was in full bloom. University of Utah's James Brophy, critical of MIT and the national press, told the *Deseret News*, "There was no peer review and they produced a press release with no information. It is worse than what we did, and I wonder if anyone is damning them for it." And note Professor Walling's reaction in the *Salt Lake Tribune* to Hagelstein's theory that mentioned ^{4}He, "It sounds from what you're telling me that he's as smart as we are." Or from the University of Utah College of Mines, Dean Milton Wadsworth, "I know these guys. They're not any different . . . your average Joe Professor isn't any better than who we have." On the other hand, Brophy expressed gratitude that at least *someone* at MIT found reason to take cold fusion seriously.

On the financial front, Pittsburgh-based Westinghouse Electric Corporation, a well-known producer of nuclear power technology, was reported to have signed a nondisclosure agreement with the University of Utah to evaluate its cold fusion work. Without public explanation, though it may have had something to do with "attached strings," Pons shocked the DOE by announcing that he was turning down its proffered 18-month $322,000 grant. Instead, he would continue to receive money from the Office of Naval Research, which had been funding his other non-cold fusion work. This, on top of the Utah money that would soon come.

On April 14, 1989, Governor Bangerter announced the nine individuals who would participate in the newly legislated Fusion Energy Advisory Council, the body that would control the spigot of Utah's cold fusion funding. The Council members included only two scientists, Professor Wilford Hansen (professor of physics and chemistry at Utah State University) and Karen W. Morse (Dean of the College of Science at Utah State University), as well as state science advisor Randy Moon.

The other members were four present and former corporate executives, an attorney, and an accountant.

At this time the University of Utah and Los Alamos National Laboratory were negotiating to come to an agreement on experimental cooperation. The process dragged on for weeks and never was fruitful, because legal considerations on the patent front apparently got in the way. By now, Utah had filed for five patents and was preparing to launch four more applications. This did not stop Stan Pons from traveling to Los Alamos to confer with scientists there. Los Alamos would have been a prized collaborator, given its great expertise in neutron and tritium measurements. There were unconfirmed reports surfacing in the press that Los Alamos researchers had, in fact, verified the Fleischmann and Pons experiment. They had not done so by this time, though in later months Los Alamos would produce some of the most compelling evidence for nuclear effects in cold fusion.

Other less well known organizations were getting into the patent business too. Researchers at Washington State University mimicked the "big boys" and announced that they too had filed a patent application. Like MIT's, it was said to cover technology based on an unverified cold fusion theory of Associate Professor of Physics Gary S. Collins. But Utah was still king in this realm, having earmarked about $500,000 to "paper the file" with patents.

Back at his lab, Pons had already set up many new experimental cells and was letting the word out that heat production was now perhaps 10 times larger than had been reported earlier. Two of Pons's University of Utah professorial colleagues, Milton E. Wadsworth, Dean of the College of Mines, and Richard W. Grow of electrical engineering were trying to duplicate the Fleischmann and Pons work. Also involved in their own separate effort were University of Utah metallurgy professors, J. Gerald Byrne and Sivaraman Guruswamy.

✱ On the Defensive

More bad news for Fleischmann and Pons came when *Nature* announced on April 19 that it would publish Jones's paper in its April 27 issue but would not be publishing their work. The paper that Jones submitted required "very big changes," according to *Nature*, modifications that Jones did in fact accede to. Pons and Fleischmann, on the other hand, were unwilling or unable to provide the additional words and data that *Nature* demanded, and so withdrew their paper on April 15. Though they held open the possibility of submitting future manuscripts, *Nature's* editorial position would quickly grow so negative about reports of excess heat, that future publication would become a moot point.

On April 20, investigators at Drexel University in Philadelphia, Michel Barsoum and Roger Doherty, dropped another piece of negativity into the public stew, saying that they had attempted electrochemical cold fusion with both light water and heavy water and found no difference between the two experiments.

A report coming out of China, a country that within two months would explode in the violence at Tinanmen Square, said that investigators at the Chinese Academy of Sciences were doubtful about cold fusion after trying and failing to get excess heat. A Duke University physics professor had satisfied himself that it must be an unknown chemical effect. North Carolina State also reported negatively.

Counterbalancing this downturn, however, was an intriguing report coming from Stanford University on April 18. Professor of Materials Science Robert Huggins had done the Fleischmann-Pons experiment with both heavy water and light water and found that the heavy water cell produced excess heat, but the light water cell did not. "While what we've seen is not chemical, it is evidently not a conventional fusion reaction either, but we don't know what it could be," Huggins told the press. There were more positive reports of various kinds: from the University of Washington, a Czechoslovakian group, a South Korea team, and Portland State University. At the University of Florida at Gainesville, Professors Glen Schoessow and John Wethington claimed to have produced significant levels of excess heat and tritium as well. Their work for a long time remained cloaked in secrecy, but they steadfastly maintained that they had achieved high power production.

On April 18, physicist Francesco Scaramuzzi and his colleagues at the Italian National Agency for Alternative Energy at Frascati announced a dramatic new finding. They had measured low-intensity bursts of neutrons coming from a metal chamber pressurized with deuterium gas and filled with titanium metal fragments. The neutrons appeared after the chamber had been cooled to the temperature of liquid nitrogen and was then allowed to warm to room temperature. This would emerge as one of the most compelling kinds of cold fusion experiments, because of the high quality of the detectors and control experiments arrayed to verify these neutrons. Others have now reproduced the Frascati results many times.

The proliferating fusion confusion simply became too much to ignore for Admiral Watkins, who met in late April with President Bush and Nobel laureate chemist Glenn Seaborg to discuss cold fusion. On April 24, Watkins issued marching orders to the 10 DOE-operated national laboratories: Find out what all this cold fusion talk was about. And he set up an illustrious panel of scientists and engineers to find out whether there was anything cooking. To that end, Los Alamos National Laboratory (LANL) announced that it would hold a cold fusion workshop in Santa Fe, New Mexico, May 23–25, 1989. Despite numerous unanswered questions (*unanswered* by the panel's own admission—even *if* cold fusion was a mere mistake or misinterpretation), the recommendation by the panel in July, ratified in November, was to spend almost nothing to chase the ephemeral cold fusion phenomena. Indeed, at most one or two million dollars annually spread over many hundreds of eager researchers around the country would be very little.

At about the same time, the MIT chemistry-plasma fusion team was beginning to think about pulling the plug on its efforts to find evidence of electrochemically induced cold fusion (the experiments terminated in late May). Other groups had just about thrown in the towel too. Steve Jones was saying to the press that he felt a "moral responsibility" to raise public skepticism about reports of excess heat. Dr. Paul-Henri Rebut, the director of the hot fusion JET tokamak at Culham Laboratory in England told the *Daily Telegraph*, "I am not God, and I don't claim to know everything in the universe. But one thing I'm absolutely certain of is that you cannot get a fusion reaction from the methods described by Martin Fleischmann and Stanley Pons. . . ."

Even a motley crew of environmental and antinuclear activists, who logically should have been instant fans of cold fusion, attacked it. Larry Tye of the *Boston Globe* elicited some of their comments: Amory Lovins said, "Most of the costs of fusion will be the stuff you wrap around it to get electricity, from the turbine, to the plant site, to the health physicists and other cleanup services you need, all of which will make it at least as expensive as fission. The right place for a fusion reactor is where we have one—in the Sun, 93 million miles away." Biologist Barry Commoner of Queens College said, "Putting [a cold fusion device] in the basement or car is nonsensical. . . . As long as radiation is involved, you need major controls." Extreme antitechnologist author Jeremy Rifkin gasped, "The fusion findings are the worst news that ever happened. Right when we are beginning to develop a global awareness of the problems of global society, here come some scientists saying we don't have to deal with these problems."

There were so many conflicting rumors and reports at this time that it was hard to know what to believe. A report surfaced that the Gandhi Center for Atomic Research in India had had some success in mimicking the Fleischmann and Pons work about mid-April, using a titanium rather than a palladium cathode. Pons was talking to the press about being able to produce a scaled-up "device" that would produce substantial power—a unit that was to be about six inches in diameter and a foot long. Pons told the *Salt Lake Tribune* that he was finding an elevated amount of helium-4 in his cells at a level a million times over what might have been present initially as natural contamination. This encouraged him to think along the lines of the Walling-Simons theory to explain his results. He was emphatic about the excess heat being real: "You could consume the palladium, the platinum, the glass, all the water, everything . . ." and still not come up with as much excess energy as they had observed. The contrast between the contentions of Jones and Pons could not have been greater.

✳ Cold Fusion Goes to Washington

Congressionally speaking, Utah came to Washington, rather than the other way around. Washington would take its time and wait until mid-May, 1989, to visit Utah, putting off its earlier plans to come in late April. Utah's other senator, Orin Hatch, had just visited with Pons in his lab and had brimmed with enthusiasm, hyperbolically saying that the research center would make Utah "the intellectual idea center of the world." He said cold fusion "could make Utah one of the wealthiest states in the world." (*Deseret News*, April 24, 1989.) On Thursday, April 26, Pons and Fleischmann arrived at the imposing congressional hearing chamber to testify before the House Science, Space, and Technology Committee, chaired by Robert A. Roe (D-NJ). Fleischmann had come from his European campaign, and Pons straight from his laboratory. Fleischmann had just been voted a full-fledged research professor in the University of Utah Chemistry Department.

The intrepid duo were accompanied on this bold venture by University of Utah President Chase Peterson, James Brophy, and other state functionaries. Also along on the Washington expedition was an experienced business consultant from Rhode Island, Ira Magaziner, who played up the age-old theme of "preserving U.S. international competitiveness." Their objective: Convince Congress that cold fusion was real and warranted immediate funding and the setting up of a national research facility at the University of Utah. Utah wanted $25 million "seed money" to begin an effort that was expected to capture a complementary $75 million from individuals and companies. Utah Congressman Owens was planning to introduce a bill to fund the center.

Surely the federal government could spare a few "crumbs" of financial aid for a promising new source of energy—one that might even be, shall we say, infinite. Of course, the money didn't all have to go to Utah, even though that was what was being requested. It could have been spread around the country—there was no dearth of eager investigators at major universities and companies. Wasn't Washington, D.C., the place whence came billions of dollars for invisible B-2 stealth bombers at $500 to $900 million per copy and billions of dollars annually for SDI (Strategic Defense Initiative) research? Had not Washington been funding the hot fusion program to the tune of about $500 million per year, a surely far more inspiring quest than even the latter glamorous

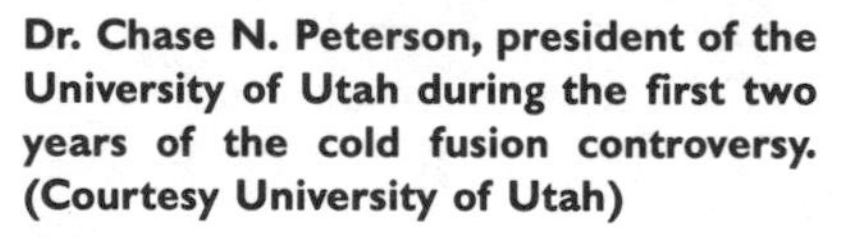

Dr. Chase N. Peterson, president of the University of Utah during the first two years of the cold fusion controversy. (Courtesy University of Utah)

Dr. James J. Brophy, vice president for research of the University of Utah when cold fusion was announced. (Courtesy University of Utah)

items? If distant Utah was thinking of spending $5 million for cold fusion, might not Uncle Sam come up with, say, $25 million—the balance of up to $100 million to be raised by an alert and vigorous private sector? Mind you, the Office of Naval Research had already begun to extend its funding of Pons to cover cold fusion work, and he had actually turned down the DOE on its offering of $322,000. You may know the tale—the $25 million was not to be.

At 9:45 A.M., Room 2318 in the Rayburn House Office Building was packed. Chairman Roe in his prefacing remarks held up the promise of a golden age: "Today, we may be poised on the threshold of a new era. It is possible that we may be witnessing the cold fusion revolution, so to speak. If so, Man will be unshackled from his dependence on finite energy resources." The ranking Republican member of the Committee, Robert Walker of Pennsylvania, sang the tune of small science, which Pons and Fleischmann had come to personify: "If this discovery is fully proven, it will show once again the importance of supporting a vigorous small science enterprise in this period of large engineering and science projects. . . . If the initial results are verified, it is essential that we do everything we can do to develop the promise of cold fusion." He chilled the hot fusion scientists who were present: ". . . I was pleased that the committee's Energy Research and Development Subcommittee accepted my amendment during its April 6 markup authorizing that $5 million be redirected from the Magnetic Fusion Program into the Basic Energy Sciences activity, specifically for room temperature fusion." He said he would move in the direction of upping that to $25 million. These high stakes were really making the hot fusion people sweat, but despite their skepticism, they could not be absolutely sure that cold fusion was a mistake or an artifact—a misinterpretation of something much more prosaic.

Congressman Owens introduced Fleischmann and Pons with soaring words: "The event, the possible achievement of solid state fusion, or the so-called cold fusion, is nothing less than a miracle with all the elements of a miracle—surprise, exaltation, disbelief, and skepticism." Pons led the cold fusion charge and recounted the tale of their amazing discovery. He had brought a mock-up of their experiment to show the congressmen and held forth with technical slides to buttress his case, explaining the science and technology of their special electrochemistry in some detail. He explained the "competition" of two paths for the deuterium—either release as D_2 gas bubbles at the palladium rod surface or deep penetration into the palladium. Pons drew a mental picture to prompt associations with hot fusion, ". . . we end up having a low-temperature plasma inside the metal instead of atoms or molecules of deuterium." Palladium can dissolve within its structure a staggering amount of deuterium (or hydrogen) without forming gas bubbles within the lattice. With the palladium lattice, nature supposedly performs a wonder that our most sophisticated technology is incapable of doing otherwise. Pons said, "If you were to try to obtain that same voltage by the compression of hydrogen gas to get that same chemical potential of 0.8 volts, you would have to exert a hydrostatic pressure of a billion, billion, billion atmospheres. . . ." Pons estimated the effective confinement time for the deuterium atoms to be 600 years! Ergo, he and Fleischmann had tamed fusion.

Then came the neutron evidence, from measurements of gamma rays from neutrons hitting a surrounding light water bath, data on tritium, and the heat measurements, calorimetry. The bottom line: ". . . the excess heat liberated is of such a magnitude that it cannot be explained by any chemical reaction." But even more wondrous: "The heat generation continues indefinitely until the cell is turned off. . . ." This time he fully owned up to the mystery of cold fusion: The amount of energy coming out was about a billion times more than could be explained with conventional d-d fusion giving those same measured levels of neutrons and tritium. "So apparently there is another nuclear reaction or another branch of the deuterium-deuterium fusion reaction that heretofore has not been considered, and it is that that we propose is, indeed, the mechanism of the excess heat generation."

Fleischmann took over and posed the conundrum with which experimenters all over the world were now wrestling, ". . . the experiment . . . is superficially simple, but it is actually quite difficult to carry out, because you have to go through a process of optimizing the experiment such that you will make a significant observation." He addressed what was foremost on everyone's mind; how could the process be scaled up? It was clever "hand waving," as engineers politely say about projections or calculations that are somewhat shaky: "A four-millimeter rod would give about eight times as much energy out as we put in." If the hot fusioneers could talk about scaling up their fusion donuts amid uncertainty, Fleischmann felt equally justified in scaling his fusion rods. And just as the hot fusioneers did, Fleischmann touted the possible benefits of using tritium and deuterium mixtures instead of deuterium alone. As he said, even though it was admittedly speculation, they were "guided by the conventional approach to nuclear fusion. . . . Our work was not just a shot into the dark, as people believe. We were guided by reasonable theoretical formulations of what might be taking place." Fleischmann proposed a goal, which to this day remains elusive for heavy water cells, but which may be getting closer: "A bench-top demonstration of a device which gives out very much more energy than you put into it." (An analogous electrochemical cold fusion cell, based on molten materials, seems now to produce spectacular multiwatt levels of excess power. See Chapter 15.)

To critics who questioned the method of announcing cold fusion to the world, Fleischmann said that their initial paper was in the nature of a preliminary note to a journal—a commonly accepted form. He agreed that there were perhaps insufficient experimental details, but they could be had by requests via fax, phone, and so on. Easier said than done for the thousands who were clamoring for information. Pons sought to contrast the methods used by chemists to disclose their latest findings, with those of physicists. He maintained that early published announcements of important results that were intended to appear later in more refined form were standard practice in chemistry. Physicists, he said, required a broader peer review process first.

Fleischmann was adamant that they had not told the media that *Nature* would also be receiving a shortened version of their *Journal of Electroanalytical Chemistry* paper. (Jerry Bishop of *The Wall Street Journal* told me that it was his early discussion with others about forthcoming possible publication in *Nature* that led to the erroneous disclosure.) Hitting back at *Nature* for suggesting that they had not replied to reviewers' criticisms, and asserting that they had sent a 19-page rejoinder back to the magazine, Fleischmann explained, "We felt that we had reached the stage where there was no point in writing a short paper on this subject, that we really had to write an extensive paper—extensive set of papers, on the different parts of our work." *Nature* was not the proper venue for these, he claimed. (True enough, because in July 1990 when the *Journal of Electroanalytical Chemistry* published their comprehensive paper on heat measurements, it ran some 56 pages—much too long for *Nature* to publish.)

To questions about reproducibility, Fleischmann claimed that many people were setting up cells with dimensions and parameters that—not surprising to him—did not yield excess heat. Just to be sure no one could question their veracity, Pons announced that researchers from Los Alamos would soon be coming to the University of Utah to measure a working cell and would take it back with them for testing (after it was charged up with deuterium). He also said that other groups in the past week had come to the university and had been satisfied with what they saw.

It would be 10 to 20 years to reach the stage of a commercial product, Fleischmann told the representatives. Neutron emission (if it was real) seemed to be low enough not to warrant concern. Tritium production (if it was real) didn't appear to be much of a drawback to implementation, since it came at such modest levels. Helium-4 would be of course no environmental concern, if it were there at all, but on this point Fleischmann was evasive. Their palladium rods were just then undergoing testing for helium content by their palladium supplier. "I don't want to discuss helium too much at this stage." (Much later it was learned that little evidence exists for helium being found in "active" electrodes at levels needed to explain the excess power.)

They were supremely confident of their results. Pons told the committee: "For five and a half years I think we were our most severe critics, and we are still as sure as sure can be. We produce our data and we believe what we are seeing." Fleischmann was only a shade more circumspect: "I do not know how to interpret our results in any other way than that we have observed a fusion phenomenon. So I'm still totally convinced about our work. But naturally, we shall have to look at everybody else's work as well, including all unsuccessful experiments, and only time will show whether we are correct or not." Fleischmann based this claim, he said, on adding up the total energy coming out of a typical cell over a period of 100 hours or more. The bottom line: The excess energy per cubic centimeter of material in the palladium electrode (5 megajoules per cubic centimeter; alternatively, 1.4 kilowatt-hours per cc) was over a 100 times "any conceivable chemical reaction in the system." Fleischmann did not doubt that if the experiment were run 10 times longer, the total excess would be 1,000 times higher than a chemical process.

It was time to get down to the real business of the committee, money—how much and how soon. Pons and Fleischmann had discreetly left these coarse matters to the more experienced, President Peterson and Ira Magaziner. Peterson, had been president of the University of Utah for about as long as Fleischmann and Pons had pursued cold fusion—5½ years—and was a native of Utah who had a taste of the East during his education at Harvard and the Harvard Medical School. Magaziner, an ivy-leaguer himself (valedictorian of the class of 1969, Brown University) was president of Rhode Island-based Telesis, USA, Inc., "an international consulting firm specializing in corporate strategy for companies and economic development policy for governments, industry associations, and unions." His was not exactly the kind of advice that other scientists looked forward to hearing. Scientists wanted to know about the truth or falsity of cold fusion, not about financial development plans, but the audience today was congressmen and the need was money. Congressman Owens announced after the testimony of Magaziner and Peterson, that Magaziner had "elected to provide his services without charge, a rather unusual circumstance for a person of his caliber and fees."

Magaziner offered the familiar and predictable litany of "Americans invented it, foreigners capitalized on it" stories—television sets, video recorders, microwave ovens, and so on. He proposed that cold fusion be a fresh start for America, "I have come here to ask you for the sake of my children and all of America's next generation, to have America do it right this time." The best way to avoid past losses, he said, would be to have the basic science proceed in parallel with applied research, product development, market research, even manufacturing. If the United States didn't do this, the Japanese would. In fact, they were already off and running, he asserted. "Fusion fever" had gripped Japan with "over a hundred companies" already involved, and Europe was beginning to stir too. He had heard that top people at the powerful Japanese consortium, MITI, were almost literally burning the midnight oil in their quest for a head start in the cold fusion business.

Magaziner wanted money to come from universities, the state, corporations, and Uncle Sam (certainly not necessarily in that order). If Utah had put up $5 million conditionally, the federal government should put up more money. He raised a specter: "If we dawdle and wait until the science is proven, and if we wait for economists to hold symposia on whether Adam Smith would approve of putting public money into it, or if we move cautiously and invest only in basic research, or only in defense applications, and wait for the spin-off, we're going to be much slower off the blocks than our Japanese and European competitors. . . ." He admitted that the whole notion of cold fusion could be wrong and that "we could all wind up with egg on our face," but to do nothing would be worse.

Peterson talked competitiveness too, surely one of the most common buzzwords in 1989. His metaphor recalled Magaziner's thrust: "We may be obliged to build the first floor of commercial development, as well as the second floor of engineering developments, while we're still building the basement of scientific confirmation and enlargement." In short, 'We need money fast.' In his view, Utah was just the kind of place to do this work, and he told of the fresh frontier spirit, perspective, and open spaces it offered. The "remote regions" that he alluded to, "only 25 minutes from the university," might have been interpreted as safe zones for possibly hazardous nuclear experiments. Los Alamos, circa 1945, revisited.

"Twenty-five million would allow us to start the 'onion' growing, with state and private sources," Peterson said. The remaining $75 million would come later. The alarmed skeptical physicists attending the session might have smelled an onion, or was it a skunk, or a rat? The entire University of Utah sponsored research budget was already $100 million (about the 30th ranked university in the country in terms of that funding), so this was to be a doubling—all for cold fusion.

A good deal of the panel's time was spent in banter about patent issues, financial arrangements, and so forth. It was scientist-turned-congressman Bruce A. Morrison (D-CT) who reminded his colleagues of the paramount need to verify cold fusion technically, so the session returned to the science. Professor Huggins was soon to get on a plane to San Diego. There he would present some of his latest heat measurement results to the Meeting of the Materials Research Society. He was next to speak and gave the panel a capsule summary of his findings—even before Steve Jones spoke. Huggins was a convincing and polished speaker.

Though Huggins was well connected to the materials science field from years of experience as founder and director of Stanford's Center for Materials Science, cold fusion had taken him completely by surprise. Huggins noted that he was already working on projects involving hydrogen storage in metals, an electrolytic process related to the newly discovered cold fusion techniques. Huggins launched right into the nub of the argument that Harold Furth had made in Dallas: Any difference in heat evolution between a cell with light water and a presumptively identical one with heavy water might be critical evidence for a nuclear process. Viva la difference! Huggins and his colleagues in his department's "Solid State Ionics Laboratory" had found it.

He put the magic conclusion directly up front, "The results that we have obtained lend credence to the Fleischmann and Pons contention that a significant amount of thermal energy is evolved when deuterium is inserted into palladium, and that this phenomenon is quite different from the behavior of the otherwise analogous hydrogen-palladium system." Not being experienced in nuclear measurements, however, he had no neutron or other radiation data to give further support to this remarkable result.

Huggins' group was claiming about 14 percent excess heat in one of the experiments. The excess power generated in the deuterium-containing cell increased continuously, while the power curve of the light water cell remained essentially flat. Huggins concluded that there must be "an appreciable internal heat generation effect in the case of the deuterium-palladium system, regardless of the presence of any chemical or thermal effects in both systems." The group had observed the phenomenon, he said, in more than one sample, on several occasions, and with different types of calorimeters. Moreover, Huggins asserted, "The magnitudes of the observed effects are comparable to those reported earlier by Fleischmann and Pons and lend strong support to the validity of their results." Huggins told the panel that the now very puzzling reproducibility problem in his view had to do with the preparation of electrode materials, a theme he would consistently echo in weeks to come. He cautiously avoided a firm conclusion that the phenomenon was a nuclear effect, though he did cite the Walling-Simons theory (the helium-4 production branch of d-d fusion) as a possible explanation.

Huggins speculated in a direction that more than a year later would bear fruit in the remarkable elevated temperature cold fusion experiments of one of his former students, Professor Bruce Liebert of the University of Hawaii (Chapter 15). ". . . there is a further important issue," Huggins told them, "whether the same phenomenon can occur at higher temperatures. High-temperature, high-quality heat, is much more useful from a commercial technological standpoint than room-temperature heat. We believe, from our experience in related matters, that this is a very strong possibility. . . ." Higher temperatures mean greater efficiency, thermodynamically speaking.

Next the claims of excess heat production were leavened by the contingent from BYU, with Steve Jones in the lead. Jones's paper was going to appear in the following day's issue of *Nature* and everyone knew that, so he was speaking with a certain degree of approval by mainstream science. He traced his involvement in electrochemical cold fusion work to his pioneering experimental activities in muon-catalyzed fusion since 1982 (Chapter 6)—the "other" kind of cold fusion. Jones made an oblique but transparent attack on the March 23 Fleischmann-Pons press conference when he held up his own earlier hesitancy to tout publicly his 1982 "scientific breakeven" achievement in muon-catalyzed cold fusion. "If we had announced it to the world, I'm afraid the public would have expected commercial power around the corner," Jones said. He also contrasted his reluctance then to start "a cold nuclear fusion center" pending scientific confirmation of what his group had found.

However, Jones grievously underestimated the time required to confirm or deny the reality of heat-producing electrochemical cold fusion, suggesting that only a few months should be needed. Like most physicists, he had very little feel for the practical problems of calorimetry. But Jones was certain of his own present research: "After years of painstaking work, we have been able to prove that fusion in metals does occur at very low levels by measuring the energy of neutrons produced." More than a few hot fusion physicists to this day, however, are not convinced that Jones and his colleagues have demonstrated in their electrochemical cells fusion neutrons above background level. Jones would find out just how much skepticism there was about his own supposedly solid work—"bona fide fusions" was his term—when he attended the Workshop on Cold Fusion in Santa Fe a month later.

Jones hammered home his central theme that the fusion-associated energy production in his group's experiments (the power one could calculate coming from that number of measured neutrons) was only a trillionth of a watt. He spoke in terms with which congressmen were familiar: "This is nothing to get excited about. . . . Yes, a new door has been opened. But the gap between the bona fide fusion yield and energy production by fusion is roughly equivalent to that which separates the dollar bill from the Federal national debt, a factor about a trillion to one. That is an enormous gap." Then he went through the common physics wisdom about why, if conventionally understood fusion was occurring, his detector would have discovered it. Truthful as Jones was, he could not admit the two other possibilities: (1) The physical parameters of his cell did not permit him to get the Fleischmann-Pons effect, whatever it was, or (2) A heretofore unknown nuclear process might be at work in the Fleischmann-Pons cells. What they were measuring was a chemical reaction, Jones suggested. He also threw a bone to his hot fusion colleagues: ". . . magnetic and inertial approaches currently represent the best paths to achieving controlled fusion energy." "I would also add that I believe that funding for cold nuclear fusion should come by peer reviews from such organizations as the Department of Energy and NSF. . . ." That would be fine for Steve Jones. He would continue to receive funding for his work and so would some of his like-minded skeptical colleagues, but the tide of official scorn would virtually cut off chances for adequate federal cold fusion research funding in the coming year—even for some skeptics who wanted more solid evidence that cold fusion was *not* real.

Jones used a prop; he held up a pathetic withered green plant in a jar and said, "This is a tender shoot, as you can tell. It is difficult to say what it will become. Some think and suggest strongly that this is a tree, and it will grow up very quickly and provide us enough wood for all our energy needs for generations. I do not think it is. Let's give it a chance to grow. I think adding too much fertilizer at this stage will be detrimental."

Daniel L. Decker, chairman of the BYU Department of Astronomy and Physics, joined Jones in being completely negative in assessing the Fleischmann and Pons excess heat results. He suggested that chemical, not nuclear reactions, were the likely ghost in the fusion machine. He urged the panel not to give equal credence to the University of Utah and the Jones work, though he felt that there was enough evidence to continue an investigation at least at a low level of support. Until a chemical explanation was convincingly ruled out, he said, "We shall not know if the Fleischmann-Pons effect represents a new source of energy for an energy-hungry world or is just a fantastic battery."

Though he was part of the crowd decrying "science by press conference," he was not above quoting an ephemeral report to emphasize his point: ". . . Just yesterday I read on the electronic mail that the University of Berlin had repeated the energy measurements of the University of Utah and claimed they could definitely show it to be a purely chemical process." To be fair, he balanced this by citing Huggins's testimony that the Fleischmann-Pons effect did not seem to be purely chemical in origin. But Decker said he would rather see the chemists change their understanding of chemical reactions 100-fold [to explain the effect as a chemical process] than have physicists revise their theories a trillionfold.

Standing squarely in the middle of pro and con arguments was Professor George Miley, a hot fusion researcher from the University of Illinois and editor of the respected journal, *Fusion Technology*. The following autumn, his technical journal would begin an excellent series of papers on cold fusion. By July 1990, his views had moved so far in the direction of belief in cold fusion, he could editorialize that evidence presented at the March 1990 conference on cold fusion in Salt Lake City had ". . . reinforced my conviction that nuclear reactions can take place in a solid." (Miley's written statement to the Committee was already quite positive and included the sentence, "I am personally convinced that solid-state catalyzed cold fusion occurs and this is an unexpected and very important new regime of physics.) His openness to new ideas about fusion was evident at the hearing. He described himself as a "proponent of fusion in any form," of the "search for so-called advanced fuels or aneutronic fusion," and of the quest for "alternate confinement concepts, which this [cold fusion] certainly is."

Miley was well aware of the need for verification with the "simultaneous measurement of reaction products from whatever this mysterious reaction is." But he took the unusual step of assuming, for the sake of argument, that heat-producing cold fusion was real. If so, he said, practical systems would require cheaper materials, higher temperatures for improved efficiency, longer lifetime electrodes, perhaps even a new way of getting deuterium into a matrix of metal atoms. What other fusion reactions could be excited in the same way, Miley wondered? He supported the Utah proposal to carry out scientific studies in parallel with technology development.

This was no orthodox hot fusioneer speaking. The caution he was recommending in cold fusion had more to do with avoiding adverse public reaction if the whole concept "crashed." He reminded the representatives of how the British hot fusion program was set back (in his view) by the premature announcement in 1958 of success with the Zeta (Zero-Energy Thermonuclear Assembly) "reverse field pinch" plasma fusion experiment. At a press conference for hundreds of reporters on January 24, 1958, Nobel laureate Sir John Cockcroft, director of the Harwell Laboratory, had announced with high confidence that Zeta had achieved thermonuclear fusion reactions at a temperature of 5 million °C. The British press went wild with headlines about "everlasting energy" from "H-power." But within six months the laboratory had to retract its claims, because fusion was not occurring in "mighty Zeta," as many physicists had almost immediately suspected. It was an unfortunate misinterpretation of data that had led to the public relations nightmare for hot fusion in Britain.

Miley's overall message advocated diversity in fusion research—the antithesis of the present direction of the financially strapped hot fusion program. "It's very important to the research activities of this nation that we have seed money to allow smaller groups to do the exploratory research," he said. "I'm a lifelong proponent of fusion. It has so many possibilities, that one of the difficulties of the moment is there is no room for funding for innovative research. Something has to be done." He was absolutely right.

Naturally and quite properly, the advocates weren't the only folks to speak at the hearing. Harold Furth, primed by his Dallas experience, would have something to say about cold fusion, as would Michael Saltmarsh, a top magnetic fusion scientist from the Oak Ridge National Laboratory in Tennessee, and Ronald Ballinger, an MIT professor of Nuclear Engineering and Materials Science, who was involved with MIT's plasma fusion effort.

Michael Saltmarsh reported that four separate groups were working at Oak Ridge to verify the cold fusion claims, but that they were having difficulty because of uncertainty about what parameters in electrochemical cells were most important for success. Three of the four groups were using calorimetry to measure heat, but all results so far were negative—no heat, no radiation found—not yet at least. He judged that both the United States and international efforts at duplication were largely failing, but, "It would be a real mistake to draw firm conclusions at this point."

But Saltmarsh did not want Utah to get a cold fusion research center to centralize the verification process. He was in favor of the university-national laboratory complex carrying out the work.

The afternoon session was wearing on. Furth reiterated the Jones distinction between the two kinds of electrochemical cold fusions—the Utah and BYU work. It was an energy difference comparable to the difference between "your personal lunch money and feeding the whole human race." For a major figure in the hot fusion community, Furth was very analytical and one might even say open-minded about the whole affair. At one point he asserted, ". . . if one should fail to confirm the reality of the nuclear evidence, that wouldn't damage necessarily the reality of the calorimetric [heat measuring] evidence. Because at the moment, the argument is that the excess heat is not produced by known fusion reactions at all, but it is produced by a new kind of fusion reaction not previously known which does not have visible nuclear reaction products." Contrariwise, if a chemical explanation were found for the Fleischmann-Pons effect, this would not necessarily rule out Jones-level cold fusion being real. Still, Furth wanted the community to resolve at least one major question: whether any helium reaction product could be found inside palladium rods that had successfully produced excess heat. And he wondered about, as he put it, "The Case of the Missing Control Experiment"—why had Fleischmann been so averse to talking about the results of his experiment with a cell using light water?

Nuclear engineer Ronald Ballinger's fusilade was less forgiving and seemingly meant to stop cold fusion in its tracks. He told the committee that the round-the-clock MIT experimenters had so far been unable to verify any cold fusion claims. "To my knowledge," he said, "with the possible exception of the people at Stanford [University], and the results from Europe and the USSR, of which I have no personal knowledge, we have not had a single confirmation, scientific confirmation of the reported neutron emissions from the experiment, nor the excess heat. I want to be careful when I say *scientifically verified*." He maintained that the MIT radiation detection measurements were at least 10 times more sensitive than the University of Utah measurements yet had found no neutrons. He also claimed that MIT's calorimetry was also "probably about ten times more sensitive."* Blasting Fleischmann and Pons's style of scientific communication, he said, ". . . the scientific community has been left to attempt to reproduce and verify a major scientific break-

---

*If Professor Ballinger at this time had had the advantage of being able to see the detailed work that Fleischmann and Pons would finally complete for publication in the spring and summer of 1990, he might not have been so free to make a claim about more sensitive MIT calorimetry. But Professor Ballinger's hostility toward Fleischmann and Pons is evident in his public discussions of cold fusion.

through while getting its experimental details from *The Wall Street Journal* and other news publications." He urged the committee to support verification experiments but not to commit major funds before that process had worked its way through.

The hearing lasted until 3:40 in the afternoon. Representative Owens thought that the Utah team had "scored a home run." But whatever merits other members of the committee might have ascribed to the pleas for money, the House rejected a supplemental spending bill that could have financed the cold fusion effort. A decision to fund cold fusion would have to be put off. Representative Walker, one of those reacting favorably, had just proposed diverting $5 million of existing fusion funding to cold fusion and planned to request $40 million for the coming fiscal year (FY '90). The magnetic fusion community feared that the situation was moving perilously close to being out of control. They knew that their own shaky budget might be the bank account from which cold fusion money would be withdrawn.

✳ The Approaching Storm

Reports of other confirmations were coming in, but it had to be troubling to the cold fusioneers that most of the more "prestigious" universities and laboratories were having trouble verifying the claims. Cold fusion was floating on a sea of rumors: that in Japan cold fusion investigations were shifting into high gear with more than 100 scientists being assigned to the job; that two U.S. national laboratories had confirmed parts of the Fleischmann-Pons experiment, but were withholding their results pending formal publication; and so on.

A group at Case Western Reserve University had confirmed at least some of the Fleischmann-Pons experiment. That was good news. And Fleischmann and Pons were soon expecting to have an audience in Washington with the technically savvy engineer John Sununu. In the sheltered gulf of Utah, the seas were relatively calm for the cold fusion crew, but very stormy weather was approaching.

A team of MIT scientists was preparing a report that it planned to release in early May at the meeting of the American Physical Society in Baltimore. It wouldn't adequately address the excess heat issue, but it would devastate Fleischmann and Pons's claim to have detected significant numbers of neutrons coming from their electrochemical cells. Other distressful signs appeared days after the supposed triumphal march to and from Washington.

Most disturbing was the first major attack by a scientific publication. It came in an editorial in the venerable publication *Nature*, whose every turn and twist has been closely watched by scientists the world over for the past 120 years. *Nature's* editor John Maddox wrote in the

April 27 issue, "The Utah phenomenon is literally unsupported by the evidence, could be an artifact, and given its improbability, is most likely to be one." And further, ". . . the likelihood of replication fades as the days go by." Maddox had apparently forgotten the nugget of wisdom his journal had published in an editorial not a month earlier, " 'Cold (con)fusion': It is the rare piece of research indeed that both flies in the face of accepted wisdom and is so compellingly correct that its significance is instantly recognized."* To the obvious dismay of Fleischmann and Pons, the editorial appeared in the same issue that carried rival and heat-skeptic Steve Jones's article, "Observation of cold fusion in condensed matter." Maddeningly, Jones and coauthors ended their article with a statement that was certainly more optimistic about cold fusion's utility than Jones's public pronouncements: "Although the fusion rates observed so far are small, the discovery of cold nuclear fusion in condensed matter opens the possibility, at least, of a new path to fusion energy."

Maddox's editorial also attacked the researchers for having neglected to perform the supposedly ideal control experiment—using an identical cell with light water in place of heavy water. Maddox wrote, "This glaring lapse from accepted practice is another casualty of people's need to be first with reports of discovery and with the patents that follow." He did however have two "good" things to say about the duo, although the sentiments were not taken as compliments, to be sure: "Fleischmann and Pons have done at least one great service for the common cause: they have kindled public curiosity in science to a degree unknown since the Apollo landings on the Moon. . . . It is remarkable that so many people are willing to accept that experimental observations, and the inferences drawn from them, acquire validity only by replication. Has what used to be called 'the scientific method' now become widely understood?"

Nature had slowly mounted its assault on Fleischmann and Pons, accusing them in its March 30 issue of having falsely claimed that their work would soon appear in the magazine. Fleischmann denied this to Congress in the week earlier and he denied it again vehemently days later to JoAnn Jacobsen-Wells in the *Deseret News* (April 30, 1989): "At the press conference we stated it was quite incorrect to disclose the name of the journal we were submitting to because it would restrict the freedom of action of the editor. We have never disclosed to anyone that we were submitting an article to *Nature.* It could have only come from other people submitting papers, or from the reviewers, or from the staff of *Nature* itself. I think it is an impertinence that *Nature* feels itself

---

**Nature*, Vol. 338, 30 March 1989: 361–362.

entitled to repeat an allegation which is untrue." Pons had even more stinging words for *Nature*: "This is sensationalism of the worst type. I have never in my life seen such a scandal by a scientific journal to discredit two people. . . . I think what you are seeing here is the development of a very nasty piece of sensationalism by the editors."

In their view, the character of their initial 1,000-word brief note to *Nature* was to inform the readers of the "character of the results we obtained and the general methodology we used." *Nature* had wanted a more complete article and on rather short notice, so the two had simply withdrawn their submission. Fleischmann and Pons had, however, sent replies to the reviewers of their article. They contend that *Nature* erroneously, or deliberately, did not pass along their replies to these reviewers. Thus, they were angered when *Nature* implied that they had not responded (April 20, 1989). Fleischmann was so mad that he characterized the *Nature* suggestion that they had not replied as "libelous."

Nature and much of the rest of the scientific community were in many ways holding Fleischmann and Pons to a much higher standard than they had any reason to, given the puzzling character of what the two had found. Instead of having the forum of *Nature* to speak within, they were relegated to the pages of the *Deseret News*—a fine publication in Utah, but hardly a suitable means to communicate with fellow scientists. About their disclosure, they made these key points in the *Deseret News* (April 30, 1989):

Fleischmann: "We presented this whole thing as an experimental observation with a minimal amount of interpretation. We regard it still as an experimental observation which needs to be verified. But rather than people criticizing it, they should go and do some of the experiments themselves." And that is precisely, of course, what hundreds of scientists were doing. Whatever its shortcomings—and there were many, the March press conference had at least accomplished that.

Pons: "If solid state fusion had been discovered in an industrial laboratory, the first time you would have heard about it would have been when somebody sold you a generator. We were certainly open about our results in the paper. It's a pity that the wide readership of *Nature* now only has access to information at second hand—that information coming from our critics." *Nature* was portraying the two as being locked up in a fortress of patent attorneys, but Fleischmann and Pons contended that they were talking to and seeing many scientists, setting up collaborations, and giving out their data.

As there is need for a calm before the storm, it is well to take refuge in a temporary respite from the fury of the cold fusion controversy. Before resuming tales of battle, let us step back in time for a brief remembrance of cooler days in the strange prehistory of cold fusion.

6 The Prehistory of Cold Fusion

> We had the short but exhilarating experience when we thought we had solved all the fuel problems of mankind for the rest of time.
>
> Luis W. Alvarez
> Nobel Prize acceptance speech, 1968

> It is possible that Fleischmann and Pons have rediscovered a 150-year-old German cigarette lighter.
>
> George Chapline
> Lawrence Livermore National Laboratory, May 1989

THE REMARKABLE PREHISTORY OF COLD FUSION recalls a stereotypical scene in a horror movie. A point comes in the movie drama when the solution to a mystery is at hand. Heroic figures have just pieced the whole scary business together and have explained the origin and weaknesses of the dangerous creature—by consulting ancient texts, by looking through books from the musty back rooms of the local library, or by examining decaying records at town hall. So it is with cold fusion if you trace scientific history far enough back; you keep coming up with eerie foreshadowing.

✳ An Amazing Element

Cold fusion phenomena are not reported exclusively in palladium—witness apparent neutron-producing cold fusion experiments that researchers have done with titanium. But historically, palladium plays the key role in the emerging awareness of the phenomenon. So the prehistory of cold fusion begins with the remarkable properties of palladium, an element whose curious behavior science has known for a long time.

Silvery lustered palladium ranks 46 among chemical elements according to increasing number of protons in the nucleus—the factor that fixes an element's identity. It is a precious metal and was first identified in 1804 by the English chemist and physicist, William Hyde Wollaston (1766–1828). In experimenting with platinum (element 78), which in the late 18th century was regarded more preciously than gold, Wollaston isolated palladium. He named it after the large asteroid Pallas (itself named after Greek goddess of wisdom Pallas or Athena) that German astronomer Heinrich Olbers had discovered in 1802. Pallas remains the second largest known asteroid, 480 kilometers in diameter, in whose massive body there may be millions of tons of palladium.*

Pallas's naming was in the tradition of the alchemists of old, who gave metals the names of planets. Closer to Wollaston's time in 1789, however, the German chemist Martin Klaproth had named his own newly discovered element uranium after Uranus—the planet discovered only eight years earlier (the first planet that had to be "discovered," because it wasn't visible to the naked eye). There may be a strange irony here, because uranium is identified, of course, with *fission* and palladium is now tied to a strange new kind of *fusion.* Also, the discovery of uranium occurred in 1789, 200 years before the 1989 announcement of cold fusion in palladium, the same year that the Voyager 2 spacecraft whizzed by the planet Uranus. Fusion, whether cold, lukewarm, or hot, is also somewhat like alchemy; its final products are often elements that were not originally present. Strange that palladium was named in an alchemical tradition!†

Palladium is a by-product of platinum mining and metallurgical refining. Its uses are mostly industrial: contacts in electrical equipment, catalysts in the petroleum and chemical industries, automobile exhaust catalytic converters, and a component in alloys for jewelry and dental materials. Supplying the world with most of its palladium is the Soviet Union (1.6 million ounces/year), then South Africa (1.1 million onces/ year). A mine in Stillwater county in southern Montana contributes an additional few hundred thousand ounces to the U.S. supply. (The county government of Stillwater—adjacent to Carbon county, no less!—has yet to consider a name change to Heavywater.)

Palladium is better known today as a precious metal in which commodities traders deal. Just before March 23, 1989, the going rate for

---

*Metal composition asteroids are thought to have 10 to 20 times the concentration of palladium as Earth's crust. Reacting to the news of cold fusion, some space exploration buffs immediately drew up plans to mine asteroids for palladium.

†Go back even further into the dusty past. More than two millennia before anyone knew about palladium, the element, a carved wooden image of the goddess Pallas (or Athena), called a Palladium, was a widely used religious symbol for the ancient Greeks. Zeus, the chief deity, supposedly had cast the Palladium from heaven at the founding of Troy; according to legend, the thievery of this "sacred stone" of the ancients by Odysseus made the fall of that city inevitable.

palladium was $145.60 per ounce on the New York Mercantile Exchange, but to no one's surprise shot up to over $170 an ounce within two weeks of the Utah announcement. Palladium's all-time high was about $335 per ounce in 1980. In the 1989 rush there were reports that the market was boosted not only by traditional investors, but by significant numbers of newcomers, not a few of whom were scientists and engineers, "in the know" or merely deluded, who had caught the cold fusion bug. Palladium had been substantially lower six months earlier, but because Ford had announced in December 1988 that it would no longer use platinum in its automobile catalytic converters, a switch to palladium in the air led to the price climbing above $120 per ounce.

✳ Of Airships and Cigar Lighters

In 1823, not long after palladium's discovery, a German chemist, Johan Dobereiner, found that the metal could catalyze the burning of hydrogen. He used this property to make a device that he called a Feuerzeug—a lighter for tobacco smoking. That was the heat-producing very early prelude to the cold fusion controversy, one that some skeptics are fond of recalling.

In the early decades of the 20th century, we find another curious palladium story. It was the 1920s, the age of the great German airships. Chemically inert helium, abundant in the Sun but much rarer on Earth, was an eagerly sought substitute for the flammable hydrogen gas that filled the transoceanic zeppelins. After years of experimenting, German scientists Fritz Paneth and Kurt Peters at the Chemical Institute of the University of Berlin reported in the scientific literature in 1926 that they had used a palladium catalyst to convert hydrogen to helium.*

It was an unproved hypothesis at the time that hydrogen atoms could combine to form helium in nuclear reactions that powered the Sun, but the researchers offered this as an explanation for their hydrogen-to-helium discovery. Unmentioned by them, however, was the even grander hypothesis of English chemist and physiologist William Prout (1785–1850). Prout had suggested in 1815 that atoms of all the elements were composed of different numbers of hydrogen atoms—a hypothesis that turned out to be not very far from the truth when the facts about the atomic nucleus became known a century later.

Scottish physical chemist Thomas Graham (1805–1869) had studied the remarkable absorption of huge volumes of hydrogen into palladium; this had become the element's most famous characteristic. So it was natural for anyone bent on manipulating hydrogen to tinker with

---

*_Berichte der Deutschen Chemischen Gesellschaft_, Vol. 59, 2039, 1926.

palladium. Paneth and Peters had tried a number of techniques to produce helium: passing hydrogen gas through heated palladium capillary tubes, and flowing hydrogen over asbestos fibers coated with a film of palladium. They sought evidence of helium production by looking for spectroscopic lines characteristic of helium in the light coming from the gas that emerged from their experiment (after the gas was heated to high temperature). After taking care that they had not mistaken their product for helium that was already in the system at the outset, they convinced themselves that they had cooked up a tiny amount of helium—about 10^{-8} cubic centimeters of the gas. A 1926 *Nature* magazine report on the Paneth-Peters work said that the researchers had calculated the expected amount of heat release from this nuclear conversion, 0.28 calorie—far too small to measure under the circumstances, the researchers said.

Unfortunately, Paneth and Peters later discovered a source of error that they had neglected, and so retracted the 1926 claims in April 1927. To their regret, the glass walls of their apparatus were letting loose tiny amounts of helium when it was heated in the presence of hydrogen—helium that had surely been in the glass from the beginning. The asbestos too turned out to be a source of helium. Yet it is interesting that in a note of retraction to *Nature*, Paneth still held out some hope that transmutation to helium might be occurring at a low level. He wrote, "By avoiding all heating of the apparatus, we shall endeavour to decide whether a transmutation of hydrogen into helium of the order of 10^{-9} cubic centimeters or less takes place." Probably it was no more than coincidence, but Paneth and a much younger Martin Fleischmann found themselves in the chemistry department at the University of Durham in the 1950s.

That would have been the end of this early and naïve cold fusion experiment, but John Tandberg, a creative genius working in Stockholm, Sweden, got wind of the Paneth-Peters results (prior to their retraction) and did some experiments of his own. Tandberg, who was employed at the Electrolux Corporation laboratory, had an idea that foreshadowed the work of Fleischmann and Pons years later: Why not use electrolysis with a palladium electrode to increase the hydrogen concentration and likelihood of a reaction? Operating at high pressure would make the process even more efficient, so he reasoned. In fact, in February 1927, Tandberg applied for a Swedish patent for just such a device to produce helium, an application that was rejected later that year.

Remember, all this work was occurring before the neutron was discovered by James Chadwick in 1932, so no one really knew what kind of nuclear reactions were even hypothetically going on. The helium nucleus, for example, was then thought to consist of four protons. The American chemist Harold Urey's 1932 discovery of heavy hydrogen (deuterium, with a proton and neutron in its nucleus) prompted further thoughts by Tandberg: Deuterium might be more amenable to fusing than hydrogen. Tandberg's biographer and former collaborator, Torsten Wilner (*Our Alchemist in Tomegraend*, by Torsten Wilner), recalled how Tandberg quickly decided to try to produce nuclear fusion in a wire of palladium that had previously been saturated with deuterium through electrolysis. On many occasions ending with deafening electrical discharges, Tandberg attached a source of high voltage to "deuterated" palladium wire in attempts to shock the deuterium into fusing. Tandberg was well aware of the hazards. Had all the deuterium in his wire fused to helium, he calculated, the explosion would have released the energy of a ton of dynamite. Fortunately (or sadly!) the deuterium did not fuse explosively. Tandberg died in 1968, never anticipating the astounding 1989 independent revival of his work.

If cold fusion circa 1991 really enters the pantheon of useful energy sources, it is interesting to speculate what would have been the course of science and technology had these early attempts at table-top fusion aroused much wider curiosity. If cold fusion had been confirmed and successfully developed in the 1930s, it is possible that no one would have taken up the daunting challenge of hot fusion.

✳ Fleischmann's Dream

About two decades after this early toying with the possibility of hydrogen fusion in palladium, Martin Fleischmann was working toward his doctorate at Imperial College, London, in the period 1947 to 1950. In a curious foreshadowing of his later thinking, Fleischmann's doctoral thesis focused on how hydrogen diffused through platinum. He used platinum membranes only 0.1 to 0.2 mm thick to study the electrochemical transport of the hydrogen.

There was little chance that Fleischmann's roving mind had turned to the possibility of electrochemical fusion in those days, the period 1944 to 1950. The information on nuclear matters circulating in the Chemistry Department at the time was of a very poor sort recalls Professor B. E. Conway, then a graduate student colleague of Fleischmann's. Today, Conway is a Professor of Electrochemistry at the University of Ottawa in Canada. The chemistry students had scant knowledge of what was happening then even on the exciting frontier of fission, which the nuclear physics community was pursuing. Conway says that even then Fleischmann had the reputation of "being very fluent with thousands of ideas."

The electrochemists formed a very close-knit group, and it remains close in that these greats of the field know each other personally and stay in touch. John Bockris, later to be a professor at Texas A&M University and a key figure in cold fusion studies, was a lecturer at Imperial College and taught Fleischmann. He once joined the band of student electrochemists (Fleischmann, Conway, Rex Watson, and Edmund Potter) on a mock "ghost-hunting" mission at the ruins of an ancient abbey north of London. One Halloween, the group was intent on playing a practical joke on the locals at a favorite pub near the abbey; they constructed a fake Rube-Goldberg apparatus, which they called a Geisterphone. They claimed it would help them detect the headless ghost of a nun who was said to walk uneasily among the ruins. On a typical foggy English night, the group brought their Geisterphone to the pub on the way to the abbey, then spent a half-hour in a mock search for the headless nun. The locals were largely duped by this ghost-hunting episode.

This ghostly 1948 event, which cold fusion skeptics might say bears some resemblance to "cold fusion" circa 1990, was not, of course, the serious flirtation with cold fusion that Fleischmann might have had in the early 1970s when he wrote a joint paper with his University of South Hampton graduate student, B. Dandapandi. In their 1972 paper, "Electrolytic Separation Factors of Palladium,"* lay the seeds of future pondering of the mysteries that lurked within a palladium lattice laced with deuterium. In the April 1989 Fleischmann-Pons-Hawkins paper, the authors say that one of the features revealed in that early 1970s study prompted their later cold fusion work. Fleischmann may have begun to wonder whether some of the peculiar behavior of the deuterium ions (D^+) in palladium would make the lattice suitable for near collisions and possible conventional fusion reactions. The 1989 paper concludes that it is "necessary to reconsider the quantum mechanics of electrons and deuterons in such host lattices."

At the First Annual Conference on Cold Fusion held in Salt Lake City (March 1990), Fleischmann presented an overview of the cold fusion phenomenon and alluded to other influences that had motivated their work. A "key element," he said, was his awareness of the mid-1970s projects in the United States and in the Soviet Union in forming metallic hydrogen and deuterium at extreme compressions. Also provocative was the "conundrum" of why was the "diffusion coefficient" for deuterium in palladium—how rapidly it moves through the lattice—higher than for either ordinary hydrogen or tritium. Fleischmann suggested that he and his colleagues were wondering whether large numbers of particles within the Pd lattice could work cooperatively to bring about fusion reactions. "We, for our part, would not have started this investigation, if we had accepted the view that nuclear reactions in host lattices could not be affected by coherent processes," he said. Their

---

***The Journal of Electroanalytical Chemistry*, Vol. 39, 1972: 323–332.

discovery of excess heat was a "great surprise." All they were hoping to find were neutrons and tritium, but they really struck pay dirt in finding what they considered to be clear evidence of anomalous excess heat. Cold fusion crept up on them.

✳ The Other Cold Fusion

Consider some bold claims: "It is now conceivable that cold fusion may become an economically viable method of generating energy"; "The process may one day become a commercial energy source"; and "[A] commercial cold fusion reactor could be built with existing technology." It may surprise you that these were not pronouncements by Drs. Pons or Fleischmann in 1989 or excerpts from a state of Utah development brochure. They appear in an otherwise staid article in *Scientific American* in July 1987, authored by Johann Rafelski and Steven E. Jones, who two years later would be a prominent skeptic on the likelihood of cold fusion applications. (Remember what Steve Jones has been fond of saying, "Twenty years to never.") But in that article titled "Cold Nuclear Fusion," Jones and Rafelski weren't talking about fusion in a glass jar with electrodes, they were discussing muon-catalyzed fusion—the "other" cold fusion, which uses exotic subatomic particles called muons to bring about fusion in hydrogen isotopes.

The headline on the front page of the *New York Times* was impressive: "Atomic Energy Produced By New, Simpler Method: Coast Scientists Achieve Reaction Without Uranium or Intense Heat—Practical Use Hinges on Further Tests." Following the main article was a brief sidebar titled, "Cold Fusion of Hydrogen Atoms," a brief explanation of the new process. The date of this world-shaking news was December 29, 1956, cold fusion making headlines not just two years but 33 years before its 1989 incarnation! Then also the subject was muon-catalyzed fusion, not electrochemical fusion. Twelve scientists at the University of California at Berkeley, led by a future Nobel laureate, Professor Louis W. Alvarez, had by sheer accident *rediscovered* cold fusion and were announcing their findings before a meeting of the American Physical Society.

Rediscovered? Yes! The birth of muon-catalyzed fusion originates even further back, beginning with the speculative ideas of the late Soviet physicist Andrei D. Sakharov and the British professor F. C. Frank, who independently conceived the notion in the late 1940s. Their idea was elegant, ingenious, and workable and demonstrates the remarkable prescience of theoretical physics: First a negatively charged muon generated by a powerful particle accelerator temporarily replaces an electron orbiting the nucleus of an atom of a hydrogen isotope such as deuterium. Since a muon is 207 times more massive than an electron, it orbits nearer to the nucleus than the former electron—about 200 times closer. Because this "muoatom" is much more compact, it binds much tighter to other hydrogen isotopes, including other muoatoms produced with additional muons. The closer proximity of two hydrogen isotope nuclei increases the probability (enhances the quantum tunneling probability) that the strong nuclear forces between their constituents will make them fuse.

Working in 1956 with a nuclear particle detector called a "bubble chamber" on a completely unrelated matter, by sheer accident the Berkeley team observed muon-catalyzed fusions. At first they didn't realize what they had seen, being unaware of Sakharov's and Frank's ideas. In fact, Edward Teller was helpful to the group in identifying the unusual subatomic events as muon-catalyzed fusions. You can imagine how the unsuspecting researchers felt when they realized what they had found—a possible breakthrough in controlled fusion, at that time still a very young industry. Hence, the dramatic announcement in 1956 and the recollection of feelings in Alvarez's 1968 Nobel acceptance speech.

Since muons not deriving from randomly occurring cosmic rays that bombard Earth must be produced artificially in high-energy collisions in particle accelerators, how, in principle, might we apply this exotic phenomenon? Begin by bombarding a chamber filled with hydrogen isotopes with accelerator-generated muons. The deuterium or tritium, or a mixture of the same could be at room temperature, much colder than room temperature, or even considerably above—say 900°C. Fortunately, a single muon in the course of its exceedingly brief lifetime before it disintegrates (a few millionths of a second) has the opportunity to catalyze more than a single fusion between nuclei. The trick would be to make each muon catalyze enough fusions so that the net energy production would be greater than that required to operate the accelerator that creates the muon beam. Like a dance partner gone mad, a muon can swing from one hydrogen isotope nucleus to the next in a fraction of its lifetime, all the while catalyzing more and more fusions. The dance can be very complex and can even result in the formation of molecules based on muoatoms of deuterium or tritium. Maximize the number of catalyzed fusions per muon before it dies to optimize the process. Perhaps we could even reach the Holy Grail of energy breakeven, but this is easier said than done.

The Alvarez group's work was with hydrogen-deuterium fusion, but it happens that deuterium-tritium is a much better mixture, and Soviet researchers later discovered experimentally that the process works better at elevated temperatures. Prompted by encouraging theoretical work on muon-catalyzed fusion done by others in the 1960s and 70s, Steve Jones and his colleagues in 1982 undertook a major experimental effort in that area. They worked at the Los Alamos Meson

Physics Facility, under the sponsorship of DOE's Ryszard Gajewski (who would later play a controversial role in funding cold fusion studies in 1989–90). The group found much better than expected performance—on the order of 150 fusions per muon—but still not enough for energy breakeven at least by a factor of 10 to 20. Through what Steve Jones aptly has termed "a gift of God," certain kinds of vibration resonances within "muomolecules" may be enhancing the fusion rate. But there are negative complications too, such as muons "sticking" to alpha particles (helium nuclei) produced in some of the reactions.

"It's a difficult thing to get from where we are [in muon-catalyzed fusion] to commercial energy," Jones said in 1989 at Columbia University. In his testimony to Congress in April 1989, he said that a tenfold improvement in fusion yields would be required to begin to make muon-catalyzed fusion practical. "It is not at all clear that we can bridge that gap, even though our yields achieved to date exceed those seen by Alvarez by a factor of several hundred," he told the Committee on Science, Space, and Technology. In order to get the energy out of muon-catalyzed fusion and make it produce useful heat or electricity, the neutrons from the fusion reactions would have to be absorbed in a surrounding blanket of lithium, thus producing heat and tritium, the latter being fed back to the reactor as fuel. This is similar to plans for using the fusion neutrons in hot fusion reactors (Chapter 2).

The field of muon-catalyzed fusion is full of promise; many new ideas have already come from it and will undoubtedly continue to do so. Jones achieved "scientific breakeven" with the method in 1982, but in a commercially practical system the energy cost of generating the muons in an accelerator would have to be factored in. Muon-catalyzed fusion helped to catalyze Jones's thinking about piezonuclear fusion; that, in turn, prompted his May 1986 thoughts about electrochemical fusion. There may even be more exotic subatomic particles that could catalyze fusion,* so given the history of unexpected turns and twists in this field, one should not rule out a future breakthrough in catalyzed cold fusion.

✳ Fusion in the Earth?

In 1986, Steve Jones teamed with Clint Van Siclen of the Idaho National Engineering Laboratory and published a paper, "Piezonuclear Fusion

---

*George Zweig of Caltech has suggested "free quarks"—exotic constituents of protons and neutrons that few physicists really expect to occur independent of those particles. So have Luciano Fonda and Gordon L. Shaw in "Fluctuations and Nonreproducibility in Cold Fusion from Free Quark Catalysis," Proceedings of Anomalous Nuclear Effects in Deuterium/Solid Systems, BYU, October 22–24, 1990.

in Isotopic Hydrogen Molecules" (*Journal of Physics G: Nuclear Physics*, Vol. 12, 1986), which speculated that the vanishingly small fusion rates (essentially nonexistent) in mixtures of hydrogen isotopes at everyday pressure might be enhanced simply by squeezing the molecular compounds to much higher densities. The paper was contemporaneous with the beginning of Jones's electrochemical fusion work that led to the seminal and still controversial 1989 paper in *Nature*, "Observation of Cold Nuclear Fusion in Condensed Matter" (April 27, 1989). Jones and his colleagues had progressed from muon-catalyzed fusion, through theoretical studies of piezonuclear fusion, and on to experimental investigations of piezonuclear fusion through electrochemistry. The lineage was all the same.

Van Siclen and Jones suggested in their 1986 paper that very high pressures could reduce the internuclear separation of hydrogen isotopes in a way that was similar to, though not as extreme as the effect in muon-catalyzed fusion. At ordinary atmospheric pressure, the separation of deuterium nuclei in a D_2 molecule is about 100,000 times the diameter of a typical atomic nucleus. Jones and Van Siclen estimated crudely that in solid deuterium at about 200,000 atmospheres pressure, the internuclear separation between the deuterium nuclei would decrease only about 5 percent from this—not much closer together for all that pressuring!

Since squeezing was said to produce the same kind of geometric effect as do muons, Jones and Van Siclen wondered what would happen if enough were applied to cut the internuclear separation between deuterons even more, say halving it (at a pressure of over 100 million atmospheres). This would lead, they found, to a fusion rate of about one fusion per minute in a kilogram of deuterium—something that would have a chance of being measured. The halving of the separation distance would be equivalent, they said, to the intervention of a negatively charged "fictional binding particle"—akin to a muon, but with only twice the mass of an electron.

Departing dramatically from their calculations, they questioned "whether piezonuclear fusion within the liquid metallic core of Jupiter can account for the excess heat radiated from the planet." It is well known that Jupiter radiates excess heat amounting to about twice the solar radiation that reaches that giant world. After a few more calculations, they answered their own query and said, "The piezonuclear fusion rate in Jupiter is evidently many orders of magnitude too small to be a significant source of energy." Bear in mind, however, that the pressure at the center of the hydrogen core of that planet is still an incredible 60 million atmospheres.

Though piezonuclear fusion might not be able to explain the enormous excess radiation coming from Jupiter, this doesn't mean that cold fusion does not play a role in the internal dynamics of that world. The effect may simply be smaller and contribute only a part of Jupiter's excess energy. But by the time Jones and his colleagues wrote their 1989 *Nature* paper, they had progressed to consider the possibility of cold fusion occurring within Earth too!

Professor Paul Palmer of BYU proved instrumental in developing these ideas, because he was aware that in hot places on the Earth, particularly at volcanoes, there is a thousandfold increase in the ratio of the helium isotopes, $^3He/^4He$. On March 12, 1986, at BYU, Palmer's thinking was catalyzed by Steve Jones's Physics Colloquium on piezonuclear fusion. In places like Hawaii, the tiny ratio is 10^{-5} but it is usually 10^{-8} elsewhere. The combination of piezonuclear idea and the geological ^3He anomaly led Palmer to suggest the possibility that geological cold fusion was occurring and might be the source of volcanic heat. Jones has said, "That was our start into this field–independent of any work elsewhere, in 1986 March." The idea: Deuterium-containing water (all water, of course) is carried along with a continental plate that creeps slowly deep into the Earth, where melting occurs. There, both direct fusion production of ^3He from the H-D reaction, the D-D reaction, and the radioactive decay of fusion-produced tritium (T) to ^3He would presumably occur in some unforeseen way. Jones and his colleagues discussed this with Harmon Craig (University of San Diego and Scripps Oceanographic Institute), who had noticed the helium anomaly years earlier. On March 13, 1986, a Palmer notebook entry suggested that oxygen in rocks could possibly catalyze the fusion reaction.

In fact, at the outset of the 1989 paper they say that certain observed anomalies in the prevalence of naturally occurring helium-3 (^3He) on the Earth led them to their laboratory investigations of cold fusion in condensed matter. Jones was also motivated by Russian physicist B. A. Mamyrin's 1978 observations of anomalies in ^3He and ^4He concentrations in metals and in diamonds. It is not certain, however, that they were first to consider this possibility. Another group of researchers at the University of Arizona, seems to have conceived the idea independently. Their thinking was triggered by the Fleischmann-Pons announcement, and they too submitted a paper to *Nature*,* which the magazine turned down.

Is it possible that cold fusion going on inside the Earth could account for at least some of our world's internal heat generation? Just possibly, but in all likelihood not much of it. Earth's internal heating is thought to be due largely to the radioactive decay of the elements

---

*McHargue, L.R., P.E. Damon, H.P. Dart, III, in a paper, "Cold Nuclear Fusion, Helium Isotopes, and Terrestrial Heat Production," written after March 23, 1989, and subsequently rejected by the *Nature* in April 1989.

uranium, thorium, and to some degree potassium, processes that lead to the production of large amounts of ^4He. Of all the startling implications of cold fusion, none is perhaps so bizarre and intriguing as the possibility that cold fusion occurs naturally inside the Earth. After all, if the Earth can have natural fission reactors in geological formations—evidence for at least one has been discovered in Africa*—why not a naturally occurring cold fusion reactor? It is even more enticing and literally far out to think that the giant outer planets of the Solar System—Jupiter, Saturn, Uranus, and Neptune—might have internal fusion reactors of the cold variety. Speculating about such remarkable possibilities gives an inkling of how topsy-turvy the world of science might become if cold fusion were proved to the satisfaction of all.

*Cowan, George A., "A Natural Fission Reactor," *Scientific American,* Vol. 235, July 1976: 36–47.

7 The Beginning of Wisdom

> Can Wisdom be put in a silver rod?
>
> William Blake
> *The Book of Thel*, 1789

> It has been said that the art of living lies in knowing where to stop, and going a little further.
>
> Arthur C. Clarke
> *Profiles of the Future*, 1963

> Someone has compared this to setting off a hand grenade in a hayloft and expecting the hay to change the nature of the hand grenade explosion. People are having trouble seeing how this would work.
>
> Harold Furth
> Congressional testimony, April 26, 1989

PARALLELING THE FRENZY OF EXPERIMENTS to verify cold fusion claims, theories about solid-state fusion began to proliferate. At the time of the initial reports about cold fusion there was literally *no* convincing theory that would lead anyone to believe that the phenomenon could occur. Experiments can only go so far without theoretical support.

There was a general feeling that the palladium lattice must have something to do with either Fleischmann-Pons or Jones-level "fusion." Could some unforeseen mechanism hidden in the atomic structure of the electrode account for reports of cold fusion? Walling and Simons seemed to have thought so with their theory, and K. Birgitta Whaley had put forth her outline of a mechanism at the ACS meeting. But these were relatively insubstantial frameworks on which to hang weighty belief. Despite their courageous intent to explain the "impossible," these ideas had an ad hoc quality that is not the hallmark of a really novel and comprehensive physical theory. Furthermore, the Walling-Simons theory was unrealistic in that to explain the excess heat it required a rate of "internal conversion" to energetic electrons—after fusion of deuterium to produce ^4He—much higher than was possible.

There were many other speculators, however, who had immediately taken up the challenge to explain cold fusion. Physics Nobel laureate Julian Schwinger was one of the earliest to join the fray, as was Peter Hagelstein, physicist Robert T. Bush at California State Polytechnique, a few theoreticians in India, an Italian group of theorists whose most outspoken member was Giuliano Preparata, nuclear physicist Yeong E. Kim from Purdue University, T. A. Chubb from the Naval Research Laboratory, and perhaps dozens more scattered around the world. As experiments sharpened the description of the erratic phenomenon, the theories evolved too, but what an uphill battle it was to be! To design a theory for a secret that nature had so cleverly conspired to conceal was extraordinarily difficult. Without tantalizing experiments to egg them on, none of these people would have deigned to try, so patently "absurd" was the notion of significant cold fusion in palladium or titanium. So some erstwhile theorists preferred to try to establish the *unlikelihood* of cold fusion, a much easier task. If the history of science (and technology) has shown anything, it is that some prominent scientists will inevitably try to prove that a spectacular advance is impossible. This serves as a very important spur to theorists who think the advance is possible via some other route. With cold fusion it would be no different.

✳ A Skeptical Theorist

Cold fusion seemed to be a pushover for the erudite and brilliant skeptic, physicist Richard L. Garwin of IBM's Thomas J. Watson Research Center. Garwin was a well-known debunker of many exotic technical matters, but technological, not scientific debunking, had been his forte. He could assault the wisdom of particular ground-based antiballistic missile systems or orbiting space shields like SDI, but could he bring low an extraordinary scientific claim? Immediately following the Utah announcement, Garwin had apparently done a lot of head scratching. He later collected and recounted his thoughts in a piece for *Nature*, (April 20, 1989) really a report on his visit to the April 12, 1989, cold fusion conference at Erice, Sicily, which Fleischmann, Jones, Steve Koonin of Caltech, and others had attended.

Any theorist was facing formidable odds. The immediate temptation was to concentrate on possible fusion between deuterium nuclei, because reports indicated that deuterium was the magic ingredient that brought out the strange effects. The probability that two deuterium at-

Dr. Peter L. Hagelstein, Associate Professor of Electrical Engineering and Computer Science at MIT, explains his theory of "coherent fusion reactions" to an audience of MIT scientists on April 14, 1989. (MIT photo by Donna Coveney)

Dr. Steven E. Koonin, Professor of Theoretical Physics at California Institute of Technology, an outspoken critic of cold fusion claims, played a leading role in convincing many scientists that Fleischmann and Pons were wrong. (Photo by Robert Paz)

oms will fuse near room temperature is vanishingly small. The positive charges on each of two deuterium nuclei present a horrendous barrier to the nuclei coming close enough together to fuse; they must climb an energy "mountain" of some 600,000 electron volts (eV). Now a particle in a relatively cool plasma at 12,000 K has only a single electron volt (eV) of energy, far below the energy of motion that could readily overcome that 600,000 eV mountain. Thus, to expect copious room temperature fusion is on its face ludicrous.

But in the strange world of the microcosm where quantum mechanics reigns, matters aren't always what they seem. Sometimes they are quite crazy. It is possible for a charged deuterium nucleus to spontaneously tunnel right through that energy barrier—an effect known as quantum tunneling. Still, the higher the barrier, the lower the probability that any given pair of deuterons will tunnel toward each other and fuse. At the Sicily meeting, Koonin reported new calculations (later published in *Nature**), which increased those quantum tunneling probabilities significantly for d-d fusion, but still nowhere near a level making it seem very likely to occur visibly at room temperature. He had actually found an astounding error in previous calculations, and by correcting it he

*Koonin, S.E., and M. Nauenberg, "Calculated Fusion Rates in Isotopic Hydrogen Molecules," *Nature*, Vol. 339, 1989: 690.

raised the probability of d-d fusion 10-billionfold. Koonin concluded that the d-d fusion rate at room temperature was less than 10^{-63} per deuteron per second (1.0 divided by 1 followed by 63 zeros). Translated, this means that in a volume of room temperature deuterium atoms about the mass of the Sun, only one fusion was likely to occur per year! Koonin discovered bare protons would fuse with deuterium—the p-d reaction—at a rate almost a billionfold greater than the d-d reaction rate, still far below what was needed to explain cold fusion by quantum tunneling.

To enhance tunneling, one might have imagined that electrons hovering near deuterium nuclei had in an unknown way become effectively more massive, and like the muons in muon-catalyzed fusion would act to bring deuterium nuclei and/or protons closer together, thereby increasing the probability of tunneling. But Koonin's work showed that the electrons would have to acquire unreasonably high "effective masses" to make cold fusion likely.

There were other possibilities, but none seemingly very likely. Koonin had suggested looking for "dynamic effects" that might augment the tunneling—vibration mechanisms that would perhaps cause deuterons to accelerate toward one another. Grasping at straws, physicist L. Ponomarev of the Soviet Union suggested at Sicily that "coherent" [i.e., moving together] waves of electrons might entrap deuterons and bang them together at unusually high velocities.

Surveying the mysteries of the early experiments, Garwin was baffled, above all, by the dearth of neutrons reported to be coming from the Fleischmann-Pons cells—only some 40,000 per cubic centimeter per second, about a billionfold less than would be expected from d-d fusion generating the amount of heat the experimenters were claiming. Garwin wondered how the roughly equal occurrence of the two dominant conventional reaction outcomes of d-d fusion could be so suppressed or distorted to leave such pathetically few neutron stragglers. He could offer no convincing possibilities. Garwin mused about the chance that the helium-4 outcome of d-d fusion could be depositing its energy throughout the lattice—the prospect offered by the Walling-Simons theory. This would generate heat but no potent gamma rays that occur, albeit rarely, in hot fusion (about one in ten million d-d reactions). Garwin could not believe this for a minute. He suggested that such a mechanism would have to be a billion times faster than the other possible particle outcomes. Magic words spilled out as Garwin noted the necessity of "a totally new physical phenomenon" to explain what he clearly thought was impossible, but he was unwilling to take any steps in that direction. Even so imaginative a physicist as Garwin could more readily believe that experimental errors or misinterpretations were afoot, than agree to seek new theoretical grounds for the experimental findings.

Garwin concluded with two marvelous quotes that he had brought home from Sicily, which seemed to express perfectly the coexistent wonder and disbelief that cold fusion had engendered: "Somebody is going to have to eat his hat," said L. Maiani; "We are also human, and need miracles, and hope they exist," said L. Ponomarev of Moscow. The question was—and still is—*who* is going to have to eat their hat?

✳ Burning Midnight Oil

Occasionally in science the uninitiated gain more ground than those attached to old ways of viewing a problem. It seems that in the cold fusion theory business, there was some advantage in not being too steeped in conventional wisdom about either electrochemistry or nuclear reactions. Peter L. Hagelstein might be such a person, or he might be a brilliant but hapless victim who fell into the swirling vortex surrounding the black hole of "cold fusion." As for theorizing about cold fusion, few could match in the sheer volume of speculations and calculations the work of Hagelstein. His theoretical work on cold fusion in the spring of 1989 caused an immediate storm of controversy, both outside and within the university. Many physicists and hot fusioneers at MIT became very antagonistic to his involvement with cold fusion. On the other hand, there were others in his Department of Electrical Engineering and Computer Science and in other departments who admired his bold thinking. Hagelstein's ideas bear some similarities to those of other theorists (particularly Julian Schwinger's independent work), but his simultaneous strength as an experimentalist and a careful observer of the experiments of others seems to make him stand out, even if his theories ultimately prove unable to give a workable mechanism for cold fusion (assuming, of course, that solid-state fusion is real).

Hagelstein's enthusiasm for high-risk, high-payoff scientific quests was ignited by the reports of cold fusion. But he did not begin speculating about it of his own volition—in fact, he was initially leery of the strange claims. After Utah announced cold fusion, Hagelstein was away from MIT at his old stomping ground, the Lawrence Livermore National Laboratory (LLNL) in California, where his abrupt 1986 departure and return to MIT had enveloped him in controversy about his work on the Star Wars X-ray laser, "Excalibur." He was consulting with the Laboratory on other matters, but his contact, Lowell Wood, urged him to put his assignment on the back burner and try to find a mechanism to explain cold fusion.* From then on Hagelstein was hooked and became a "believer," as critics would say.

Hagelstein, who grew up in Los Angeles and graduated from MIT, had returned to his alma mater directly from Livermore. He had received all three of his degrees in electrical engineering and computer science from MIT—gaining his Ph.D. in 1981. As an undergraduate, Hagelstein whizzed through the Institute in less than three years, yet still had time for athletics (track and swimming), playing the violin and piano, and a bit of musical composition and performance.

Hagelstein's Livermore affiliation had begun with a student summer job in 1975; he professes that on signing up he did not know that LLNL was the nation's premier nuclear weapons laboratory. There he had built his reputation for brilliance in laser physics and became known as one of the chief contributors to the concept of the X-ray laser, a project that he fell into apparently after some arm-twisting. He was the only one deemed agile enough to solve a thorny problem in testing the concept. Hagelstein's brilliant insight and suggestion to modify a competitor's (George Chapline's) formulation led to a highly successful underground test of the concept in late 1980. Hagelstein had bet a "million-to-one" against success, but a "concatenation of miracles," he says, made it work and this became literally his nightmare. Despite his intellectual growth in tackling difficult computational problems, Hagelstein wrestled with his conscience about his nuclear weapons work. He called it a "Faustian bargain" to preserve his soft-X-ray laser research.†

The effectiveness of a possible high-powered X-ray laser (which was to be energized by a nuclear bomb explosion) had been a key issue for a time in the debate over the direction of SDI. Hagelstein had returned to MIT in his department's Research Laboratory of Electronics to develop a table-top low-energy or "soft" X-ray laser for more benign applications. When he left Livermore after 11 years, some there commented that his departure would be a severe blow to the X-ray laser program. Of his experience at LLNL, he told the *San Jose Mercury* in 1987, "My own record of published scientific papers is not as strong as it might have been had I not engaged in classified research." But above all it seems that Hagelstein preferred campus life to the dualistic "management style" of the weapons laboratory.

Edward Teller had disparagingly told Hagelstein that he was a "pharmacist," his metaphor for someone who cops out of weapons work against evil empires—either the Nazis or the Communists. The lanky

---

*It is quite possible that LLNL had found out about Fleischmann and Pons's work in advance of the public announcement, because scientists there may have reviewed their cold fusion research proposal to DOE.

†An interview with Peter Hagelstein in *Omni*, May 1989, by Bill Moseley.

34-year-old professor seemed, indeed, an unlikely pharmacist to fall into the cold fusion controversy. Those who know Hagelstein well know him as a shy person and are fond of his boyish, soft-spoken manner and his insistence on scrupulous fairness. Having been burned a bit by some descriptions of the reasons behind his return to MIT from Livermore, he is media-shy and generally does not speak to the press. On the other hand, Hagelstein acknowledges that the universe should know what is going on in cold fusion.

Much controversy surrounded the announcement that Hagelstein was working on a cold fusion theory. Some scientists within and outside MIT were offended that anyone would waste time taking cold fusion seriously, thereby giving it some credibility. Yet Hagelstein believes that criticism—even severe criticism—is absolutely essential to good science and he relishes receiving suggestions on where he might be going wrong. He is also a man of good humor, much of which is self-effacing. An electrical engineer, not a physicist, he still feels quite comfortable treading on the sacred ground of theoreticians, which provokes some people in that community.

After nearly continuous round-the-clock work, as is Hagelstein's custom when he gets excited about something, between the 5th and 12th of April, 1989, he submitted four papers to *Physical Review Letters*, which proposed a unified theory to explain the phenomena reported in cold fusion experiments. Each of the four articles treated a different aspect of the theory. To the many instant cold fusion skeptics, this appeared to be exceedingly rash, given that the experimental support for cold fusion was then so very tentative and shaky. After all, why waste time on theories when the very existence of the phenomenon in question was in great doubt?

Seeing the possible enormous implications of his work, and for reasons having absolutely nothing to do with a personal desire for profit, Hagelstein immediately contacted the MIT Technology Licensing Office so that the MIT-supported developments could be protected for the benefit of the university. This was a perfectly natural thing to do and would not have raised a single eyebrow in connection with virtually any other field, but in the atmosphere of the brewing cold fusion war, it was almost a life-threatening act in academia! MIT did, indeed, immediately file several patent applications for technological devices connected with Hagelstein's theoretical analysis. These patent applications are still pending, along with more than 60 other filings for patents connected with cold fusion that have been submitted by many universities and individuals both here and abroad.

When word of these MIT patent requests emerged, there was no end of confusion—vaguely resembling the stir made by the University of Utah. In those frenzied times, the decision not to conceal the filings emerged from a strained, symbiotic consensus among the MIT News Office, the Technology Licensing Office, and a professor in the Department of Electrical Engineering and Computer Science. Immediately, people began calling the MIT News Office asking how MIT could presume to patent a *theory*. The Hagelstein-initiated patents deal with *technological applications.* No physical theory or phenomenon of nature can be patented. The specific technology governed by the MIT patents remains today a still undisclosed matter as the logjam of cold fusion patent applications winds through the mill.

The MIT News Office issued a press release that was absolutely clear: Theories were not being patented. Moreover, top officials at MIT had approved the contents of the release. Then MIT Provost John Deutch, a physical chemist, with the review and approval of electrical engineering professors Richard Adler (now deceased) and Louis Smullin, issued a general statement as part of the news release: "MIT is a place where creative individuals are encouraged to address scientific subjects of the greatest significance. We are pleased to see Professor Hagelstein proposing an explanation for 'cold fusion' and we are encouraging investigators both here and at other research institutions to continue their work on this most surprising phenomenon, which may have enormous consequences." How could anyone have disagreed with that? But this was cold fusion, and they did!

Long after the ensuing storm, Provost Deutch told me that he had reason for second thoughts about the wisdom of the MIT news release and wondered how the momentum for it had started. Frank Press, president of the National Academy of Sciences had ribbed him good-naturedly about the patents, when Press spoke at a colloquium at MIT at which Deutch was not present. Given the highly charged atmosphere surrounding cold fusion, almost anything that one could say on the subject was fair game for humor or an indignant assault.

The first version of the press release about Hagelstein's work went out on Wednesday, April 12, 1989, and told the world almost nothing about Hagelstein's theory other than that it dealt with "both quantum, collective, and coherent effects." Reporters were justifiably baffled. This unfortunate absence of information came about because an exhausted Hagelstein had rushed off to Logan airport to catch a plane to attend a meeting. He maintained his press silence. He was still working feverishly through many nights to complete his papers. With evident pride, he said he was putting finishing touches on a fourth one. On the morning of April 12, key members of the media were alerted that a press release on Hagelstein's theory would probably be forthcoming that day, pending hearing exactly what he wanted to say about his work. But at the last minute, Hagelstein left the News Office with instructions to issue only the following statements:

✳ "Papers describing a speculative theory on the new cold fusion have been written by Professor Peter L. Hagelstein, Massachusetts Institute of Technology, Research Laboratory of Electronics, Cambridge, Massachusetts, and submitted to the appropriate journals for scientific peer review. The papers were submitted between 4/5/89 and 4/12/89."

✳ "The model involves quantum, collective, and coherent effects."

✳ "Further details will be provided upon acceptance of the papers."

Members of the press were sorely disappointed with the few crumbs that the News Office was able to give them. The announcement was titled "MIT Professor Has a Theory to Explain 'Cold Fusion.' " It began, "An MIT professor, Dr. Peter L. Hagelstein, has submitted four papers to a scientific journal, each of which addresses a possible theoretical basis for the 'cold fusion' experiments that have been reported recently by researchers at the University of Utah, Brigham Young University, Texas A&M University, and Georgia Institute of Technology. MIT has filed patent applications in connection with this theoretical analysis. The News Office cannot discuss the specific technology prior to the publication of the journal articles." The journal was to be *Physical Review Letters*, but even that information was off limits.

A subsequently updated press release on April 14 said that Hagelstein acknowledged that his theory was "highly speculative, that little hard experimental evidence currently exists to support the claim of [cold] fusion, and that no support for any aspects of the present model have been demonstrated experimentally." The release quoted Hagelstein quite clearly, saying that his model "does not predict that [cold fusion] should or should not exist, but if it did, the model proposes how it might work." Hagelstein was not presuming to pronounce final judgment on the validity of reported cold fusion experiments. Far from it, he was only developing a theory, a theory that was to radically transform itself over the following eight or nine months as new information came in from the experimentalists.

The most extraordinary fact of the affair, however, was that few were spending much time wondering about the scientific implications of Hagelstein's theory. That there was any theory at all coming from such a noted scientist was the main news. All the energy was going into questions about the procedure and "etiquette" of scientific communication. This trend was enhanced by the difficulty of comprehending Hagelstein's ideas about "coherent fusion" in a palladium lattice—concepts that were still evolving. Even though scientists were getting much of their information about cold fusion from daily newspaper accounts, there was a strong disposition by some (mostly hot fusion physicists) against announcing developments in the press. This was definitely related to antagonism over the March 23 Utah press conference.

The focus of Hagelstein's April 1989 analysis was the possibility that within the confines of a metallic lattice, two deuterium nuclei might be able to fuse to form a helium-4 nucleus (garden variety helium) and then with a very exotic mechanism transfer the liberated energy of fusion (23.8 MeV) into organized vibrations of the lattice that would become heat. The exotic reaction is just like the rare one in plasma fusion, except the energy goes into the lattice:

$$D + D \rightarrow {}^4He + 23.8 \text{ MeV (lattice energy)}$$

This means that two deuterons (deuterium nuclei each containing one proton and one neutron) react to form a helium-4 nucleus (having two protons and two neutrons) and an amount of energy (23.8 million electron volts) that goes *not* into a dangerous, highly penetrating gamma ray, but into energy that stays in the palladium lattice and eventually appears as heat. By contrast, in high-temperature plasma fusion experiments, two deuterons combining to form helium-4 occurs very infrequently and results in a 23.8 MeV gamma ray (only in one instance out of every 10 million d-d fusions). Most of the d-d reactions result in either production of tritium plus a proton or helium-3 plus a neutron.

Though there was no good evidence for significant helium-4 production in cold fusion cells, such atoms—"ashes" of the reaction—might linger in the palladium rods, or so many people began to believe. Hagelstein at that time considered this to be the likely dominant reaction in cold fusion, just as Walling and Simons had at the University of Utah and Julian Schwinger at UCLA. He would later come to completely disagree with that initial conclusion (Chapter 12) and would propose an even more exotic mechanism, but one that still embodied the general idea of what he continues to call "coherent fusion."

Other theorists too were then speculating that if cold fusion were real—producing significant amounts of net power—the helium-4 route was a good possibility. But the energy would have to be dumped into the palladium lattice or into the electrolyte in liquid solution. That was the essence of what Walling and Simons had proposed, but their mechanism for coupling the energy of the reaction to the palladium lattice with high-energy electrons had problems. The pioneering notion that Hagelstein was developing—coherent fusion—was a concept that emerged naturally from his familiarity with laser physics. Lasers are *coherent* optical devices—light waves move lockstep with one another—and through stimulation of surrounding atoms intensify the ultimate effects, giving us bright, narrowly collimated beams of light. The coherent processes that might be going on in cold fusion had to do with something akin to a falling-domino kind of chain reaction. Hagelstein's calculations suggested that the laserlike mechanism of coherent fusion provided "a considerable enhancement of the fusion [reaction] rates at low temperature."

The analysis involved complex mathematical relations that describe the quantum mechanical properties of the metallic palladium crystal lattice, although Hagelstein allowed that cold fusion might also be occurring in liquid solution too. His first unpublished paper was titled, "A Simple Model for Coherent DD Fusion in the Presence of a Lattice." It was anything but simple to comprehend, and few people were taking the time to even try.

The coherent process would still involve so-called quantum tunneling, whereby the reacting deuterium nuclei could fuse and overcome the electrical forces of repulsion that normally keep them apart. Coherent fusion would enhance tunneling. The coherent fusion process would involve long chains of individual fusions, but it would have to be triggered initially by a cosmic ray or other energetic nuclear particle—an alpha particle, for example, possibly present in the local background radiation. Hagelstein offered the latter as a possible explanation of why the reaction often took weeks or longer to begin.

The fusion is sustained, he suggested, by *optical phonons* (high-frequency soundlike vibrations), "possible stimulation by electric current," and what he termed "coherent transitions between degenerate levels." He explained the presence of low numbers of neutrons and tritium reported in some experiments as "incoherent by-products" of the main coherent fusion reaction. In other words, the neutrons and the tritium were just from ordinary fusion reactions that occurred randomly now and then—secondary emissions from the main cold fusion reaction leading to helium-4. These ideas went into papers two and three, respectively: "Dephasing in Coherent DD Fusion and the Long Chain Model" and "Rates for Neutron and Tritium Production in Coherent DD Fusion."

In his fourth paper, "Phonon Interactions in Coherent Fusion," Hagelstein postulated that it may be possible to "couple the coherent fusion energy into electrical energy with some efficiency." In other words, go directly from the coherent fusion process to electricity, bypassing the need for heat to create steam.

✳ The Press Conference That Wasn't

The eleventh commandment in the cold fusion controversy had become, "Thou shalt not hold a press conference." Heaven forbid that the public should hear the rancorous deliberations of scientists—that some people were actually taking cold fusion very seriously, while others were rejecting it. There would, indeed, be no public discussion of Hagelstein's theories at MIT, contrary to reports that he had held a press conference to promote his ideas. But on Friday afternoon, April 14, Hagelstein did explain himself at a closed-door presentation to about two hundred specially invited faculty members, researchers, and students.

Professor Louis Smullin, a mentor of Hagelstein, introduced him. In his remarks, he linked the novelty of Hagelstein's theory, with the concepts of the laser and the maser that Nobel laureate Charles Townes had pioneered decades earlier. Hagelstein began in earnest quiet tones, but he was absolutely clear. "Thanks for your introduction. You're much too kind. If part of this is right, I hope it helps people. If it's not right, then next Monday I'll be looking for a job, I think." The audience laughed.

"I want to emphasize that this is a *theory.* I mean, some theories are right, some theories are wrong. These are some ideas. If you like, we've got the problem that we've got a tough thing to try to explain. I'm not going to claim that I can explain it, but I think I have some ideas of some things which *might* be going on. . . . Also, let me make one other thing clear: In this presentation, my goal is not to *convince*, my goal is to try to *explain.* If all of you walk out completely unconvinced, that's fine by me. But at least if you understand what it is I've tried to say, then I will have achieved what I was after.

"I want to start by explaining why cold fusion is absolutely impossible," Hagelstein said rhetorically. He showed what everyone knew to be depressingly true—that the reaction rate for low-temperature fusion of deuterium, calculated conventionally, is so fantastically small, one would not expect to be able to measure it. Then he made a logical slip that Plasma Fusion Center director Ronald Parker was quick to pick up on. There was a bit of scientific one-upsmanship:

Peter: "So what are the possibilities? One is that Pons and Fleischmann made a mistake. If so, so have a number of other groups and groups who haven't reported yet."
Ron, interrupting: "That's not true. I mean, it doesn't follow that BYU made a mistake, if Pons and Fleischmann did."
Peter: "Oh, I'm sorry. What I mean—I know that I'm in front of a critical audience. . . ."
Ron: "Georgia Tech already retracted, so we know they were wrong. What is the consequence of what you're saying? Why?"
Peter: "Once again, I'm taking an optimistic side of this. I have not heard of Georgia Tech's retraction and I have not heard of the retractions from the other places yet. In fact, my point of view on this is: If in fact it is true, how could it possibly work? O.K., is that more clear? If you like, this is not a proof. . . ."
Ron, calmly: Just throw away that view graph because it's wrong. It doesn't follow."

Completely composed, but bent on using a disarming gesture, Hagelstein walked ceremoniously over to the nearest circular file and dropped the "offending" slide in. The group loved it.

> **Peter continued:** "The first possibility, given that the alternate first possibility has been retracted, is that chemical energy can explain Pons and Fleischmann's result. And if you like, I think that's controversial. And the argument for is that potentially you can't explain sustained fusion power as being reported by Fleischmann and Pons. And in addition it's hard to explain neutrons from chemical energy. . . ."
> **Ron:** "You can explain neutrons by saying that what they measured weren't neutrons. You are taking a leap of faith here by suggesting that their results are, in fact, correct."
> **Peter:** "The whole talk is presupposing that they are right. Suppose they're right. If they are right, how in the world can you possibly explain it?"

Hagelstein resumed: "Another possibility is that there is something new. . . . Then the question is how might it work? In the first paper I propose two new ideas in this business. One is that the fusion, if it works, conceivably might be a coherent process and a coherent process in the sense of coherent processes in lasers. The second thing is that it might be a collective interaction between a lattice—mediated by the nuclear force, and mediated in a way which, in principle, you can be pretty quantitative about. . . .

"The second paper is a proposal that if you accept the possibility that the alleged fusion could be coherent, and if there are neutrons and/or protons and/or tritium observed, then you might be able to get production of these—I'll call them incoherent products—as a by-product of the initial coherent product. . . .

"I'm also proposing that given that you've initiated the fusion reaction and you're going coherently, once you get started you keep on sailing. It's conceivable that the total number of [fusions] based on one initiation could be a very large number. It could be 10^{10} or 10^{15} or something like that, because transitions between degenerate states—if there's something driving them—can just keep on going. You've got no mechanism to stop. . . ."

But even if the chain of fusions were to start, Hagelstein was faced with explaining how it could transfer its energy "benignly" to the palladium lattice, not through potent gamma rays: ". . . How are you going to unload 23.8 MeV," he asked? "That's not chicken feed. There are no 23.8 MeV phonons [extremely high energy vibrations in the lattice] running around. . . . the idea is that the coupling between the nuclear energy and the phonons occurs—one quantum of nuclear energy gets spread over something like 10^9 phonons." In other words, the energy of a single d-d fusion would be spread over a billion smaller vibrations in the lattice.

Hagelstein was thinking aloud that if there were any vibrations already present in the lattice—*thermal phonons*—they would stimulate the process. "If your sample happens to be warm or even hot," he said, "then . . . there are going to be more phonons around and it conceivably might work better. . . . I've got a scenario—if you like, a proposal for a mechanism which is kind of a concatenation of pieces and parts. You stack them one upon the other, and it's kind of like a house of cards or something. I'm not going to say that it can account for [cold fusion]. . . . If you've been up too many hours or had too much to drink or something, you might be able to rationalize how this works."

Then Hagelstein spoke about the mechanism to climb the "coulomb mountain" that was so daunting to Garwin and others. Hagelstein thought that energetic particles present in showers of cosmic rays, or even alpha particle radiation coming from radioactive elements in tiny amounts in a lattice, would do the trick. A particle would hit a deuteron and cause it to begin a cascade of fusions.

Next, something would have to stimulate the process into more dramatic action. Perhaps vibrations at high frequency—optical phonons—would do? Then after much calculation: "I look at it, I scratch my head and say, 'Oh, this is a stimulated process.' " It had to work something like a laser, Hagelstein deeply believed, and continues to believe. If heat-producing cold fusion is real, he thinks it has to be from a coherent process that no one ever suspected could happen, which is why no one had ever taken the time to carry out these wild computations and speculations. It was too bizarre, too outrageous even to contemplate without that kick in the rear that Fleischmann, Pons, and Jones had given the scientific world.

Coherent fusion requires a falling domino kind of chain reaction. "You start out with a thermal state and a cosmic ray brings you up to a state where there is at least a finite chance that you could begin a fusion chain," Hagelstein said. "Once you've gotten up into this state—if in fact you can fuse—then you are going to go into the next state. If the energy is given to the lattice—and shortly I'm going to argue that the lattice isn't the only place where energy could go—you have 23.8 MeV of energy to dump. You dump most of it into the lattice." The process bootstraps itself along by heating the vast array of atoms, pumping energy into them but saving just enough to provoke subsequent fusions in the chain of deuterons.

There could be "un-fusion" too. The reaction, being coherent, could theoretically reverse itself, Hagelstein claimed. Some would say Hagelstein had stretched people's imagination to the breaking point, that he had "gone off the deep end" with this. It would be unlikely, he said, but a helium-4 nucleus within the lattice could disassemble itself and become two deuterons—absorbing 23.8 MeV of lattice energy in the process.

Hagelstein suggested that the energy might also unload itself directly into the excitation of electrons and "electron holes." He imagined that if fusion were going on and if the fusion is actually managing to couple to the current going through a palladium cathode, then the cold fusion process might become in some sense an amplifier of electrical power.

In closing his talk, he remarked, "Hopefully there are enough details in these papers, so if there is anything that I didn't explain well, it would be obvious where I made my mistakes in the paper. In years from now when I'm thinking back fondly at my time spent at MIT, I'll pore over how it is that I went wrong and made a mistake."

Then onto questions. Someone asked the obvious shocker, "Once it starts by a triggering event, which could be caused by a cosmic ray, is there any mechanism to stop it from continuing indefinitely?" In less polite terms, could this turn into some kind of bomb? Hagelstein's assessment was "no."

Someone asked, "Have people tried to initiate these things by taking palladium and exposing it to some high-energy radiation of particles?" Hagelstein said that he had proposed that to some researchers, who were going to try it.* Professor Philip Morrison remarked, somewhat facetiously, "I hope they work on a small scale!" There was laughter at this. Comic relief. But this was uncharted territory, and who could be sure it was safe? Fleischmann and Pons had more than once issued stern warnings about the possibility of fusion "ignition," after having seen a partially vaporized palladium electrode.

Hagelstein said, "Given this is optimistic—believing that this happens at all—then you have to *optimistically* believe that all this is going to work to *pessimistically* believe that they are going to end up hurting themselves." Then the swan song, "Any other questions or comments? Well, let me prevail upon you to help me find a job next week." Whether Hagelstein had given anyone additional reasons to take cold fusion more seriously could not be read in the polite applause that followed.

✳ Theories Cooking

Hagelstein's papers were never published in *Physical Review Letters*, but they did not fade away—certainly the ideas didn't. There were many requests for preprints, strong evidence that people—physicists, engineers, and entrepreneurs—were curious about theoretical and other pos-

*Researchers at Yale looking for neutrons from cold fusion cells tried it.

sibilities in the field. The peer review process at some journals had sandbagged many proponents. In Hagelstein's case, no matter how the papers were changed, they would not pass muster—some physicists would say because they were technically flawed. Hagelstein kept revising them as new information filtered back from the experimentalists. But it was not unusual at this time for papers supporting the possibility of cold fusion to be rejected by certain mainline journals like *Nature*. Other journals were beginning to publish the works of other theorists as well as experimentalists. In December, Hagelstein would deliver a substantially revised version of his theory to a San Francisco meeting of the American Society of Mechanical Engineers, and the comprehensive paper was published in the meeting's proceedings (Chapter 12).

Schwinger, who shared the Nobel Prize for physics in 1965 with Richard P. Feynmann and Sin Itiro Tomanaga, was working on a theory, which, like Haglestein's, relied on transferring energy to the lattice. Schwinger was one of the earliest theorists to take very seriously the possibility that a new path to fusion was now open, and he dedicated himself to understanding it (see Chapter 13). He also had *his* cold fusion papers rejected by *Physical Review Letters.* Schwinger later wrote to me, "Although I anticipated rejection I was staggered by the heights (depths?) to which the calumny reached. My only recourse was to resign from the American Physical Society." Theorist Giuliano Preparata and his colleagues in Italy even thought they had arrived at a rough way to calculate power levels in cold fusion ("First Steps Toward an Understanding of "Cold" Nuclear Fusion," *Il Nuovo Cimento*, Vol. 101 A, May 1989: 845–849.) Robert T. Bush entered the theoretical fray early on with his "transmission resonance model" of cold fusion. There were many others, and some who may never have gone public for fear of being ridiculed.

Some theorists initially thought they could "explain away" electrochemical cold fusion. Their attempts were valiant and interesting. George Chapline and two of his colleagues at LLNL proposed that neutron bursts were coming from heretofore unsuspected chains of deuterium fusions that were catalyzed by cosmic ray muons in the confines of a palladium lattice.* Linus Pauling, Nobel laureate in chemistry (1954), suggested that the electrolysis-induced pressure of deuterium within the palladium lattice was giving rise to a "higher deuteride"—possibly PdD_2. The alloy's decomposition later during electrolysis might be responsible, he thought, for the apparent excess heat and the destruc-

*Guinan, M.W., G.F. Chapline, and R.W. Moir, "Catalysis of Deuterium Fusion in Metal Hydrides by Cosmic Ray Muons," Lawrence Livermore National Laboratory, UCRL-100881 Preprint, April 7, 1989.

tive meltdown of an electrode that Fleischmann and Pons had observed (*Nature*, Vol. 339, 1989).

Within MIT itself, Professor Keith H. Johnson of the Department of Materials Science and Engineering suggested, at about the time that Hagelstein was coming out with his theory, that cold fusion in a palladium crystal lattice might be related to the metal's well-known superconducting properties when it was saturated with deuterium or hydrogen. He believed that most of the energy being reported in cold fusion experiments was of an unusual chemical nature, which he and a graduate student Dennis Clougherty explained in a published article.* Johnson and Clougherty did not rule out that nuclear fusion reactions could be occurring concurrently at a relatively low level in such experiments, but would add negligibly to the observed power production.

Johnson came to believe that the vibrations or *optical phonons*—the term used by Hagelstein, albeit in a different context—along these chemical bonds either: (1) promote electron pairing necessary for superconductivity at low temperature, or (2) are the main source of heat production in cold fusion experiments. They had estimated the total chemical energy of the optical phonons in a palladium lattice that is saturated with deuterium. Regarding the possibility of fusion reactions, he claimed that electrical "screening" of deuterium nuclei by some of the palladium atoms' electrons would increase the probability of quantum tunneling by reducing the electrical repulsion barrier between the nuclei. This might facilitate either nuclear fusion or the decay of the deuterium nuclei. But Johnson's theory did not agree with the tremendous excess energies that at least some workers were beginning to see. His theory had come out at a time when many of the heat experiments were still very much in a state of flux.

The role of experiments is to inform theory. When things were proceeding rationally in the cold fusion controversy, that was happening. When arguments verged on the irrational, theory was too much in the driver's seat and experiments were being ignored. If heat was not being generated anywhere else, it was certainly coming from the inelastic collision of blind theory with an array of baffling experiments.

---

*Johnson, K.H., and D.P. Clougherty, "Hydrogen-Hydrogen/Deuterium-Deuterium Bonding in Palladium and the Superconducting/Electrochemical Properties of PdH_x/PdD_x," *Modern Physics Letters B*. Vol. 3, October 1989: 795–803.

8 Yes, We Have No Neutrons

We too often forget that not only is there a "soul of goodness in things evil," but very generally a soul of truth in things erroneous.

Herbert Spencer
First Principles, 1861

When you have eliminated the impossible, whatever remains, however improbable, must be the truth.

Sir Arthur Conan Doyle
The Sign of Four, 1890

✳ The Closing Vise

UNTIL EARLY MAY 1989, investigations of cold fusion suffered from a serious paradox: both too few and too many neutrons being observed. Not enough neutrons to explain the excess heat (trillions per second were needed), but still too many neutrons to be credible. Neutrons and neutron bursts were being seen in other experiments, but no one but Fleischmann and Pons was finding their vast numbers. This was the Achilles heel of cold fusion.

Researchers at MIT's Plasma Fusion Center realized almost immediately—as had physicists elsewhere—that Fleischmann and Pons's neutron measurements were suspect. While an interdisciplinary team of investigators tackled the question of excess heat, another group worked to demolish the neutron claims of the Utah pair. They succeeded, and in the process cleared the air of much confusion surrounding cold fusion. The MIT neutron-busters may have thought that they had dug cold fusion's grave, but what they actually accomplished was to prevent further wild goose chases after copious neutrons and to make cold fusion even more appealing to some investigators than it had originally seemed. Wonder of wonders, electrochemical cold fusion, if it was nuclear at all, was largely an *aneutronic* process (producing few if any neutrons)—the dream of some hot fusion researchers like George Miley and others.

Fleischmann and Pons had claimed to have detected neutrons via telltale gamma rays produced by neutron-proton fusions; it was one of their supposedly convincing pieces of evidence for a nuclear process. In their experiment, they had deliberately placed their cell within a tank of water. One of the reasons for that water had to do with measuring heat—they needed to stabilize the temperature external to the cell. But the other purpose of the ordinary water bath surrounding the test cell was to measure possible neutrons that might emanate from the unknown nuclear process.

Neutrons appearing with possibly very high energy would slam into the vat of light water molecules and would be slowed down or thermalized, as nuclear physicists say. In this process, most often an individual neutron fuses with a proton (the nucleus of a hydrogen atom in an H_2O molecule) and generates a gamma ray—one with about 2.22 MeV (million electron volts of energy).* The gamma rays from the neutron bombardment of the water tank would go flying off every which way; some would be captured in the gamma-ray detector that Fleischmann and Pons set up. They had a detector with an efficiency such that about one out of 100 gamma rays actually entered it. (It is possible to measure neutrons *directly* with a sophisticated detector, but measuring gamma rays as evidence of neutrons is a perfectly good way too.)

It was the forementioned 2.22 MeV gamma ray that they claimed to have detected, inferring from the calculated inefficiency of their detector that their test cell must have produced about 40,000 neutrons per second. As evidence that they had detected gamma rays of that particular energy, they showed how the energy was distributed into different energy channels of their detector—the statistical spectrum of the gamma-ray energies. (Not every gamma comes into the detector with precisely 2.22 of MeV energy. There is a broadening of what might otherwise be a sharp spike on a graph.)

When the Fleischmann-Pons paper was received at the Plasma Fusion Center about a week after the announcement, the gamma evidence looked superficially compelling. But a closer look at the spectrum revealed three very troubling aspects of the data: (1) In the plot of gamma ray counts versus energy, the curve was too narrow for the predicted ability of the detector to resolve that sharp a feature. There were technical reasons to believe that the peak couldn't be narrower than about twice what Fleischmann and Pons showed; (2) More important, some-

*The reaction is: $p + n \rightarrow d + \gamma$ (2.22 MeV). A proton joins with a neutron to form a deuteron, and energy comes off as a very energetic gamma ray that has an energy of about 2.22 MeV.

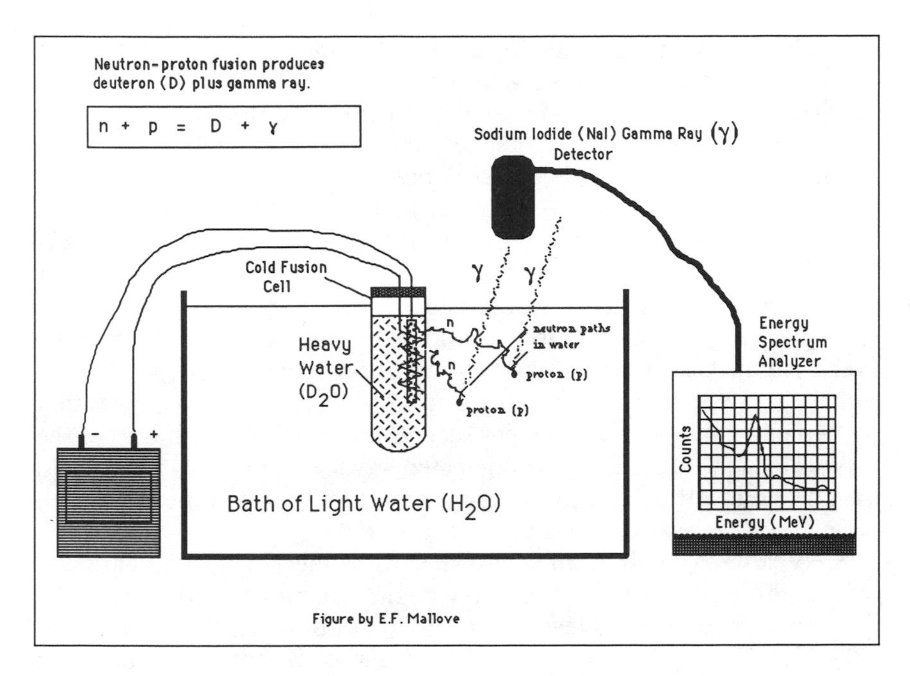

A schematic view of Fleischmann and Pons's experimental setup.

thing else was missing that had to be there. Absent was the so-called *Compton edge*—a feature that comes from the basic process by which this kind of detector operates; and (3) The maximum count rate for the neutrons inferred by Fleischmann and Pons was inconsistent with other data reported by them. Moreover, Fleischmann and Pons had not even shown the "bigger picture" of the entire gamma-ray spectrum across the full range of energies, which would have shown where this particular peak fit with respect to other peaks caused by natural background radiation.

The reality of the gamma rays, hence that of the neutrons, was immediately suspect. But the MIT group had to be sure. They set out at once to demonstrate conclusively the errors in the neutron counting.

The team did an experiment to simulate how genuine neutrons would bombard water and affect a detector functionally similar to the one Fleischmann and Pons had used. The crew took a plastic garbage can, filled it with water, and put a neutron source in it, called a "Pu/Be" source, because it contains the artificially made radioactive element plutonium and the light element beryllium. The Plasma Fusion Center group got their gamma peak in the spectrum, just where they assumed they would see it, and they saw the all-important Compton edge. They didn't have to waste any time trying to generate *possible* neutrons with an electrochemical cell; these were *real* neutrons.

The peak was at 2.22 MeV, but unlike the Fleishmann-Pons peak, it was as wide as theory said it should be; it also had the required Compton edge. It had other additional features—a so-called "escape peak" and so forth; these are all the properties that have to appear when one detects gamma rays with this kind of detector. All this was made clear in a technical paper that the group sent to *Nature.* The paper suggested that the spectrum obtained by Fleischmann and Pons had some unknown electronic origin—that it was an experimental artifact. It also charged that "the purported γ-ray line actually resides at 2.5 MeV rather than 2.22 MeV." The spectrum most surely did not suggest that anything like 40,000 neutrons per second could be emanating from the Utah cell. If the Utah group had detected any neutrons at all, they would have to be below a level of 400 per second (still quite a healthy rate), or so the MIT group calculated.

To calibrate their gamma-ray spectrum, the MIT group carefully measured the background radiation in their laboratory by having the detector turned on to detect "everything"—gammas from cosmic rays as well as from natural radioactivity in the environment. Like a road map being laid out, this careful measurement gave essentially a calibration of all the gamma-ray spectrum lines; it showed the precise electronic channels of their detector where peaks of gamma rays from certain radioactive disintegrations would be found.

Here is where the modern "miracle" of television enters. On some of the initial television broadcasts, the probing eyes of the TV cameras captured the details of the gamma-ray background spectrum being measured in the University of Utah lab; the glowing display screen of the electronic spectrum analyzer was shot straight on. They knew that Fleischmann and Pons had to obtain this background spectrum in order to find out exactly where their gamma-ray signal was. So in discussing the Fleischmann-Pons gamma-ray data in their May 18, 1989, scientific correspondence in *Nature*,* the group actually published one of the TV images below the PFC's own measurement curve of the natural background of gammas. Team leader Richard Petrasso describes it as "the first time in the history of television that data was taken from it and actually published in a scientific journal."

The MIT scientists did place calls to the University of Utah and spoke, not directly to either Fleischmann or Pons, but to some of the researchers in their lab (Marvin Hawkins and R. Hoffmann), who told

---

*Petrasso, R.D., X. Chen, K.W. Wenzel, R.R. Parker, C.K. Li, and C. Fiore, "Problems with the γ-ray Spectrum in the Fleischmann et al. Experiments," *Nature*, Vol. 339, May 18, 1989: 183–185.

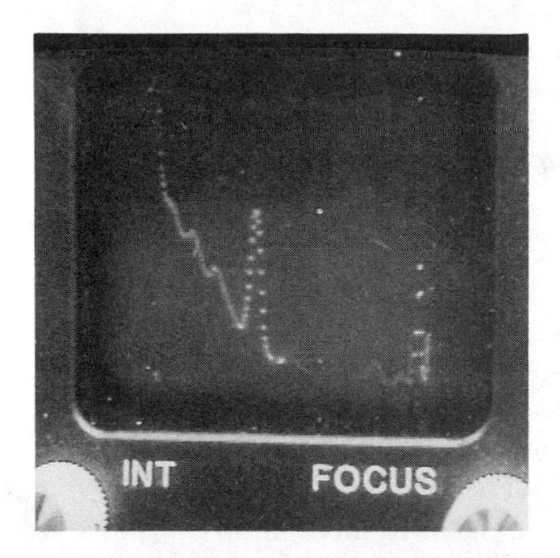

A television image that the MIT group asserted was the spectrum of the gamma-ray background obtained by Fleischmann and Pons. (Courtesy KSL-TV, Salt Lake City, Utah)

them that the TV-portrayed spectrum was a real background spectrum taken in the lab. This strengthened the belief at MIT that the Utah gamma-ray data could not be interpreted safely as evidence of fusion neutron emissions. The bottom line of the MIT analysis published in *Nature* was: "We can offer no plausible explanation for the feature, other than it is possibly an instrumental artifact [an electronic "glitch"] with no relation to a γ-ray interaction."

A month later on the scientific correspondence pages of *Nature* there followed a technical exchange on the gamma-ray question—a somewhat heated debate between Fleischmann and Pons and the MIT group. In their rebuttal, Fleischmann, Pons, and Hawkins made a number of assertions contradicting the original MIT group's statements: (1) They wrote that Hawkins "did not state that the quoted television spectrum was made in these laboratories, as it most certainly was not." (2) They criticized as innuendo part of one of the captions to the MIT-published TV display of the Utah background spectrum. The caption had noted a "curious structure at about 2.5 MeV." The Utah people asserted that the "curious structure" was nothing more than the electronic cursor—like the blinking cursor on a computer screen. [Indeed, in retrospect, that is what it seems to be.] (3) They asserted that what they really had meant to say was that their data revealed a gamma-ray peak at 2.496 MeV, whose origin they could not identify, but it was certainly not, they agreed, the conventionally understood 2.22 MeV peak of fusion neutrons smashing into hydrogen in water. They concluded with a bold assertion: "In spite of the problems underlying the interpretation of these spectra, we consider that the measurements show the emission of gamma rays from the cell environment; removal of the cells leads to the removal of the signal peak."

For their part, the MIT group stood its ground and brought up even heavier artillery, virtually accusing the Utah group of dangerously loose interpretation of data. The MIT group said that Fleischmann and others were now changing their story on the location of the peak. First it was supposed to have been conveniently at 2.22 MeV (where fusion neutron evidence would be expected). Now, Fleischmann and Pons were claiming to have found a peak at 2.496 MeV. (The matter was even more confusing, because some evidence suggests that Fleischmann and Pons initially thought that their gamma-ray peak was at 2.5 MeV, but that they hastily and unjustifiably moved its location down to 2.22 MeV when reminded by knowledgeable physicists in the first week after the announcement where it should be.) The MIT people wrote bitingly: "Although they inexplicably no longer claim to have observed the 2.22 MeV neutron-capture γ-ray, they now contend that their γ-ray signal line is a true γ-ray of energy 2.496 MeV and, most importantly, that this signal is evidence for nuclear reactions in their cell. They make this claim despite their inability to identify the nuclear process associated with their purported 2.496 MeV γ-ray or to account for its distinctly unphysical lineshape." Though Fleischmann and Pons scored some points in the debate, for example, the "curious cursor" gambit, it was clear that the argument was lost. Whatever the case, Martin Fleischmann told me in 1991, "I'm absolutely 100 percent sure that there was a difference in the gamma-ray spectra, between blank and measured, in our measurements. I'm sure that is correct. But *why* that was so is not clear." As far as why they may have initially put their peak at 2.5 MeV, Fleischmann said, "It was a straight mistake." He implied that the error was made because they were thinking of the 2.45 MeV fusion neutron that produces the 2.22 MeV gamma ray.

By lining up the MIT background peaks with the newly revealed Fleischmann and Pons peaks, the MIT group concluded that Utah had completely miscalibrated its curve—the molehill of gamma evidence was actually at 2.8 MeV—not 2.22, not 2.496, but 2.8 MeV. What a mess, how could cold fusion survive this disaster? Yet it did.

✳ An MIT "Bombshell"

Richard Petrasso was preparing to present the MIT neutron analysis before a meeting of the American Physical Society in Baltimore. Appropriately enough for the impending distress of cold fusion, the session of the meeting devoted to the subject was to occur on May Day. But before Petrasso could give his technical paper, news of the MIT work leaked out—how embarrassing! Only Fleischmann and Pons supposedly stood accused of the "crime" of the artful scientific leak. In truth, it came from a bit of innocence in dealing with a reporter. This is not unusual for scientists who have to deal with investigative journalists, for rarely in the course of normal scientific work does such an intense and sensitive issue arise.

Director of the MIT Plasma Fusion Center Ronald Parker and his colleague Professor Ronald Ballinger had given an exclusive interview on Friday (April 28, 1989) to science reporter Nick Tate of the *Boston Herald.*

They were very frank in their discussions with Tate, leaving little doubt in Tate's view that they felt Fleischmann and Pons's work—the neutron results, in particular—could then hardly be taken seriously, because it was so unreliable. What came out on Monday morning was a banner headline in the *Herald* that blared "MIT Bombshell Knocks Fusion 'Breakthrough' Cold." The article said that Parker had accused Fleischmann and Pons of having engaged in "scientific schlock" that was possibly "fraud." But on Sunday evening before Monday's *Herald* hit the news stands, the piece went out over a wire service from the *Herald.* CBS Television news had picked up on it and called Parker at home to get an additional response from him. When the wire service account was read over the phone, Parker was flabbergasted to hear the words "fraud" and "scientific schlock" in their prominent, and in his view, out of context form.

The two professors well knew that they had been very harsh on Fleischmann and Pons in the interview with Tate, but the way the words came out were not what they said they had intended. On the other hand, in reviewing what Tate heard from the two,* it was clear that the article was close to the mark in its general assessment that Parker and Ballinger believed that Fleischmann and Pons were trying to short-circuit the scientific review process to get federal funding as quickly as possible. Tate has said that an unfortunate editorial change in the lead sentence of the article also presented the word "fraud" in an incorrect manner, making it seem that the MIT scientific paper had employed that word, when it had only appeared in the interview as the worst possible characterization that could be applied to the Utah "misinterpretation" of data.

A distraught Parker called me at my home in Bow, New Hampshire, very late Sunday evening. He was upset about what was to come out in the paper and wanted to know what could be done. I listened in disbelief at what clearly seemed to me at the time a serious possible distortion of Parker's views. In the early A.M. hours over the telephone, Parker and I prepared a statement for a number of wire services in an attempt to cancel the effect of the *Herald* story—in particular, denying

*Partial transcript of the interview, *Boston Herald*, May 2, 1989: 4, as well as the original tape.

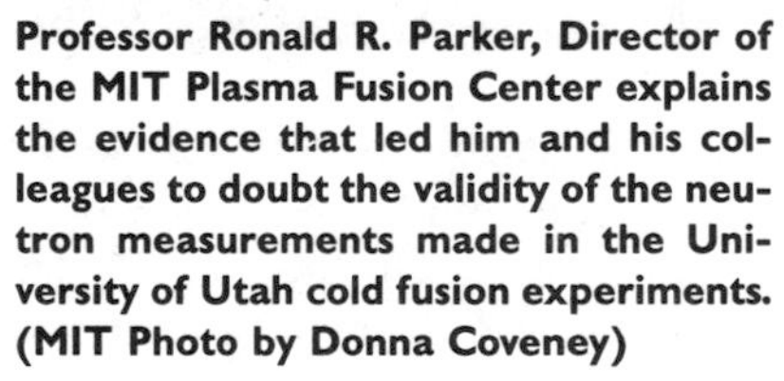
Professor Ronald R. Parker, Director of the MIT Plasma Fusion Center explains the evidence that led him and his colleagues to doubt the validity of the neutron measurements made in the University of Utah cold fusion experiments. (MIT Photo by Donna Coveney)

Dr. Nathan Lewis, Associate Professor of Chemistry at the California Institute of Technology and one of the more strident critics of the heat measurements performed by Fleischmann and Pons. (Courtesy California Institute of Technology)

the specific use of the word and phrase "fraud" and "scientific schlock," and saying that the article seriously misquoted him and gave a largely incorrect view of his discussions with Tate. In the news release to the wire services Parker said, "The [*Herald*] article erroneously characterizes remarks that I made regarding the cold fusion experiments done at the University of Utah. Specifically, I did not: (1) Deride the University of Utah experiments as 'scientific schlock' or (2) Accuse Drs. Fleischmann and Pons of 'misrepresentation and maybe fraud'." Of course the *Herald* article came out, and the wire services did report that there was some disagreement as to what Parker had said.

That morning at the MIT News Office was one of the the most hectic and hair-raising that I can recall. My colleagues and I determined quite early on, with Parker's agreement, that it would be advisable to hold a press conference to clear up the matter. Tate arrived at the Plasma Fusion Center, urging that the two scientists hear the tape of their interview before talking about its substance at the news conference. Parker and Ballinger listened politely behind closed doors (I was not present) and were still satisfied that their interpretation of what they had said was correct. But Tate still felt he had a good case.

Whatever the true meaning of harsh words said in or out of context, the well-attended news conference held that morning diffused the explosive situation. Parker did not dwell extensively on the matter of the words that appeared in the *Herald.* Instead, he delivered to the press the substance of what Richard Petrasso would have to say in Baltimore that night about the gamma-ray spectrum and the lack of neutrons. Thus, by his own admission, out of necessity, he had done exactly what he had determined *not* to do–engage in "science by press conference."

During the news conference, Parker explained in detail his team's efforts to investigate the Fleischmann-Pons work. Most of his remarks addressed the "glitch," as he phrased it, or "some sort of interference or malfunction of equipment"–that the team concluded had put into question the Fleischmann-Pons gamma-ray data. He said that if any neutrons at all were present in the Utah experiment, they would have been at a level at least 100 times less than the 40,000 per second suggested by Drs. Fleischmann and Pons–a level even if accepted was still 100 million times less than might be consistent with their reported level of energy production (according to *conventional* fusion reactions). "We're asserting that their neutron emission was below what they thought it was, including the possibility that it could have been none at all," he said in a telephone interview with the Associated Press earlier in the day. At the news conference, Parker was eminently fair to Fleischmann and Pons in acknowledging that other exotic, though in his view unlikely reactions not producing neutrons, could still be occurring. He said, "If fusion is taking place at all, it's a very strange kind of fusion. . . . It would require suspension of disbelief."

He ended his remarks on an upbeat note, "In a way, what's good about this and the media attention being given to it, is that there is a heightened awareness of the capability of fusion. For one who has spent his whole career on fusion, I find this very rewarding. The idea of turning deuterium into helium is a very, very attractive fuel cycle. It does away with many of the problems that we worry about–CO_2 emissions, oil spills, problems of waste disposal. So I hope what this leads to is a good feeling about fusion and perhaps a renewed sense of commitment to develop fusion.

"My own outlook is that there are at least three ways now that fusion might work: magnetic, inertial–laser-produced fusion–and now let's say even cold fusion–who knows? My objective in my career is to make any *one* of them work. I'd be delighted if there were a breakthrough of this type in cold fusion or in magnetic or laser fusion. We in the fusion community hold that big goal out there of wanting to make it happen and are gratified to see the public respond.

"But there could be a down side as well. There could be a negative reaction and there has been to some degree. When this [cold fusion] first came out, there was more than one individual saying, 'There go those fusion guys again. It must be Congressional hearing time again.' It hurts because we in the magnetic fusion area and the inertial area are really working very hard and believe very much in what we're doing."

As for his remarks about "scientific schlock," Parker said, "I regret coining that phrase, but when I said it I meant it to describe the whole process (of) somebody avoiding or short-circuiting the whole (scientific peer-review) process." He had not said specifically that Fleischmann and Pons had committed fraud, but there was an implied suggestion that they had asked for money for research in advance of reasonable confirmation. It was clear that the PFC people, including Parker and Ballinger, were very angry that Fleischmann and Pons had sought funding for cold fusion from the federal government, before the reality of the phenomenon had been demonstrated to the scientific community. The proposed cold fusion research money was to have come out of the hot fusion budget—a plan that congressmen, certainly not Pons or Fleischmann, had suggested.

Parker offered an olive branch of sorts to Fleischmann and Pons: "As you know, Professors Fleischmann and Pons are chemists and I believe they have very little experience with these systems—neutron systems. And so one can easily explain perhaps, let's say, confusion that might have resulted in resolving some of these points. It's very easy to get carried away in doing experiments and perhaps see things that sometimes you don't challenge enough, and perhaps that's what happened in this case."

Nick Tate left the news conference happy that some degree of peace had been made, even though he still stood by the substance of his article. Ultimately, Fleischmann and Pons seemed to consider the matter closed. The incident was one of the first public indications of how deeply emotions were running in the effort to come to grips with the astounding reports of cold fusion.

✳ Nuclear War in Baltimore

Though relative peace had been declared in Cambridge, Massachusetts, a nuclear war of sorts broke out that evening in Baltimore at the meeting of the APS—a war over claims about nuclear phenomena. The highly skeptical group of physicists had gathered to listen to their colleagues explain over and over again why cold fusion couldn't possibly be real.

In fairness, there were some papers that attempted to rationalize cold fusion as a possible nuclear phenomenon, but not many. The sentiment was heavily against Fleischmann and Pons. Of the 19 papers presented, none offered experimental evidence to support Fleischmann and Pons's claims.

The largest organization of physicists in the United States, the APS has 40,000 members. Its 1989 Baltimore meeting drew some 1,500 members, more than 1,000 of whom attended the specially arranged evening sessions on cold fusion. Originally only one evening session had been scheduled, but with enough interest and sufficient speakers, they filled up two sessions, each lasting until midnight. The targets of the angry physicists, Fleischmann and Pons, had been invited to the meeting but both initially declined. They said that they were preparing for the upcoming Los Angeles meeting (May 8) of the Electrochemical Society. Had the duo gone to Baltimore, it would have been like sheep wandering into a lion's den, though Fleischmann's agile defensive wit might have been a strong shield. Actually, Pons had told the chairman of the meeting that he would try to come, but just before the meeting began, he called and said he was too involved in discussing cold fusion with members of the congressional committee. Many participants speculated that the duo had stayed away largely for fear of a scientific "lynching."

Petrasso reiterated the results of the MIT neutron analysis in his talk. Some have suggested that the story in the *Herald* had preconditioned the audience and participants for their assault on Fleischmann and Pons. Others have suggested that the MIT attack was a serious emotional turning point, for Pons in particular—pushing him to adopt an increasingly defensive stance toward critics. In a companion paper also delivered at the Baltimore meeting, Dr. Stanley Luckhardt of MIT's Plasma Fusion Center presented the results of the cold fusion group's efforts to reproduce the University of Utah's reported phenomena—thermal effects, tritium, neutrons, and all. In these experiments, conducted over a five-week period (and still continuing at that time), the researchers used various sizes and metallurgical treatments of palladium electrodes in a number of electrolysis cells, containing both light and heavy water.

The MIT team used sensitive, calibrated calorimetry, to measure possible excess heat from the cells, and sensitive radiation monitors to detect possible fusion by-products. The results had revealed, the team concluded, neither excess heat—heat that could not be explained by ordinary resistive heating in the electrolysis bath—nor radiation statistically above the natural background. Yet an expert electrochemist visiting the United States from Bulgaria, Dr. Vesco C. Noninski, would later take issue with the interpretation of the MIT calorimetry and believes that the MIT experiment may indicate the presence of an unknown heat source (see Chapter 9).

Caltech contributed the other side of the jaws of the vise that was squeezing Fleischmann and Pons. Chemist Nathan Lewis delivered his biting attack on the Utah heat measurements. "One of the main things we've learned during the course of these experiments is just how easy it is to fool oneself into thinking that there is an effect when there actually is none," or so Lewis thought. He didn't know it then, but his own heat measurements and methodology would ultimately come under the intense scrutiny of others, who believe that an inappropriate calibration was made and who found reason to believe that Lewis and others might have measured a mysterious excess heat after all (see Chapter 9).

Lewis said that his group had found no excess heat in their experiments with cold fusion cells. He found fault everywhere in the methodology of Fleischmann and Pons. His biggest complaint was that the solution in the generic Utah cell was not properly stirred, so temperature measurements intended to reflect the condition of the entire medium only bore on a local and perhaps excessively heated region, thereby giving false results. He said, "These problems may lead to errors large enough to cast serious doubts on published determinations of excess heat. When these errors are avoided, we obtain no evidence for excess heat production. . . . We have no reason to invoke fusion to explain any of their results. There is no evidence of any unknown nuclear process. At this point, we can find no evidence of anything except conventional chemistry." He mocked their cold fusion cells as "great fusion refrigerators."

Lewis's big point was the allegedly neglected stirring, and this became the simple-minded canard that was widely touted in the media. It was easy for journalists to buy that line—even the best of them—because it seemed so straightforward and obvious. It was much simpler than all this complex talk surrounding gas-recombination, thermoneutral potentials, and so on. Lewis was sufficiently stirred up at the meeting to deal Fleischmann and Pons what some might consider the ultimate scientific insult: that their device "violates the first law of thermodynamics," that is, the conservation of energy or, as is often said, "the universe offers no free lunch."

Lewis claimed his techniques were 10,000 times more sensitive than Utah's. He had carefully measured tritium, gammas, helium, and the all-important excess heat. None of these measurements showed anything unusual. The small amounts of helium that Pons had alluded to were probably caused, he said, by small quantities of the gas in laboratory air.

Lewis was fighting mad. "If they are going to have publication by press conference," he averred, "I want to have peer review by press conference." Assaulting the calculations that Fleischmann and Pons had carried out in figuring their excess heat, Lewis made it appear that the duo had sheepishly admitted scientific sins: "We confronted them with this information privately, and they acknowledged that this is the way they did their calculation."

Lewis's theoretical physicist colleague at Caltech, Steve Koonin, blasted Fleischmann and Pons with the weapon of scientific orthodoxy: "If fusion were taking place, we would see radiation in one form or another, and you would simply not be able to hide that radiation." Koonin told *New York Times* reporter Malcolm Browne at the time of the meeting, "It's all very well to theorize about how cold fusion in a palladium cathode might take place . . . one could also theorize about how pigs would behave if they had wings. But pigs don't have wings."

Koonin raised the possibility that the gamma-ray spectrum Fleischmann and Pons had obtained was the result of radon decay in their lab. His line, "I don't know how much radon they have in their lab, but I do know they mine uranium in Utah," drew a good laugh. On the other hand, the MIT-analysis contradicted Koonin's claim that radon was responsible for the Fleischmann-Pons gamma-ray error. Koonin charged, nonetheless, that Fleischmann and Pons's claims "had not been proven by the usual standards we expect in scientific discourse." Then to sustained applause (and a weak smattering of boos) he administered his immortal coup de grace: "My conclusion is that the experiments are just wrong and that we are suffering from the incompetence and delusion of Doctors Pons and Fleischmann."

Nor did Koonin let Steve Jones off the hook: "Our theoretical studies indicate that the BYU results are quite improbable, but perhaps not impossible. However, we know of no way of accounting for the University of Utah results." Koonin told Warren Strobel of the *Washington Times*: "I think it's going to be a wake for the claims of Messrs. Pons and Fleischmann."

Jones presented his by now well-known work and continued his litany against claims for excess heat. "There is no shortcut, no royal road to fusion energy in my work. . . . My reaction to theirs is like the ratio of a $1 bill to the entire national debt." Concluding his talk, he said, "Is this a shortcut to fusion energy? Read my lips: No!"

A team at Oak Ridge National Laboratory reported detecting no neutrons above background so far, since beginning their effort in late March. Ditto for a Brookhaven-Yale collaboration. Arch-skeptical physicist Douglas Morrison of CERN (European Center for Nuclear Research) in Geneva reported that essentially all Western European experiments had failed to duplicate Fleischmann and Pons's work, which he characterized as "pathological science."

Dr. Walter E. Meyerhof, a professor of physics at Stanford University found Huggins's earlier reported heat-measuring experiment flawed for the same reason that Caltech's Lewis faulted Utah's work. Meyerhof recited a disparaging poem:

"Tens of millions of dollars at
stake, Dear Brother,
Because some scientist put a thermometer
At one place and not another."

For his part, Huggins was unmoved. He told the media, "We stand convinced. We don't see anything wrong with our experiments."

More unfriendly fire: Physics Nobel laureate Leon Lederman of Fermilab said that Fleischmann and Pons should be given a "public spanking" and their university's president fired for misleading the world. His assessment of the duo: "First they were just sloppy, but then they became venal."

Finally, "science by press conference" occurred again, degenerating even further into "science by poll." At a news conference on the second day of the Baltimore cold fusion fest, Steve Jones asked for an impromptu "straw poll." He asked nine of the session's leading speakers whether they were at least 95 percent confident that the University of Utah claim to have generated heat by fusion could be ruled out. Eight answered "yes" and one, Rafelski, Jones's colleague, wisely withheld judgment. Rafelski commented, "This should not be taken as the matter is settled." However, Yale physicist Moshe Gai said of his group's work, "Our results exclude without any doubt the Pons and Fleischmann results." The panel voted more favorably on whether the claim that neutrons were being seen in a number of cold fusion experiments could be ruled out—three of nine kept an open mind.

University of Utah's James Brophy fired back at all of this. He told the press that Fleischmann and Pons had "spent five-and-a-half years on this research and now people spending four weeks on crude experiments are criticizing their work. Four weeks is not a long time. It's unfortunate." If the physicists had seen Pons's critique of all this in the *Deseret News*, they would have been gape-mouthed: "The absence of neutrons doesn't concern us in the slightest. We couldn't be happier. We and other scientists will soon tell them why this is so."

The chorus of criticism at the meeting came shortly before Pons was to meet with aides to President Bush. Apparently the noise was heard at the White House; Pons never did meet with any administration officials, as he was supposed to on May 4. Utah Governor Bangerter called the apparent snub "shabby treatment." The University of Utah work had become the laughingstock of science following the Baltimore meeting; the state of Utah was thrown into the pot for good measure. *Science* magazine recalled the 1972 claim by three scientists at the University of Utah to have made an X-ray laser that didn't turn out to work as advertised. These false alarms, including cold fusion, were being identified as the "Utah Effect." *The New York Times* editorialized that the University of Utah ". . . may now claim credit for the artificial heart horror show and the cold fusion circus, two milestones at least in the history of entertainment, if not science." The editorial also said, "Given the present state of evidence for cold fusion, the government would do better to put the money on a horse."* The editorial admitted, however, that there might be something to the Utah work.

One certain effect of the Baltimore meeting: Palladium prices plummeted on the commodities exchange. Then weeks and months afterward, readers of the *Bulletin of the American Physical Society* were regaled with talk of cold fusion's demise. Robert Park, not an official spokesman for the APS, but whose "What's New" news-opinion columns appear in the organization's *Bulletin* wrote: "The corpse of cold fusion will probably continue to twitch for awhile, even after two nights of unrelenting assaults at the APS Baltimore Meeting" (May 5, 1989); "Alas, experiments conducted by Sandia scientists, using multiple neutron detectors in a deep underground laboratory, would seem to bury the Brigham Young reports of cold fusion right alongside the more extravagant claims of Pons and Fleischmann."

✳ The Death of Cold Fusion: Greatly Exaggerated

Heaping scorn on scientific research, calling it "pathological science," being so self-assured as not to admit the possibility of error in one's own experiments that seemed to show others to be in error; these were some of the many thrusts used by critics of cold fusion to "kill it." The stench of the death metaphor was in the air. But cold fusion was good at playing possum. Tinkerers and scientists around the world kept plugging away with their electrochemical cells, despite skepticism by "the scientific establishment" and by a barrage of ridicule in the media. Enough money was flowing from the major electric utility research consortium, the California-based Electric Power Research Institute (EPRI), to keep some researchers afloat, a bit of leftovers from DOE, Utah money, some private efforts, and rampant bootlegging of tinkering time from defense and other research contracts. Case Western Reserve University, Stanford University, Texas A&M, and the Bhabha Atomic Research Institute in India had gotten positive results by now. There was now reason to believe and hope.

**New York Times*, April 30, 1989. But the *New York Times* editorial page was not necessarily a font of scientific wisdom. Many years earlier a *Times* editorial had excoriated the now-accepted ideas of the great American rocket pioneer, Robert H. Goddard—an editorial that they symbolically withdrew after the first manned Moon landing in 1969.

Yet the cold fusion situation was deeply mysterious. Some laboratories weren't reproducing cold fusion phenomena, while others ostensibly doing the same experiments were. Soon after the Baltimore meeting, a Duke University group gave up in frustration; University of Michigan researchers announced that they too had reached a dead end; ditto for North Carolina State University in Pons's native state. There was a pervasive feeling that Fleischmann and Pons were consciously not revealing enough particulars for others to be successful in their experiments. There were even hints that the two felt almost relieved by the skeptical reaction in Baltimore. With less laboratories striving to reproduce their work, they could continue their studies at a more deliberate pace—as they had originally intended. And they did. Also there were patent considerations, as University of Utah lawyers tried to keep a lid on information flow.

Everyone was awaiting May 8, when at the special cold fusion session of the Electrochemical Society spring meeting in Los Angeles, Fleischmann and Pons were supposed to present a "thorough, clean analysis" of the thermal aspects of their experiment. Pons told Jacobsen-Wells of the *Deseret News*, "We are going to supply all the information that we can. People evidently are misunderstanding a lot about calorimetry. A lot of people are making calorimetric measurements with instruments that may not be suitable for these experiments."

The meeting began with controversy over the relative absence of critical scientists; had it been arranged to be a celebration of only positive results? Lewis of Caltech was present at least as a token skeptic. As he had done in Baltimore, he proclaimed his numerous permutations and combinations of materials and conditions, all of which had failed to show excess power or nuclear products. "I'd be happy to say this is fusion as soon as somebody shows that it is," a self-assured Lewis told the 1,600 assembled. Fleischmann and Pons were having no trouble. Now they were claiming to get bursts of heat lasting a few days up to 50 times the power input to their cell—the claim was even more extreme than before! Was this a tip-off that they were really onto something, or that they had completely gone off the deep end? To rebut Lewis, they showed a brief film clip of a bubbling cell in which they had injected red dye. Within 20 seconds the dye had spread uniformly through the cell, intuitively giving the lie to Lewis's accusation about improper stirring.

Concerning their neutron results, Fleischmann and Pons backed off a bit, acknowledging reluctantly that their measurements were deficient and were the "least satisfactory" part of their research. They said that they would rerun their experiment with a new detector. More disturbing was their withholding of the long-awaited and promised ^{4}He measurements. There was an emerging feeling (not necessarily a correct one) that if there were no copious neutrons, there had to be helium-4 to make the claim for a nuclear process. The Fleischmann-Pons rods were being analyzed for helium by Johnson-Matthey Corporation, the 170-year-old British precious metals supplier, under an agreement of exclusivity with the company. This was the presumed reason for the turning down of many other offers to do the rod "autopsy." Fleischmann had admitted at the meeting that if no helium were to turn up, "it would eliminate a very strong part of our understanding of the experiment."

Bockris from Texas A&M, Huggins from Stanford, and Uziel Landau from Case Western all backed up the Utah duo with positive heat measurements. At a press conference Huggins said, ". . . It's fair to say that something very unusual and large is happening. There is conclusive evidence there is a lot of heat generated here—much larger than the proposed chemical reactions that people suggest might be happening." A thinly veiled criticism of physicists by a Society official, Dr. Bruce Deal, drew applause: "Unlike other societies, we do not attempt to solve complex technical problems by a show of hands." But not every electrochemist left the meeting convinced. The experiments were subtle, apparently difficult to reproduce consistently, and of course totally unexplained. Steve Jones again reiterated his faith in his neutrons and disbelief on the question of heat—at least in cold fusion cells. Cold fusion might still be partly responsible, he thought, for the hellish conditions inside the planet.

Soon cold fusion would face increasingly acid opposition. Martin Deutsch, professor of physics emeritus at MIT had told *Science News*, "In one word, it's garbage." (*Science News*, Vol. 135, May 6, 1989.) Some media had essentially written it off. Scientists who had genuinely tried to make cold fusion happen, but who for reasons still not clear could not coax their cells into working, would be joining the ranks of the opposition. They were frustrated and mad. They had wasted precious research time chasing rainbows. Enough was enough! Time to move on.

But those who believed in the tantalizing results of some experiments would not be stilled. Others who were bold enough to theorize about fantastic mechanisms to explain cold fusion did not give up either. They persevered, egged on by the serious critics.

If people were having trouble finding neutrons, perhaps the mysterious "cold fusion" was a kind of nuclear reaction that was largely neutronless—as the MIT analysis seemed to suggest. As skeptic Petrasso himself would say in January 1990 at a lecture at the PFC, "We may turn out to be the big allies of Fleischmann and Pons if they can now prove that they have fusion, because what we've demonstrated now is that they basically didn't have any neutrons at all coming from their heat-producing cell. . . . So now they can claim that they are having neutronless heat generation." If this turns out to be true, a mind-boggling technological revolution may be in store for us.

9 *New Mexico Sunrise*

There is some phenomenon, and goodness knows what it is—I daren't talk about it being fusion, but it's something—which sometimes produces a heat. After you've got it, you might look for fusion products—they might be there.

John Bockris
Santa Fe, New Mexico, May 1989

Where there's smoke, there's fire.

Folk wisdom

A foolish consistency is the hobgoblin of little minds, adored by little statesmen and philosophers and divines.

Ralph Waldo Emerson
Essays, 1841

EVEN A CASUAL WITNESS to the proceedings of the "Workshop on Cold Fusion" held in Santa Fe, New Mexico, at the end of May, would realize that reports of the demise of cold fusion were premature. A mere two months after March 23, researchers from around the world—most from the United States—gathered at what was to be a remarkable first comprehensive meeting on cold fusion. They came to report the results of their hurriedly engaged experiments and to confront enigmatic anomalies with theories and debate.

The gathering sponsored by the U.S. Department of Energy and Los Alamos National Laboratory, whose roots of course were in the Manhattan Project, drew some 500 researchers. But many more scientists and interested citizens "attended" the Sweeney Convention Center meeting via live satellite broadcasts. Neither Stan Pons nor Martin Fleischmann came to witness the proceedings firsthand.

Barely one month into the cold fusion controversy, LANL—one of the significant birthplaces of hot fusion—had decided to sponsor this cold fusion conference. Cochairmen of the meeting were Nobel laureate physicist J. Robert Schrieffer of the University of California, Santa Barbara (one of the main contributors to the accepted theory of low-temperature superconductivity), and Dr. Norman Hackerman of Rice University. During the two and a half day meeting, some 110 papers would be presented, 75 of them as poster displays, and the balance orally.

Los Alamos's director, Dr. Siegfried S. Hecker, opened the meeting with a marvelous bit of understatement: "Cold fusion, if it exists, has turned out to be very complex and most elusive." Then, Dr. Gajewski, a mover in DOE funding of cold fusion research, characterized the government's role: "Immediately following the Utah-BYU announcement, scientists at virtually all DOE laboratories spontaneously set up experiments to reproduce or otherwise verify the reported phenomena. . . . In all these activities, the Department has but one objective: to find out the scientific truth about the occurrence or non-occurrence of nuclear fusion reactions in solid materials." The frenetic activity of the federal laboratories followed the mid-April directive by U.S. DOE chief Admiral James Watkins. About a dozen federal laboratories were by then engaged in research on cold fusion, aimed at providing input to a preliminary report that would be finalized by the end of November.

✳ Some Heat, Some Light

The meeting had a breathless quality—a feeling that everyone had arrived from all-night sessions with test tube, pen, and paper, and indeed, many had. John Appleby of Texas A&M University, one of five groups there working on cold fusion research, led with his team's exciting results on excess heat. So fast were events unfolding, that some of the data were obtained as recently as the day before. The group had measured excess power coming from closed cells (cells in which no gases or liquids could enter or leave) using very fine electrodes in a precision "microcalorimeter," with a heat-measuring quality some 1,000 times better than coarser conventional calorimeters afforded. The wary and careful Appleby prefaced his talk with cautionary remarks that his results "prove nothing" and had "a number of possible explanations. . . . We do not as yet have any evidence that would stand up in court—there are loose ends all over the place." Yet his group was obtaining anywhere from 16 to 30 watts of power output per cubic centimeter of palladium—about what Fleischmann and Pons were claiming. Forming the ratio of the excess power to the volume of the palladium electrode had by now become standard in cold fusion research, even though it was uncertain exactly from where in the cell the power was coming.

Typically, it had taken some 60 hours for an Appleby cell to begin producing excess heat when the electrolyte was the lithium deuteroxide (LiOD) that Fleischmann and Pons had used. No excess heat had emerged from an 80-hour test in which the palladium electrode was replaced with a "dummy" platinum one. Moreover, ordinary lithium hydroxide (LiOH) as electrolyte—without the apparently all-important deuterium (D) of LiOD—gave no excess power during 140 hours.

Appleby casually described "a result which is rather remarkable," the quenching of the excess heat when sodium deuteroxide (NaOD) was added, followed by the excess heat coming back when lithium deuteroxide (LiOD) was reintroduced. There was slight chance, he claimed, that the excess heat could be accounted for by the much-debated chemical *recombination* of oxygen (O_2) and deuterium (D_2) gas. Those who had attempted to explain away the "excess heat" by O_2-D_2 recombination thought they had found a possible energy "bookkeeping" mistake: Part of the electrical power that went into a cell was accounted for as being consumed in breaking the D_2O apart into deuterium and oxygen gas—D_2 and O_2. If, however, the D_2 and O_2 were to chemically *recombine* anywhere within the cell—including in the space above the fluid level as the gases bubbled up—this reverse reaction would give back energy to the cell and might make it appear that more heat was being generated than was being put in via the electrical power.

Appleby said, however, that it was ". . . hard to believe that recombination occurs only with LiOD." He and many of his colleagues were saying that all evidence pointed to chemical recombination occurring only to a *negligible* extent, yet critics were using presumed recombination as a weapon against people reporting excess heat. The energy coming out (per palladium atom at least) represented some 1,000 times the available chemical bond strength. His conclusion was a dramatic prelude to what followed: ". . . chemical explanations, unless it's recombination, seem unlikely. . . . I think we can discount chemical explanations unless we have a trivial explanation. That is to say, for reasons we don't understand, recombination or partial recombination is occurring only under the conditions when we have Pd (palladium) and LiOD (lithium deuteroxide) present."

The Appleby work was far more precise, to be sure, than the heat measurements that Fleischmann and Pons had offered. It was measuring the approximately 30 milliwatt (thousandths of a watt) excess power output to an accuracy of 1 percent (three ten-thousandths of a watt). There was much less reason to doubt its significance. These electrochemical cells were closed, nothing flowing in or out except, of course, electricity and the conduction of heat. The Appleby group calorimetry later prompted even skeptical MIT researcher, Richard Crooks, to openly praise the work, "The group from A&M has shown us the finest calorimetry that we have seen so far in any of the talks surrounding cold fusion."

After Appleby, nuclear chemist Kevin Wolf, from the Cyclotron Laboratory at Texas A&M, reviewed the nuclear measurements that he had made on cells being run by the electrochemistry group. Electrochemist John Bockris had asked colleague Wolf to present the nuclear measurements, and Wolf recounted seven instances of tritium being found. Considering the startling nature of his results, he presented them calmly, almost understating them, "I would consider [them] overpowering levels compared to background levels." Seven out of ten cells had shown tritium, indisputably a radioactive product that had to have come from some nuclear process. The question already had to have been on everyone's mind: Was there some possible inadvertent contamination involved? However the tritium had emerged in the A&M cells, Wolf's careful checking had convinced him that it was, indeed, tritium. In May 1989, even ordinary elementary tritium checks could rightly be challenged, given their momentous implications of nuclear processes at work. Some of the tritium levels were 20,000 times elevated above the residual count level (about 100 tritium disintegrations per minute per milliliter of heavy water). Wolf noted that one cell had produced 3×10^{14} tritium atoms, "a significant number," he said.

Wolf had looked for but had not found gamma rays coming out of any cells, but he *had* found neutrons, he said. The neutron flux rose not more than about four times the natural background level due to cosmic rays and the like, a noise level which he put at 0.8 counts per minute. He called his own work "very strong evidence for detection of neutrons," and to this day he has not disavowed that work, despite his present doubts about the validity of his tritium measurements. Rhetorically he asked and answered, "Convincing? You bc thc judgc. Upon analyzing the data, however, we are convinced.

"Yes, there is the possibility of pre-loading of tritium in the electrodes, not a very big one, however. It would have had to have been quite an accident—of storing the palladium rods near tritiated water." But just to be sure there was no contamination, Wolf was conducting experiments, which were even then in progress, involving melting of electrodes before use to flush out impurities such as tritium.

Finally, it was the turn of an experimenter at a federal laboratory to speak out publicly. Charles D. Scott of Oak Ridge National Laboratory (ORNL) in Tennessee represented but one of several groups then working on cold fusion at Oak Ridge, the former site of the secret Manhattan Project uranium enrichment effort. His group was running Fleischmann-Pons-type cells and attempting to measure neutrons, tritium, heat, and gamma rays. Some of their tests had already run 350 to 400 hours. Scott described low levels of neutrons about four times above background—"an anomaly that we can't explain," Scott said. He

Dr. John Appleby, Professor and Director of the Center for Electrochemical Systems and Hydrogen Research at Texas A&M University, whose high-precision measurements revealed unexplained excess power in a Fleischmann-Pons-type cell. (Courtesy Texas A&M University)

Dr. John O'M. Bockris, Distinguished Professor of Chemistry at Texas A&M University, whose experiments that yielded measured excess power, and high, but irregularly occurring bursts of tritium, led him to become one of the leading proponents of cold fusion. (Courtesy Texas A&M University)

described a general increase in overall neutron emission during the test, which dropped after shutting off current to the cell. As to heat, there was a suggestion of about 10 percent excess power output at about 80 hours into a test—"certainly no greater than 10 percent over energy input," he said. He offered no convincing evidence of tritium or gamma rays.

Next, the same MIT team that had assailed Fleischmann and Pons in May offered their thoroughly negative report. Speaking for the group of 16 members who had been burning midnight oil for the past two months was chemist Richard Crooks. After 60 days of running with many different kind of cells, the MIT group had no excess heat to report, and no fusion products. They had looked for them all: neutrons, tritium, helium, all the conventional possibilities. They had found no indication of any anomaly, much less anything presumptively nuclear going on. Crooks delivered the group's message, which was to be heard over and over again at the conference: One must look for fusion products commensurate with any excess heat that is found. The MIT team had gotten neither excess heat so far as they knew, nor obvious fusion products,

Dr. Kevin L. Wolf, Professor of Chemistry at Texas A&M University, whose report on measurements of neutrons and tritium in Fleischmann-Pons-type cells electrified attendees at the Santa Fe workshop. (Courtesy Texas A&M University)

so it was a point made only to caution other groups. The MIT team had run a heavy water Fleischmann-Pons cell in parallel with one using light water. After ten days, nothing, but this was only ten days. Crooks flashed Petrasso and others' comparison of the MIT-obtained gamma-ray spectrum with Fleischmann and Pons's—the one that was to appear in *Nature.*

The MIT work did not go unchallenged, however. Physicist Yeong E. Kim questioned whether the MIT work was a fair comparison with the Fleischmann-Pons cell, given that its electric current level was less than the high range of current density (the rate of electrons flowing through each square centimeter of surface of the palladium electrode) used by the duo. Crooks replied that the group had initially chosen a middle range current used by Fleischmann and Pons. Later, they would try different conditions and higher currents, he said. But there was to be no later for the MIT group, at least in 1989 and through 1990. They had gotten discouraged or were too disbelieving to try any further.

Bockris offered a challenge to the MIT work, but one that was aimed at others as well: "What I'm saying applies to most of the papers with negative results. . . . The first thing that Martin Fleischmann said to me when I called him up after the announcement was you won't see anything under four to six millimeters [diameter of palladium rod]. If you take four to six millimeter rods you have to have a charging time of more than 72 days, so charging times alone will prevent attainment of the Pons-Fleischmann conditions. . . . It is of course quite pointless to do great numbers of experiments with very sophisticated nuclear detection apparatus and find negative results if you haven't first of all shown that you are getting heat."

Microphone in hand in the front row, Bockris was facing off directly with Crooks, "There is some phenomenon, and goodness knows what it is—I daren't talk about it being fusion but it's something—which sometimes produces a heat. After you've got it, you might look for fusion products—they might be there. But to look for them when you don't get the heat, well, something about shutting the door when the horse is gone—the horse was never there." Crooks could offer no challenge to this bit of philosophy other than to fall back on the admonition to correlate fusion by-products with heat, a perfectly reasonable position. The trouble was, however, that if this were some mysterious new phenomenon—even if one had "gotten heat," who could be so bold as to say that all possible reaction products had been searched for and found absent?

But did MIT really have no evidence for excess heat? Two expert electrochemists have examined the MIT thermal data carefully, after doing their own precision "classical" calorimetry in a cold fusion experiment that indicated excess power. Scientists from Bulgaria, Drs. Vesco C. Noninski and his father, C. I. Noninski, were impressed enough with the MIT work to evaluate it from a different perspective, and they are convinced that an MIT heavy water cell shows clear evidence of excess power—a specific excess of 1.71 watts per cubic centimeter 80 hours into the test, while the corresponding MIT light water cell shows no excess. They have submitted their work in 1990 to *The Journal of Fusion Energy* and another analysis to *Fusion Technology*. They have corresponded with the MIT cold fusion team, who remain equally convinced that the MIT experiments show no excess power "within estimated levels of accuracy." Though one cannot prove conclusively that the MIT heavy water cell gave rise to excess power, for a variety of reasons neither is it possible to say that MIT's calorimetry rules out excess power production—in its own cell or any other.

Another negative report came from Dr. J. Paquette of the Canadian Chalk River Lab in Ontario. The Canadian team had tried many kinds of cold fusion experiments, all to no avail. Kim from Purdue again objected that the Canadians had used too low a current density. Stanley Luckhardt popped up to observe, correctly enough, "So far, in all the experiments reporting excess heat, the excess evolved is not greater than the power, in [electrochemically] producing D_2." Matthijs Broer of AT&T Bell Labs also reported negatively: no neutrons whatsoever, even at a level (less than 0.1 neutron per second) below what Steve Jones had found.

Some physicists had privately observed that Caltech and MIT were engaging in a kind of "pincer" maneuver to attack the beleaguered crew at the University of Utah, both east and west coasts firing shells in the direction of the Great Salt Lake. It was Caltech's turn, and chemist

Nathan Lewis took aim. He averred that his group had never seen any neutrons above background level and had gotten absolutely no excess heat, despite the numerous kinds of rod treatments that they had attempted. It was a seemingly impressive volley. Cells at Caltech had been run for hundreds of hours in a massive collaboration between the chemistry and physics departments. No tritium had been found either, nor helium of any kind. But Lewis said that the "excess heat" being measured—on the order of 6 percent maximum—"clearly could be consistent with a small amount of [oxygen-deuterium gas] recombination." Case closed, or so was the implication. "We don't absolutely know for certain that this is not excess heat," said Lewis in reply to a questioner, "but the light water controls have also run 5 to 6 percent excess. If I got a 10 percent difference here I wouldn't think that would be evidence for anything." Caltech's negative heat results, being more extensive and elaborate than MIT's, were to be cited very often in the coming months as convincing evidence against the reality of cold fusion.

Again, the Caltech results by Lewis and others are not by any means the final word in cold fusion calorimetry, even though *Nature* magazine's editors seem to think so. The Drs. Noninski have examined the Caltech results and are convinced that because of possible errors, that Caltech may actually have recorded evidence of excess heat at about the level obtained by Fleischmann and Pons.* The Caltech calorimetry appears to have two glaring problems: (1) a faulty procedure the essence of which is that the potentially unknown heat source within the palladium electrode could cause an incorrect calibration and (2) a completely erroneous method for deciding whether a difference exists between the power production in heavy water and light water cells.

Just as he had politely hit MIT, Bockris went into action against Caltech. It sounded a bit like a line from Shakespeare's Julius Caesar—"They are all honorable men. . . ." Bockris said, "I want to make a contribution to Dr. Lewis's talk, because I want to tell all of you who are not electrochemists in this audience that Dr. Lewis is world-recognized as an extremely competent and excellent electrochemist. . . . I want to ask why it is that Dr. Lewis and all the people working with him can't replicate these results? It is becoming increasingly clear that you can or you cannot. People get it—some people do, some people don't. It is a phenomenon of great irreproducibility.

"If I knew exactly what to do," he continued, "I would be back at Texas A&M telling him exactly what to do. Because, of course, at Texas

---

*Drs. Noninski have repeatedly submitted critiques of the Caltech work to *Nature.* Editor David Lindley now says that he does not know whether Caltech or the Noninskis are correct, but he is apparently leaving it up to Nathan Lewis of Caltech to decide whether the critique will be published! Drs. Noninski have now submitted another critique of Lewis's work to *Science.*

A&M, although we've got a lot of results, many of our results are negative, like Dr. Lewis's." He suggested why many people were not getting the positive results that others did: (1) not enough deuterium was being jammed into the palladium lattice, that is, insufficient D/Pd ratio; (2) the need to have high enough current passing though each square centimeter of electrode (current density) and the need simultaneously to operate with adequate voltage conditions, known as "overpotential" by electrochemists; and (3) the critical need to eliminate hydrogen, H, from the palladium to allow sufficient D to get in.

Lewis showed impatience with this line of reasoning, saying that there was no public knowledge that Pons and Fleischmann had "poisoned" their electrodes with contaminants to create special voltage conditions. "If that's true it would be very important to know—obviously that would make a difference. . . . It would be useful for people who see effects to tell us what their H content is." Lewis grinned, somewhat nervously, being a bit on the spot as Robert Huggins also challenged him: "You annealed (preheated the palladium) at 350°C, ccrtainly not enough to get H out. . . . We found that we need to do this 8 to 12 times. . . . You need to be at 80 percent of the melting point [of palladium] to drive out the H."

Coming to the defense of Lewis, John Wacker of Pacific Northwest Laboratory threw a barb at Bockris, complaining about the transmission of information from Fleischmann and Pons via the Texan with the British accent: "Frankly—and I suspect the sentiment is shared by many here, we're getting a little tired of science being done by news conference. . . . For a while we were going through a fact of the week. . . . I think it's time we started clearing the air on these things." There was hearty applause from the audience and Lewis smiled approvingly.

An extremely revealing reply came from Bockris evidencing his own frustration, but getting to the nub of the whole affair: "I agree largely with the person who has just spoken. I've talked extensively to Fleischmann and Pons—maybe I've talked to them six or seven times, something like that. Each time I've had hedged information. They'll give me something, they won't give me something. They say some things, they contradict it. It's not a clear situation. It's my own personal opinion—and I'll go on the record as saying this, and certainly this will get back to them: I don't think they know exactly the conditions themselves. I think they are playing with it and trying to find the right conditions themselves, and I don't think that they can reproduce it. They are often very puzzled about the fact that some rods don't work and they don't know why. I think they are only two or three times better than we. So we have to find these conditions. But I think the overall point I'm making is that very many people have reported positive results. Many people have reported negative results, and we have to find out what the conditions are."

A host of speakers were not in the main line of investigation—the attempt to simply replicate the Fleischmann-Pons experiment and find excess heat or neutrons. Some spoke of attempts to use explosives to apply high pressure to metals infused with deuterium, and thereby possibly induce fusion. Attempts to observe neutrons coming out were to no avail, though it was interesting to see how creative the researchers were in arranging conditions that might arouse the cold fusion Genie. A collaboration between MIT and Brookhaven National Laboratory combined a bit of the Fleischmann-Pons-Jones palladium-deuterium approach with the "other" cold fusion—muon-catalyzed. At the Brookhaven AGS accelerator on Long Island, the team was setting up to determine the effect of a beam of muons on samples of palladium, titanium, and other metals that had been infused with deuterium.

Nate Hoffman of Energy Technology Engineering Center (ETEC), Rockwell International Corporation, described the complexity of the surfaces of the supposedly simple palladium rods that experimenters were using. Many people were sending their rods to be analyzed to his DOE-supported laboratory. Hoffman said, "I have examined specimens from all sorts of tests in cold fusion. I have always seen growths on people's specimens, the growths being platinum, sometimes chromium, and iron, sometimes products from the glass, sometimes the entire surface covered with growths that are very sharp pointed, which I'm sure would interfere with any weight measurement and might even interfere with current density and overvoltage considerations." This complexity would have to be reckoned with to determine whether any two supposedly similar experiments were truly identical.

Theorists had their opportunity too during a session on the possible physics of cold fusion reactions. G. M. Hale from LANL presented a pessimistic analysis of how deuterium-deuterium fusion reactions might be enhanced by charge-screening of electrons, and by such exotic mechanisms as the so-called Oppenheimer-Philips effect in which deuterons approached one another with their neutrons head on. The bleak conclusion: Neither screening nor the Oppenheimer-Phillips effect would do much to enhance reaction rates. Michael Danos, of the National Bureau of Standards, who was hard at work himself on a cold fusion theory, disagreed with Hale. The serious bottom line in all this, however, was that "fiddle and twiddle physics"—blending a bit of this or that "conventional" physical mechanism to explain elevated fusion rates—was not working. So ended the first day of the New Mexico conference.

✳ Steve Jones's Mother Earth Soup

Steve Jones is a likable person when he speaks. So as he attempted to address the Wednesday morning audience with a nonfunctioning microphone, one could laugh with him as he joked good-naturedly. In his gentle voice, he said, "Neutrons are very weak, as you know." Disarming his potential critics even more: "I told my sister that we had some neat results to report this morning. And she said, 'Did you say *heat* results?' And I said, 'No, not *heat* results, *neat* results.' I want to clarify that at the outset."

Jones, the one true "celebrity" speaker at the conference, was first at bat on the second day of the meeting, a session devoted to measuring elusive neutrons. His theme: "Cold Fusion in Condensed Matter: Recent Results and Open Questions." He had a lot to say, and of course he wanted to straighten out any doubts about where his ideas had come from. He traced the history of the BYU work, which began, he said, in 1985. It had become clear that muon-catalyzed fusion was faced with some fundamental difficulties. He had suggested to an associate, Clint Van Siclen, that perhaps they should look at other experimental techniques. Would it be possible, for example, to create fusion at room temperature by increasing the density of isotopes?

Jones and Van Siclen had published their piezonuclear fusion paper in March 1986 and had given a talk at BYU on muon-catalyzed and piezo-cold fusion. Professor Paul Palmer had proved instrumental in developing further ideas, because he was aware that in hot places on the Earth, particularly at volcanoes, there is a thousandfold increase in the ratio of the helium isotopes, $^{3}He/^{4}He$.

When *Nature* published Jones's work in 1989, they had asked him to reduce the emphasis on this geological fusion business. "The peer review process is a wonderful thing, as you know," laughed Jones. But significantly, the researchers had calculated that 10^{-24} fusions per deuteron pair per second could account for the anomalous terrestrial ^{3}He. If they could imitate Mother Earth's presumed cold fusion, they reasoned, they could detect this extremely low fusion rate in the laboratory by observing resulting neutrons. Jones recalled "a very exciting day," April 7, 1986. Rafelski had come to BYU where their brainstorming session occurred. To create fusion in the lab, they considered using various metals like palladium, platinum, aluminum, nickel, even lithium. "We talked about this very important guiding principle: nonequilibrium conditions. We talked about shocking the hydride, heat, and vibration. This is 1986; we outlined what we were going to do for the next few years. We took some measurements." They plunged into the quest for a new kind of cold fusion.

The group started with electrolysis; their first experiment occurred May 22, 1986. Explaining the concurrence of the BYU and University of Utah thinking, Jones said, "How are you going to get hydrogen into metals? Well, electrolysis is one of the easiest ways . . . the point is once you get this idea, which we did get in 1986 independently of any other work, then the electrolysis cell from that point is going to look similar to other electrolysis cells. Electrolysis happens to be one way of getting hydrogen into metals—deuterium in this case. We did use other ways, however, and, in particular, we have used the pressure-loading method used in Italy. We started that back in 1986 as well. But we never cooled the pressure-loaded samples. We only dealt with higher temperatures. My hat is off to Professor Scaramuzzi for trying it out." [Referring to the Italian neutron results from super-cooled and subsequently warmed-up titanium infused with deuterium. This work that had made news headlines was reported at the conference by Scaramuzzi himself.]

Jones and others called the electrolyte that they used their "Mother Earth soup," a complex concoction of chemicals that they thought might mimic the geochemical features causing cold fusion within the Earth. By contrast, Fleischmann and Pons had used a simple electrolyte, LiOD, to better evaluate what was happening.

"We saw fusions, we thought, at a small level," said Jones. "But we realized that if we were going to believe our results of the summer of 1986, we were going to have to refine our neutron counter." So Jones and company built a unique detector around two sensitive light-detecting photomultiplier tubes.*

The BYU researchers collected most of their data in January and February 1989, with a bit of it earlier in December 1988. They had gotten about 170 counts in the 2.5 MeV neutron energy region. They could barely believe their own results. Said Jones, "I asked my son, 'Do you see a peak there?' And he said, 'I think so.'" In a statistical sense, the Jones neutron curve peaks, so the group claims, at about five standard deviations—about five times the background level. The rates of neutron production varies from run to run, something that is not well understood. They were getting an average of 0.06 neutrons per second during their first eight tests, with a maximum level of 0.4 neutrons/second. Jones opened himself to criticism, however, when he said, "I must admit, that if we hadn't seen that [0.4 neutron per second] run, we probably would not have published what we did. . . . Could the 2.5 MeV neutron signal be artificial? The answer is yes." Many hot fusioneers still don't buy Jones's results. Many didn't believe him in Baltimore.

---

*Neutrons were to be slowed in liquid scintillator (light emitting) fluid by collisions with protons. It would take about six collisions to make fusion neutrons react with glass "doped" with lithium-6 (^{6}Li). The signature of a fusion neutron was a required coincidence between light pulses from the liquid and glass. The intensity of the light output was a measure of the neutron energy. Jones thought it was important to identify these as 2.5 MeV neutrons—to tell whether we have fusion reactions. "It took several man-years to develop this detector to the point where we could believe it," says Jones. Unlike Pons and Fleischmann, Jones et al. made a *direct* measurement of neutrons.

But Jones was and remains convinced that he had created fusion neutrons electrochemically. His group used control experiments with ordinary light water, as well as cells with no current turned on. These did not behave like the active ones. Fusion appeared to be happening only in the live heavy water cells. Yet, as Jones remarked, *Nature* magazine's editor, John Maddox, had claimed that Jones had not tried light water cells—an assertion at odds with Jones's own paper. If Maddox was so fast with this fact about a researcher whom he had published, could he be trusted to treat those with more impressive results fairly?

Jones pointed out that most people were getting neutron rates comparable to this, and he was firm on the importance of quantifying them. "I think that it's important that we discuss *rates*—Not simply, 'I saw neutrons, ergo, cold fusion is going to be the power source of the future.' We really have to be more scientific and quantitative than that. . . . I don't see how you can get a correlation between the heat and tritium from the data presented yesterday." As he concluded his talk, Jones put up a slide showing quotes from *Nature*: "There may be something in the Brigham Young data but that requires confirmation. . . . The [University of] Utah phenomenon is literally unsupported by the evidence, could be an artifact, and given its improbability, is most likely to be one." Neutrons yes—amazing as they were—heat from fusion, no. "I think the low neutron rates challenge very strongly these notions that the excess heat observed in the Pons-Fleischmann experiment are due to fusion," he said. "However, I do think there is a cold fusion phenomenon at very low rates—two possible explanations—piezonuclear fusion, or fracto-fusion [fusion by accelerating particles in the electric fields of microscopic cracks opening up within materials]."

✳ If It Quacks, It's a Duck!

The neutron story was far from over. An Italian physicist, A. Bertin, recounted other neutron detections coming from a collaboration between the University of Bologna and the BYU group. Deep inside a mountain 120 kilometers from Rome (the Gran Sasso Massif) the work had proceeded. In chambers adjacent to one of the longest of Europe's superhighway tunnels, the Italian-BYU team hunted their quarry—fusion neutrons from titanium metal being electrochemically infused with deuterium.

The mountain rock over the tunnel looms 1,400 meters above three underground laboratories—chambers about 20 meters high and wide and 100 meters long. They went underground because the natural radioactivity level in the tunnels is about 10 times smaller than usual and the mountain also provides the equivalent of a 4,000-meter shield of water against cosmic rays and nuclear by-products of their collisions. It reduced the particle flux by million-folds.

Running with Jones's Mother Earth soup and titanium, about one hour after operation the neutron counts had begun, peaked, then dropped off over three hours to background levels. This was observed in two other runs. When the sample was withdrawn from one detector, the counts dropped to background. Moreover, the counts in a second detector increased when it was brought near the active cell. The team had measured fusion neutrons at a rate about 900 per hour, and they stood firmly by their data. Said Bertin, "We consider the present results as further evidence of the occurrence of cold nuclear fusion in metals. The observed rate is equivalent to that observed by Jones and collaborators, and in this respect their experiment is confirmed by the present measurement with a statistics which are more than twice [as good]—and with an entirely different detection system. It is confirmed that electrochemistry plays in this type of measurement a quite significant and critical role."

What more could one want for proof that a new and heretofore unknown nuclear process was occurring? The answer: One could hope to find those wonders going on in an altogether different kind of physical system. That is indeed what physicist Howard Menlove from LANL next described—an effort to find neutrons coming from titanium and/or palladium metal chips immersed in high-pressure deuterium gas—using no electricity at all. Menlove, an ultra-cautious researcher and a cold fusion skeptic in the beginning, ordinarily speaks in serious deadpan tones, making his injection of a humorous comparison all the more appreciated. He began, "There has been a joke around the lab that looking for neutrons is like looking for ducks, so here are 'Priorities for a Neutron (Duck) Hunter': Arc there any ducks? (sensitivity); When do they come and go? (the time history); How fast are they? (energy); Do they arrive in flocks? (pulses); How many ducks are there? (absolute yield)."

Menlove asserted that his group at LANL had results to substantiate all of those ducky characteristics except the energy of the neutrons. Plenty of neutrons had come out, but not enough to get a good read on their energy. The LANL approach was to cool a cylinder containing the deuterated titanium chips down to liquid nitrogen temperature (−196°C)—literally by dipping it into a vat of the liquefied gas. As the cylinder warmed after being taken out of its frigid bath, the numerous surrounding neutron detectors began to register counts. Some neutrons were coming out randomly, but others grouped themselves in bursts. No one could accuse Menlove of being incautious or sloppy. He had four different kinds of detectors at work, one with 16 independent detection tubes surrounding the sample chamber.

Menlove tried to provoke his apparatus into false counts by exposing it to other kinds of physical assaults: gamma rays, high intensity electromagnetic radiation, temperature cycling, acoustic noise. Since the detectors were well shielded, efficient, and equipped with various kinds of discrimination apparatus, they proved impervious to these insults. But just to be sure, the work was carried out in a room three floors underground with a heavy concrete shield overhead about one meter thick. On April 28, they had their first success: "When we put the first cylinder into the counter under the pressurized gas system, we observed our first significant event. . . . It was a genuine observation that other people haven't seen yet."

He showed an amusing picture of a duck hunter's decoy resting in one of their counters. Indeed, he had his ducks in a row. To measure the environment, Menlove had even used two control counters near the primary neutron counter during the entire experimental period. His group used dummy cells intermittently with active ones—all these were negative. The control counters had been operating for over one month with none of the burst effects that his group found. The sample cylinders would usually take two to three hours to warm up, and then come the bursts of neutrons, usually at about the −30°C point. There were from 10 to 300 counts in a time interval of 100 microseconds. On repeat runs, they found that the sample had to be taken through its warm-up cycle a number of times before neutrons were registered. Possibly some material phase change was occurring at −30°C, Menlove suggested, that might have something to do with the unknown, but clearly nuclear process. Perhaps some creaking and cracking of the metal lattice was somehow provoking nuclear events.

Six different cylinders had evidenced neutrons in 15 separate bursts with an emission rate averaging 0.05 to 0.2 neutrons per second—roughly the Jones rate, albeit in a much different (it would seem) arrangement. "We don't know if it's related to cold fusion, or the hot fracture fusion through electric fields, or some other mechanism," Menlove concluded. In future work the team would try to characterize the neutron yields in different materials, determine their distribution of energy, and attempt the experiment with ordinary hydrogen and tritium gas.

Physicist Moshe Gai from Yale, an outspoken skeptic of all the neutron results, attacked Menlove's work and charged that even the Gran Sasso team was counting background gamma rays, not neutrons. "You cannot look for the effect with a background count of 100 counts per hour," he said angrily. There followed a free-for-all debate about whose neutron results were correct. Kevin Wolf criticized Moshe Gai's assertion that Wolf's detector wasn't suitable to measure the fusion neutrons.

But there were still more neutron results to back up those that came before. F. Scaramuzzi from the Frascati Energy Research Center in Italy was part of a group that had made its big splash weeks earlier with an approach that Menlove had duplicated. He reported the group's latest results. Ironically, the rugged Frascati counter had been prepared for use in a hot fusion tokamak experiment! They had cycled the sample through temperature changes with liquid nitrogen in the treatment that Menlove had followed. They had gotten neutrons every time, an average of 1,000 per hour while the background level was only two or three per hour. Sadly, in their attempt to replicate the experiment, they met with little success [long after the meeting they would again observe neutrons]. On this he remarked pointedly, "The . . . lack of reproducibility is something that I think should be considered as one of the experimental characteristics of this phenomenon. It means we are not aware of what we are doing. In particular, we know that we are working with very little amounts of deuterium in our metal."

Of the comments leveled against the Frascati work, none was more damning than that of skeptic Richard Garwin. "I've analyzed that bell-shaped curve," he said, "and it has no trace of bursts in it. In fact it's smoother than it ought to be. . . . These may not be neutrons. . . ." Scaramuzzi shot back, "What do you mean not neutrons, they could be a disturbance?" Garwin replied: "God knows what they are, but unless they come out with the same exponential decay time as in the moderator, they are not neutrons." An IBM laboratory had tried the Frascati experiment and had gotten no neutrons. Scaramuzzi delivered a message that many skeptics could well have learned from, but their eyes and ears were closed—or at least were not transmitting effectively to their brains: "This is quite a general problem. We don't quite understand yet what are the parameters with which we are working. . . . I'm aware that we are not able to repeat the experiment every time. I'm claiming that we have to learn a little more about this. I'm very uneasy when I hear people calling me from England, the United States, or Germany, and asking me what is the 'recipe.' What I tell you is what I've done, but I also tell you that it doesn't work every time. . . . In my opinion, this is one of these phenomena in which we have to learn to handle the parameters and we are very far from this."

✳ Defiance

At a large scientific meeting, invariably there is someone who becomes a self-appointed "police officer," with the assigned task of straightening everyone out on technical points. Moshe Gai was the one who took up that mantle at the Santa Fe conference over the issue of neutron measurements. At one point he even said as much, "I realize that my job in this conference is to keep everybody honest."

He and his collaborators at Yale and Brookhaven National Laboratory had built a truly impressive neutron counter, one that by appropriate electronic discrimination could create an effective background noise signal of about two neutrons per hour. The work that Gai reported at the conference was submitted to *Nature* and eventually was published in the onslaught of papers with negative results (July 6, 1989). (By contrast, Menlove's work–also submitted to *Nature*–would languish for months unpublished.) In an effort to conjure up the cold fusion phenomenon, Gai had even ripped the tiny radioactive alpha particle source out of his home smoke detector to try to get something going. He had heard about Peter Hagelstein's theory that such nuclear particles might start coherent fusion. Gai had been at least that much a believer, but no more.

Gai was supremely confident and proud of his detector. In a loud and commanding voice he boomed, "The rate we are talking about is 1.7 counts, plus or minus 0.5 counts per hour. Before you show me such kind of data I will not believe in cold fusion, and you shouldn't too." Of course, all of the Yale-Brookhaven experiments had given null results. He was suggesting that Jones's data could well be a statistical fluke.

Steve Jones, gentleman that he is, approached the front of the room to rebut what Gai had just said, first putting his arm on Gai's shoulder. But before Jones could utter one word in his own defense, Gai said in his Israeli-accented English and peculiar idiom, "I told Steve Jones that the only way I'll believe it is if he brings a cell to Yale and we measure it on my system. You are now evidence that he promised to do that. We shake hands on that. It's a contract." Jones replied, "I will show you some neutrons we did at Los Alamos, I believe we can do it at Yale." Then another pat on Gai's back, but Gai would not yield. Jones was now at the platform trying to talk, with Gai interrupting at every turn.

Jones commented that he had prepared his electrode somewhat differently from Gai's method, but Gai challenged him, "Could you explain to me why they should be fused [from metal powder]?" Jones said that he thought that the surface-to-volume ratio was important. "I'm not sure you're doing the right experiment either, Moshe." Gai's anger rose as he tried, unsuccessfully, to grab the microphone from Jones, "You're not sure I'm not doing the right experiment? I'm convinced you're not doing the right neutron counting!"

Jones gulped, "Well, it's very funny, but we worked pretty hard on our neutron counting and we did not use pulse shaped discrimination, we did not use BF_3 counters, we used a very sophisticated . . ." Gai, finally grabbing the microphone away before Jones could finish, "Maybe I should finalize this. I will not believe cold fusion before I see data of the quality I've shown to you, and the quality that will be shown to you from Bugey [European investigators] today. He will show you one count per day, and I showed you one count per hour. I think this is the kind of data we need." Jones, grabbing the microphone back, "What about the Los Alamos data? I really think the Los Alamos data is very solid, Moshe." Gai replied, "That's not what I heard ten minutes before my talk." Jones barely managed to prevent his anger from boiling over, but he did. To cool the by now very hot cold fusion proceedings, chairman Reed Jensen of LANL called a 20-minute break and said, "Let's give them a hand." Applause followed. Indeed, it was a performance that had to be seen to be believed.

✳ And More Heat . . .

Materials scientist Robert Huggins moved the forum back to the question of heat. The Stanford group was not looking for nuclear products—neutrons or tritium—just excess heat, and they had found it in a modified replication of the Fleischmann-Pons experiment. Huggins laid out the intriguing evidence, as he had partially done before Congress, with a confidence that perhaps grated on skeptics but was music to the ears of proponents. "This field has some major quandaries," he admitted. "Is anything unusual happening when deuterium is inserted? What is this reproducibility problem all about? Some people observe effects, others do not. I'm confident that after you hear what I have to say that you will agree that there are significant effects under certain conditions, and that even in the same laboratory people report that some samples show effects that are readily measurable and others do not.

"We don't really know what is important here, but there are some things that could be." He speculated that the microstructure of the palladium could be important, as could impurities within the lattice. He reminded the assembly that in open cells heavy water tends to capture light water from the atmosphere. His group had found that if they operated in the presence of air, after a while any positive indication of the heat effect would disappear, a twist that not everyone had considered. Also, cast and wrought [extruded] palladium seemed to behave differently. Perhaps dislocations in the lattice were acting as tiny traps for hydrogen. Huggins and his colleagues went to extraordinary lengths to get the gas as completely out of the palladium as possible, remelting the metal in an arc 10 or 12 times in an inert argon gas atmosphere! Huggins noted the possible importance of blocking layers on the surface, including atoms of silicon and carbon. These were issues that some of the skeptics had not addressed in carrying out their dead-end experiments.

Huggins and his colleagues had laboriously recalibrated their cell, "for every data point in every sample at every time," he said. They measured cell temperature in their precision calorimeter to 0.05°C. For some reason they had used a disc-shaped electrode, unlike the rods favored by most. They loaded their cells in an inert nitrogen gas glove box. With heavy water, the group got a steadily accumulating excess heat. The amount varied but from very recent data it looked like about 12 percent excess power. The group had done the experiment with light water too—just as the critics had demanded earlier—and had gotten no excess heat. "There is indeed a difference in the case of deuterium and the case of hydrogen. There is, indeed, real excess heat." Huggins denied that the excess could possibly be due to the favored bugaboo—recombination of the deuterium and oxygen—the by-products of the ordinary electrochemical decomposition of heavy water. He could conceive of no possible chemical explanation, but he did not utter the "F-word," fusion. When Lewis of Caltech asked whether Huggins would agree to measure a Caltech sample, Huggins unhesitatingly agreed.

D. De Maria of the Department of Chemistry, University of Rome, recounted a really hot event in his group's work that later became a bit of apocrypha in the cold fusion lore—like the Fleischmann-Pons electrode "ignition." While attempting to measure excess heat in a Fleischmann-Pons-type cell, they encountered a neutron burst over 150 times background during which there was a simultaneous elevation of the electrode temperature by an estimated 150°C. People had worried about hydrogen-oxygen explosions when the palladium cathode was inadvertently exposed to gas and the water level dropped during electrolysis—the Pd could catalyze rapid combustion or an explosion. De Maria said that about five hours before the event, the level of the solution had definitely covered the electrode, but after the event the solution had dropped below. The neutron emission had lasted about 200 seconds and triggered an automatic switching off of the current. No one had been in the lab, however, to witness this unusual event, because by chance it occurred on an Italian festival day!

Skeptical physicist Richard Garwin couldn't resist chiming in immediately: "I can imagine that the catastrophic event in which the temperature rose 100 degrees or so was not fusion at all, but simply a response to having stuffed deuterons into high energy sites in the lattice—similar to 'Wigner energy' in graphite in [nuclear] reactors. You see neutrons, but they may have nothing to do at all with the heat that caused the temperature to rise."

Argonne National Laboratory had gotten no excess heat in their attempts to reproduce the Fleischmann-Pons effect. Y. Desclais, of the

Laboratory of Particle Physics at Annecy-le-Vieux (LAPP) described the extremely low background hunt for fusion neutrons that had so excited Moshe Gai. In 300 hours near the Bugey power plant in the Frejus tunnel, the French group found no neutrons above 0.017 per hour coming from a Fleischmann-Pons cell. Nothing too from a Frascati-type gas cell experiment that they had tried. On and on went the talks, a mixed bag of positive and negative results, an unevenness mirroring the unexplained comings and goings in cold fusion cells around the world.

✳ Chemistry Won't Do

Could known chemical phenomena rescue the meeting from the cold fusion conundrum? Probably not, as John Bockris concluded in his devilishly titled talk, "Seven Explanations of the Pons-Fleischmann Effect," as though even one explanation was at hand. The much disputed neutrons, if they came at all, were coming in bursts in those experiments that were positive. The excess heat, on the other hand, typically switched on after tens of hours and lasted many more hours, though it too had burstlike behavior sometimes. Bockris asked whether any chemical phenomenon could yield anything like 10 watts excess power per cubic centimeter (10 W/cc) of palladium for a period of time like 100 hours—making the assumption that that order of excess heat had indeed been measured by a number of groups (a contention the critics were by no means ready to accept).

At the top of the Bockris list of possibilities—and most worrisome because of its possible large size—was the chemical recombination of D_2 and O_2 in the gases leaving the liquid. Recombination could amount to 18.5 W/cc. But Bockris did not believe that recombination was tenable: No one had successfully found it. At least 98 percent of the evolved gases do not recombine, he estimated.

What about chemical recombination in the liquid phase at the palladium surface? No more than 0.2 W/cm^3 were available from that. Phase changes within the metal? The metallic change in phase from the alpha to beta states—as more deuterons were packed in—would yield no more than 0.03 W/cc, obviously also negligible. What about the energy liberated when deuterium formed chemical bonds with the palladium, even if the extraordinary ratio of D/Pd of 6.0 could be achieved—really packing the deuterons in? From that, no more than 0.6 W/cc could be expected, Bockris asserted. What of the energy released in the formation of a lithium-palladium alloy at the electrode surface? No more than 0.08 W/cc from that. Linus Pauling had prescribed the alloy heat release mechanism to explain reports of excess heat. No, that wouldn't do, said Bockris. The overall conclusion: Not more than 10 percent of the posited Fleischmann-Pons effect could be accounted for by chemistry.

Ed Cecil of the Colorado School of Mines reported that his group used a beam of deuterons from an accelerator to implant them in a thin film of palladium only six microns (six millionths of a meter) thick. Obvious and expected fusion between deuterons occurred during the bombardment, but the surprise was what happened when the deuteron beam was off and an electric current was passed through the thin film—a small actual current giving a very high current density. Cecil and colleagues observed charged particles, possibly protons, coming off with an energy grouping around five million electron volts (MeV). He grasped at straws but had no explanation for whatever they were seeing. Later, his results were put on more solid footing and confirmed by others, in work that may be among the most compelling evidence for cold fusion if it can be substantiated (see Chapter 15).

An old hand at tritium measurements, J. Bigeleisen from SUNY (State University of New York) tried to throw cold water on the tritium measurements reported earlier. He explained that enrichment of the solution in tritium could occur as more heavy water was added to make up for that lost to electrolysis. Tritium, he explained, is present in extremely low concentration in heavy water, and the process being carried out by cold fusion researchers was liable to enrich its concentration three to five times—if 95 percent of the initial volume of heavy water was electrolyzed. "This range of enrichment explains quite a number of the reported tritium enrichment experiments." But he admitted, "It cannot by any means explain the tritium levels reported by Kevin Wolf."

Could reports of cold fusion really be hot fusion in disguise—a semantic quandary if ever there was? F. J. Mayer, representing a research group with members from Ford and the University of Michigan Research Laboratory, suggested that neutrons might be emerging from fusions generated by cracks in the metal lattice, "a familiar mechanism in an unfamiliar place," as he put it. He showed that as a crack opened in a metal lattice it could develop a charge imbalance that might accelerate deuterons and smash them into one another at high energy, thus producing hot fusion in tiny localized regions. Such cracking might be part and parcel of electrolysis experiments as deuterium packed the palladium or titanium lattice and distorted it. He recalled Soviet experiments with projectiles fired into targets made of lithium deuteride (LiD), in which roughly the same kinds of emissions were seen as were found in cold fusion experiments. If this were the true mechanism of cold fusion, an ominous implication for its proponents: a lower likelihood that the process could be scaled up to useful levels.

✳ Commandments for Cold Fusion Research

In the free-wheeling discussions reserved for the evening sessions and some of the question and answer periods lay gems of scientific wisdom and wit. Even biblical allusions were brought forth. Said Ed Storms of Los Alamos, who had been working with Carol Talcott to measure tritium, "I would like to suggest two commandments of cold fusion research. There'll be another eight to follow, not by me but by others, I'm sure. I would like to suggest that the first commandment is to be sure that the deuterium to palladium ratio of the cathode is near or greater than 1.0. The second commandment is make no conclusions about the possibility of cold fusion in public unless the first law is followed." This brought much applause.

The bearded and very biblical Moshe Gai matched Storms: "I would like to add to this a 'Thou shalt not. . . .' The first 'Thou shalt not' is Thou shalt not use BF_3 counters for neutron detectors."

John Bockris introduced Robert T. Bush from California State Polytechnic University, saying with not a little irony, "He can explain everything." The flamboyant Bush then launched into his theory of how deuterons could clump together in huge globules within palladium, some of which would fuse to produce helium-4 and transfer energy to the entire lattice by localized melting. A bit of everything was thrown in.

Stanley Luckhardt of the frustrated MIT group rose to offer a fundamental bit of philosophy—a commandment of sorts. In Luckhardt's view, the problem had to be approached by a systematic crossing-off of all known nuclear reactions that could occur in the brew of isotopes in a cold fusion cell. He quickly did his own crossing off and concluded, ". . . you are left with no evidence, really, of any nuclear process to account for the heat generated . . ." Cutting him off, Bockris retorted, "We don't know anything like enough to make statements of that kind." Session chairman Bockris then tried to pass on to another questioner, provoking MIT's Richard Petrasso to shout "Hold it, come on John!" to reserve some more time for colleague Luckhardt. Luckhardt continued, "I think we really need an answer. The neutron production is much too small to account for the excess heat . . . the helium-4 and so on. . . ." Bockris: "Everyone has said that right from the beginning—we all agree about that." Luckhardt: "Is there any evidence that the excess heat is due to nuclear processes?" Bockris: "No comment. I don't know."

Stanford's Robert Huggins was roundly challenged by a number of skeptics on the significance of his heat measurements, but he stood his ground and explained the enduring, baffling mystery of the excess heat: "I'm talking about the power ratio—'power out' to 'power in.' If you do it on energy, it depends on how long you run your experiment. We haven't run our experiment as long as we can to see when it dies, *if* it dies, ever. We have run experiments that are well beyond a factor of 100 percent [excess energy]. At the beginning of our experiments we see the endothermic effect [energy *consuming* effect] which you expect for the electrolysis of the water or D_2O. As the experiment with hydrogen, light water, continues, nothing changes. As the experiment with heavy water continues, the amount of heat gradually increases and after about two days you come out even.

"So the amount of heat generated somehow by some kind of process within the material is about equal to the endothermic heat of the electrolysis process. After five days, we are well beyond that and we have real, live excess heat coming out in the form of power. As of yesterday we have run five days, the total amount of energy is greater than that that went in, but the first part that went in is an investment. In the beginning you have an endothermic effect, so for a couple days you're in the negative part of that energy balance. But we're beyond that, we're in the positive side. We have no idea how much longer this thing is going to work—the longer it works the more positive the energy balance is. But the power balance is always positive. . . . The cell we're running right now is the cell that we have all these very careful calibrations on and we're seeing something like 120 percent energy out to energy in, but there is no reason to think it isn't going to get larger or smaller. We don't know."

"We don't know," perhaps as good as any summation of what the New Mexico proceedings had revealed. A tantalizing challenge to the proponents, a supreme irritant to the skeptics. The press was largely unimpressed with the prospects of cold fusion, although still open to the possibility that something curious had been discovered. Typical was the view of science reporter William Booth of the *Washington Post*, who wrote, "While a scientific consensus builds that cold fusion will probably never become the answer to the world's energy needs, even the most skeptical researchers are struggling to explain experiments that continue to produce excess heat and an enigmatic trickle of by-products that indicate some kind of fusion may be occurring."

Conference cochairman Robert Schrieffer summarized the proceedings and said, "I am personally optimistic that this [Jones-level cold fusion] is real. It will be a real shame if this goes away." Schrieffer also said that he was prepared to believe that excess heat exists—because "good people" were doing the experiments—but was chastened by the great risks of "diagnosis by elimination." Later he told the press that he believed the heat to be a chemical reaction that went "in tandem with the nuclear reaction." He warned that nonreproducibility was a crucial difficulty, but he reminded his colleagues that the development of semiconductors in the 1930s was also plagued by nonreproducibility.

Was the New Mexico workshop a sunrise or sunset for cold fusion? It seemed that the answer depended on who was watching. Was it a ticket to many more months—perhaps years—of fruitless struggle to conjure up an effect of minimal scientific importance, or was it a new door to controlled fusion power? Time alone would tell.

10 *Evidence Builds and Skeptics Dig In*

One of the problems the fusioneers have had is that their experiments last far longer than the media's attention span.

The Economist, September 30, 1989

Cold fusion has been a kind of wild party, with everyone having a giddy time. But now we have to deal with the hangover.

Harold Furth
New York Times, June 20, 1989

It is certainly not the least charm of a theory that it is refutable.

Friedrich Wilhelm Nietzsche, 1886
Beyond Good and Evil

* A Long Hot Summer

THE SPRINGTIME OF COLD FUSION was slipping away and a long hot summer was beginning, with cold fusion proponents increasingly on the defensive. In late May, Pons turned down a request by the DOE cold fusion evaluators to visit his laboratory. There were simply too many disbelievers on the panel for his work to get a fair shake, Pons charged. Matters were sufficiently smoothed, however, for the Energy Research Advisory Board (ERAB) panel cochairman John Huizenga to go there in early June. Pons insisted on barring from the visit Caltech's Steve Koonin, who was skeptical even about Jones's paltry few neutrons, though he said he was "leaning toward" belief in Jones's work.

The hoped-for collaboration between the University of Utah and LANL fell through in June, the result of failed negotiations in which legal issues, patent rights, and so forth had gotten in the way of science. Perhaps Pons's own misgivings about that initiative had interfered too. Nonetheless, many Los Alamos researchers persevered in studying cold fusion and eventually came up with some of the most convincing supporting evidence.

Then on June 15 came the supposed death knell: The British Atomic Energy Authority announced that it was abandoning efforts to replicate the Utah work and was writing up its results for *Nature.* Newspapers described this as "virtually burying hopes for cold fusion." Particular significance was attached to the bad news, because Fleischmann was a consultant to the Harwell Laboratory where the attempts at replication were conducted. (He had only briefly conferred with the Harwell team in the early spring even before the March 23 announcement, hoping thereby to make the lab's subsequent work an "independent verification.") The lab had spent nearly $500,000 and employed 10 full-time scientists and many assistants to do its work. Claiming to have tried over 125 variations of the experiment with many kinds of metallurgically worked palladium and still not to have found excess heat, neutrons, or anything else suspicious, the lab seemed to have sealed cold fusion's fate. "When brilliant people have mad ideas, it can come down on them like a ton of bricks. What's happened here is that a brilliant man, Fleischmann, has had a mad idea" was how the leader of the Harwell work, David Williams, summed up the situation at a news conference. For funding reasons, no complete internal report was ever produced by Harwell on the Williams group's work, so all one can go on for future evaluation is the summary published in *Nature.* Even in what Harwell has published, however, major problems with their experiments can be seen (see Chapter 11).

✳ Nature's Ill Wind Blows Strong

Even before the Harwell work appeared in its pages, *Nature* added fuel to the fire of the controversy by giving prominent coverage to other negative experiments. It ran a cover story for its issue of July 6 with the stark banner, "No Evidence for Cold Fusion Neutrons," under a photo of the Yale-Brookhaven National Laboratory equipment that had failed to find neutrons. John Maddox began his editorial that week: "It seems the time has come to dismiss cold fusion as an illusion of the past four months or so." An odd remark of introduction just above this contradicted not only the title but the first sentence of the editorial: "*End of cold fusion in sight*—Although the evidence now accumulating does not prove that the original observations of cold fusion were mistaken, there seems no doubt that cold fusion will never be a commercial source of energy." The conclusion about applications came completely out of the blue and was unwarranted based on the level of uncertainty at the time. Apparently, many science reporters were probably strongly influenced by that assessment.

Maddox observed further on, ". . . managers of orthodox experiments intended to replicate what happens within the Sun will be relieved." He misstated even elementary facts, such as when the Utah announcement occurred; he stated February 23, not March 23, three times! (Earlier Maddox had said that Jones had not done a light water control experiment, an error for which he later apologized—very inconspicuously.) He humored Fleischmann and Pons a bit by acknowledging gratuitously that the conventional peer review process would have been "too bland" in view of their astounding belief that they had come up with a "way of changing the world."

This kind of editorial added further obstacles in the path of those who were genuinely struggling to solve the cold fusion puzzle. Simultaneously, funding began to dry up, and it became necessary all too often to discuss scientific work behind closed doors. John Bockris recounts one such episode: "A Ph.D. from a national lab, asked [me] to step into a conference room at a meeting. After shutting the door, he pulled graphs and results from his briefcase and said, 'Don't tell anyone about this. My boss would kill me if he knew I was telling you. I have positive results, you see.' "*

That summer, Richard Oriani of the University of Minnesota in the Department of Chemical Engineering had tried and tried again to duplicate excess heat production and had finally found heat bursts. On the other side, Westinghouse Electric Corporation of Pittsburgh announced in late August that it had so far met with no success. Likewise for Oregon State University, though the group there was still continuing experiments and eventually reported positive results in heat measurements.†

In August, the National Cold Fusion Institute (NCFI) finally opened a 25,000-square-foot facility on the outskirts of the University of Utah's campus in Research Park. To run the spanking new facility, the state was looking for someone to replace Dean of the College of Science, Hugo Rossi, who had been serving as NCFI's interim director. Cold fusioneer John Bockris was being considered, but there was a perceived need for someone with more direct corporate experience. Almost unnoticed, neighboring BYU had opened its less well-endowed Center for Fusion Studies in late June, with Jones as its director.

The DOE panel was careening on its course toward an anticipated negative result—not an uncommon modus operandi for federal studies.

---

*John Bockris and Dalibor Hodko, "Is There Evidence for Cold Fusion?" *Chemistry and Industry*, November 5, 1990: 688–692.

†Lance L. Zahn et al., "Experimental Investigations of the Electrolysis of D_2O Using Palladium Cathodes and Platinum Anodes," *Journal of Electroanalytical Chemistry*, Vol. 281, March 26, 1990: 313–321.

Panel member Koonin was still intrigued by Menlove's neutron data and by Wolf and others' tritium, but he was betting on contamination in the latter. Science journalist Gary Taubes was investigating whether some of the A&M tritium might have been put there "by human hands." Word of his travels spread along the scientific grapevine; the results of his investigation appeared in print the following June (see Chapter 14).

After reaching its preliminary negative view in July, the DOE panel planned to meet again in mid-October. No matter, the final results of the "autopsy" for helium ash in Utah's palladium rods were not expected before the end of September. Poor Howard Menlove at Los Alamos, who had such compelling neutron data. He would submit his work to *Nature* at the end of the summer, and there it would languish unpublished. Though Menlove had satisfied four of five reviewers, *Nature* would not publish his work. Even when he obtained additional data to satisfy *Nature*, the magazine blocked publication again. But Menlove's results were getting better. To deny the soundness of his data would become an exercise in futility. He had clearly detected real neutrons coming from a place they were not supposed to be.

Nature was receiving more papers on cold fusion than on any other single topic. By their editor's own admission, these were roughly evenly divided between supporting and nonsupporting evidence. David Lindley thought that the negative papers were better and was claiming that the positive ones were encountering referee problems. He wanted detailed and thorough papers comparable to the negative ones. I have seen firsthand some of the correspondence between Lindley and an unpublished researcher. What Lindley was doing, it appeared, was to set up the negative papers—such as Nathan Lewis's—as a "standard" against which any positive results paper would have to stand. Even when Lindley acknowledged that he could no longer tell whether proponent or skeptic was correct, still no positive papers were published.

The East was unmoved by *Nature's* quirks and the DOE panel's maneuvering. Indian researchers at the Bhabha Atomic Research Center (BARC) hadn't given up on cold fusion. In fact, some were saying that the skeptical tones being heard in the United States were a "cover" for secret intensive work here! At the end of 1989 they would publish a compendious report, "BARC-1500," which told of their extensive cold fusion studies carried out from April through September—"the first six months of the 'cold fusion era,' " as they phrased it. The Japanese were being more discreet about their efforts, although there were reliable reports that some 40 groups with a total of 150 researchers were working on cold fusion in Japan. Referring to their purported work, Robert Huggins said at a conference at the University of Utah, "There is a deafening silence across the Pacific Ocean." In mid-September, Fleischmann and Pons went to Japan to attend the annual meeting of the

International Society of Electrochemistry to see for themselves and were not disappointed.

While the "thermodynamic duo" were away, archskeptic Douglas Morrison of CERN visited the University of Utah and received a standing ovation during his talk on "pathological science." I have talked with a prominent researcher from eastern Europe who assured me that Morrison's anti-cold fusion electronic mail spread to the upper management of many Soviet and Eastern European labs–people who typically controlled such computer communication nodes. This had the effect of squashing cold fusion investigations in many labs.

Morrison's show rolled onto campus and he made many representations, which were true at the time but later turned out to be suspect. For example, Ed Storms at Los Alamos was no longer able to get tritium in his experiments (true, but only temporarily so); an Italian group was no longer able to get neutrons in their gas-cell experiments (true, but later they got them again). Morrison spoke of his famous "regionalization of results" theory about cold fusion. His thesis: Scientists in eastern Europe and Italy were getting positive results while western Europe was seeing nothing. He noted that Florida, Texas, Utah, and Minnesota were getting positive results, while larger, "better known labs" in the northeast and California were not.

All this negativism seemed to be having a troubling effect on Rossi, the interim director of NCFI. Toward the end of September he was unnerved by the lack of results at the Institute and wondered whether it would have to close up shop even before its planned major cold fusion conference the following spring (originally scheduled for mid-February 1990). These public musings angered Pons, but he and Fleischmann weren't exactly "team players" at NCFI. Instead, they were continuing their experiments in the Chemistry Building, while acting as advisors to the Institute. Fleischmann told me, "Stan and I disapproved of [the NCFI]. . . . We were opposed to it. We said, 'It's too soon. Bricks and mortar are not the right thing even if you've got the money. Don't do it.' Nevertheless, we supported the research effort."

If heat was hard to pin down–in Utah and elsewhere–the tritium seemed secure. It was a great hope keeping the search for cold fusion alive. Kevin Wolf continued to defend its reality, and in early October was pleased with the exciting development at A&M of finding elevations in tritium levels coinciding with bursts of heat.

✳ Lukewarm Fusion

An interlude: It seemed unbelievable, but it was true. Twice in the same year chemists had stepped onto physics turf and discovered the possibility of another kind of fusion! In the September 18 issue of *Physical*

Review Letters, three chemists at Brookhaven National Laboratory announced a new way to pull off hot fusion: by using a particle accelerator to fire at a target tiny electrically charged crystals of heavy water—with up to 1,300 heavy water molecules (D_2O) in each. The laboratory immediately sought patents for this "cluster impact fusion," as it was dubbed by its three practitioners, Robert J. Beuhler, Jr., Gerhart Friedlander, and Lewis Friedman. Some people instantly began to call the discovery "lukewarm fusion," because the energies of the particles were only equivalent to hundreds of thousands of degrees (an energy of 200 to 235 eV)—not millions of degrees and certainly not room temperature. Cold fusion skeptic Steven Koonin was anything but cool to lukewarm fusion, calling the new phenomenon "a surprising new opening." Would this oddball kind of fusion be practical? Friedman certainly thought so and said, "We have a goal of trying to create microscopic stars. We think we've approached that goal."

The scientists claimed nothing physically very exotic behind their cluster-impact fusion, merely ordinary d-d fusion occurring because of the compression and heating of the clusters on impact. High-energy tritons and protons were observed blasting out. It was fusion all right. However, a 10-billionfold scale-up would be necessary to make the process practical. But because the reaction rates for lukewarm fusion were 10 billion times higher than had been predicted by theory, questions arose about how this unexpectedly high reaction rate might fit in with mechanisms that could be behind purported cold fusion. There may or may not be a link, but EPRI physicists Mario Rabinowitz and David H. Worledge have remarked about a glaring irony, "Curiously, and in sharp contrast to [cold fusion], apparently it has taken only one experiment to establish low-energy cluster-impact fusion as scientific fact." (*Fusion Technology,* March 1990)

✳ The DOE Blast

In November 1989 the panel of 23 eminent scientists sitting on the official "Cold Fusion Panel"[1] turned in a final report to the ERAB of the U.S. Department of Energy. It responded to the charge given by the authority of James D. Watkins. On April 24, Watkins had set three goals for the assessment of cold fusion:

1. Review the experiments and theory of the recent work on cold fusion.
2. Identify research that should be undertaken to determine, if possible, what physical, chemical, or other processes may be involved.

3. Finally, identify what research and development direction the DOE should pursue to fully understand these phenomena and develop the information that could lead to their practical application.

The Panel's conclusions and recommendations were reviewed by the full ERAB, which approved the report unanimously, "making only minor revisions," and sending the document to Admiral Watkins (November 26, 1989). The *New York Times* (November 11, 1989) reported the committee's negative assessment under the banner, "Panel Doubts Cold Fusion Research Will Pay Off on Energy." It quoted the Panel cochairman, Huizenga of the University of Rochester, "The present evidence for a new nuclear fusion process is just not persuasive." The article described the committee's final report as differing "only slightly from a preliminary report the committee made public in July."

Though widely touted in the media and in hot fusion circles as an exemplary document (hereafter called the "ERAB" report) that spelled the bitter end for cold fusion, the report appears to have many obvious shortcomings. Even were cold fusion to be without foundation—not even a new interesting chemical phenomenon lurking in those bubbling cells—the report is seriously flawed. Even in light of the evidence that existed at the time it was written, the report was excessively skeptical; it did not, in my opinion, objectively assess the *possibility* of a significant new phenomenon. Remarkably, even if cold fusion had been a completely erroneous quest, the stance of the Panel would have precluded determining that. The report left far too many unanswered questions.

Page after page of the 64-page report clearly shows that insuring the discovery of the truth about a possible new phenomenon was not what the Panel had in mind. Without question, the many highly respected members believed sincerely that they were participating in an objective effort, but the result fell far short of that standard. Busy scientists working for such panels are often guided by their leaders, and the leader in this case, Huizenga, was an admitted skeptic. Throughout this entire affair, he has repeatedly made statements to the press dismissing and playing down the evidence for cold fusion. This negativism, in my view, is completely inconsistent with the level of uncertainty on the matter.

In the executive summary of the ERAB report we read, ". . . the present evidence for the discovery of a new nuclear process termed cold fusion is not persuasive," but it left in a catchall "escape" clause to the effect, "The Panel also concludes that some observations attributed to cold fusion are not yet invalidated." Why then were these not pursued with a vengeance?

What were the Panel's all-important recommendations for further investigation of these "not yet invalidated" observations? "The Panel recommends against the establishment of special programs or research centers to develop cold fusion. However, there remain some unresolved issues which may have *interesting implications.* [author's italics] The Panel is, therefore, sympathetic toward modest support for carefully focused and cooperative experiments within the present funding system." Overall, a skeptical assessment, written from the perspective that all that had to be done to put the final nail into the coffin of cold fusion was to go after the slightly worrisome lingering doubts. No sense of urgency is communicated for finding out whether there might really be a new phenomenon. The "interesting implications" alluded to are never clearly spelled out. The phrase was vague, as intended, and as we are seeing more and more, the implications were, indeed, "interesting"!

To the ERAB Panel's credit, in the report's more detailed "Conclusions and Recommendations" section, there are several good and pointed recommendations. Among the suggestions: ". . . research efforts in the area of heat production focused primarily on confirming or disproving reports of excess heat"; ". . . investigations designed to check the reported observations of excess tritium in electrolytic cells . . ."; and "If the excess heat is to be attributed to fusion, such a claim should be supported by measurements of fusion products at commensurate levels."

But this seems, in retrospect, like so much window dressing to the Panel's real conclusion, namely that nothing particularly interesting, certainly nothing of a practical nature, could come out of this work. The Panel's purpose, it seems, was to give not the slightest bone of encouragement to the voices in the wilderness that were claiming the opposite.

Some perceptive souls in the media saw the Panel's report as a smoke screen. Editor Elizabeth Sullivan of the Cleveland *Plain Dealer* newspaper, being aware of the provocative work being done in her own backyard at Case Western Reserve University, ended her editorial assault on the DOE panel's conclusions with: "What prudes scientists have become if they *fail* to be intrigued by the unexplained. And how unlikely to find the answer if they fail to look."

✳ A Flawed Report

The central problem of the Panel, one that all along characterized most of the skeptics' efforts to come to grips with cold fusion, was a stubborn insistence that it was unlikely that any new physical mechanism was at work. They insisted that all mysteries had to be seen in this light, hence the incessant invoking of other supposed physical "requirements" that the new phenomenon should obey. Very rarely in the ERAB report was there even a bow to a possible new physical paradigm, and when there was, the pleasantry was quickly withdrawn with a negative remark.

The executive summary could state, with neither embarrassment nor qualification, in referring to the Jones and others and Frascati work, "Neutrons near background levels have been reported in some D_2O electrolysis and pressurized D_2 gas experiments, but at levels 10^{12} below the amounts *required to explain the experiments claiming excess heat.*" [author's italics] Implying, of course, that cold fusion would have to work by the same mechanism as known hot fusion reactions. The summary goes on to state, "Although these experiments have no apparent application to the production of useful energy, they would be of scientific interest, if confirmed." It was an echo of the earlier editorial in *Nature.* Only of "scientific interest"? It would have been far more appropriate to say "*dramatic* or *extraordinary* scientific interest," given that subtle and unexpected phenomena have a long history of becoming very useful in technology. Did anyone remember that hot fusion had pulled off a minor miracle of its own in the last two decades, climbing 10,000-fold in performance? Where was scientific imagination and creativity here?

The irony of the report was its contention that there might be some "scientific interest" in a possible new phenomenon and that reports of anomalous heat were "not persuasive"—by implication, not entirely ruled out. On the other hand, it was entirely too fond of the refrain—made several times in various forms—"no apparent application to the production of useful energy."

The report's saving grace was several assertions made in the "Preamble" to its conclusions and recommendations. However, the softening language was inserted only on the last day the Panel met, to satisfy the strong complaint of physicist Norman F. Ramsey of Harvard University, who weeks earlier had won the 1989 Nobel Prize for physics.

Attendees reported that Ramsey was so dissatisfied with the Panel's work and blanket negative conclusions (written largely by cochairman Huizenga), which earlier had been unanimously agreed on by other Panel members, that he told Huizenga that he wanted to resign if the conclusions were not modified. Huizenga was extremely angry and upset, but very reluctantly relented and allowed Ramsey to write a few sentences of softening, qualifying language that appeared in the report's preamble. This was an open public meeting with press in attendance, yet the media did not report that Ramsey had made an actual resignation threat, only that he had been dissatisfied with some of the draft report's language. Ramsey's resignation would have dealt a severe blow to the Panel's work.

Unfortunately, Ramsey had only been able to attend the first and last meetings of the Panel, and it is hard to know whether he would have been more or less negative about the report had he been present

One of the staunchest critics of cold fusion research, Dr. John R. Huizenga, Professor of Chemistry and Physics at the University of Rochester. (Courtesy University of Rochester)

Dr. Norman F. Ramsey, Professor of Physics at Harvard University and winner of the 1989 Nobel Prize for Physics, cochaired the DOE Cold Fusion Panel with Dr. John R. Huizenga. (Courtesy Harvard University)

in the middle phase. The adjusted language mollified Ramsey enough to allow him to stay on. The preamble left open the distant possibility that cold fusion might be real after all. In part it read: ". . . it is difficult convincingly to resolve all cold fusion claims since, for example, any good experiment that fails to find cold fusion can be discounted as merely not working for unknown reasons. Likewise the failure of a theory to account for cold fusion can be discounted on the grounds that the correct explanation and theory has not been provided. Consequently, with the many contradictory existing claims it is not possible at this time to state categorically that all the claims for cold fusion have been either convincingly either proved or disproved."

Cochairman John Huizenga was not happy with this "weakening" of the report, the first draft of which he was the primary author. He wanted the final knife to go in with a totally negative report. There is little doubt that the "rush to judgment" approach which had made Ramsey so uncomfortable might well have torpedoed the report's negative conclusions, if Ramsey had carried through with his threat. Federal cold fusion funding hung by that thin a thread.

Spoiling the otherwise meritorious sentiment, inserted at the 11th hour to please Ramsey, was a fifth conclusion: "Nuclear fusion at room temperature, of the type discussed in this report, would be contrary to all understanding gained of nuclear reactions in the last half century; it would require the invention of an entirely new nuclear process."

Of course! That was exactly the point that cold fusion proponents had been making all along. This conclusion was most certainly *not* being put forth as a compliment to those who were working on theories to explain cold fusion. Its intent was to reiterate the cry of the skeptics against a new scientific paradigm; the experimental evidence seemed to them overwhelmingly on their side.

Immediately preceding this conclusion is further evidence that the Panel was assuming that cold fusion had to be understood from straight extensions of known mechanisms—as though all the phenomena of nature had to be codified according to previous schema. The example of superconductivity alone should have warned them against that mistake. Nevertheless, with a faulty "appeal to authority," the Panel's fourth conclusion stated in part: "Current understanding of the very extensive literature of experimental and theoretical results for hydrogen in solids gives no support for the occurrence of cold fusion in solids. . . . The known behavior of deuterium in solids does not give any support for the supposition that the fusion probability is enhanced by the presence of the palladium, titanium, or other elements." Perilously close to saying "all that can be known about the palladium-hydrogen system is already known."

It gets worse. On page six of the ERAB report is the statement, "A third reason for skepticism is that cold fusion should not be possible based on established theory." Then follows a page of discussion that presents the well-known physical foundation of hot fusion, plus a justification for ruling out cold fusion based on little more than that the conditions hot fusion or muon-catalyzed fusion require can't possibly be present in the recent cold fusion experiments. Ergo, it is unlikely that cold fusion can be real.

✳ A Not-So-Secret Meeting

If cold fusion had "died" with the ERAB report cooked up in July and ratified in November, it was reborn in October, during a calm before a storm. A hush-hush meeting of 50 scientists came together (October 16–

18) in Washington under the auspices of the National Science Foundation and the Electric Power Research Institute. Palo Alto-based EPRI gets a tiny amount of money for advanced energy research from almost everyone who pays an electric utility bill. For two-and-a-half days, a motley crew of skeptics and proponents—mostly the latter—wrestled with the thorny question of cold fusion. Attending were notables Nathan Lewis, Steve Jones, Stanley Pons, Martin Fleischmann, Hugo Rossi, John Bockris, and John Appleby. Two scientific luminaries who hadn't really wet their feet in cold fusion were also there, physicist Edward Teller and Paul Chu of the University of Houston, who had achieved fame for his high-temperature superconductivity work.

To please all, the meeting flew as "Workshop on Anomalous Effects in Deuterated Metals"—no mention of the offensive F-word. Behind closed doors, however, some extraordinary evidence for fusion surfaced. There were discussions of persisting cases of excess energy from cells, heat bursts, neutron bursts, and tritium found where it definitely should not be. Teller went so far as to suggest that the widespread exotic effects could be due to a "yet undiscovered" neutral particle. The elusive particle acquired the name "meshuganon" from the Yiddish word meshuga, meaning "crazy."

After the meeting, cochairmen Paul Chu and John Appleby issued a provocative statement to the press: "Based on the information that we have, these effects cannot be explained as the result of artifacts, equipment, or human errors. However, the predictability and reproducibility of the occurrence of these effects and possible correlations among the various effects, which are common for accepted established scientific facts, are still lacking. Given the potential significance of the problem, further research is definitely desirable to improve the reproducibility of the effects and to unravel the mystery of the observations." Appleby said, "We are happy our results are showing there is something strange going on, and we have found that other people have confirmed those results, and those of Fleischmann and Pons. Carefully performed new experiments show that anomalous heating . . . appears to be real in many cases." Edward Teller told the press that he favored more research to determine whether the mysterious effects "are due to sophisticated difficulties in the experiments or whether a new phenomenon is involved." He advocated that uranium (^{235}U) metal be tested for its ability to support cold fusion reactions, possibly substituting beryllium nuclei for deuterium as "neutron donors" for the uranium. He recommended that "in recognition of the high class work that yielded surprising results, that the effort be supported to obtain clarification, whether the results are due to sophisticated difficulties in the experiments, or whether a new phenomenon is involved." Dr. Chu told the press, "Everyone who participated agreed more work should be

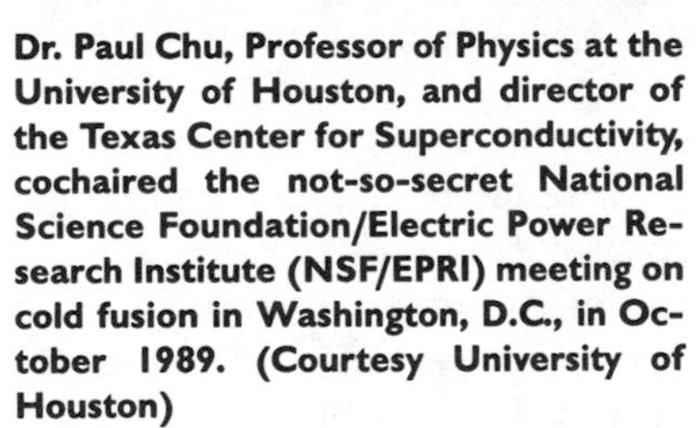

Dr. Paul Chu, Professor of Physics at the University of Houston, and director of the Texas Center for Superconductivity, cochaired the not-so-secret National Science Foundation/Electric Power Research Institute (NSF/EPRI) meeting on cold fusion in Washington, D.C., in October 1989. (Courtesy University of Houston)

Dr. Edward Teller, former director of the Lawrence Livermore National Laboratory, who has been called the "father of the U.S. hydrogen bomb," made extremely favorable public remarks about cold fusion immediately after the March 23, 1989, announcement. He made further positive comments after the NSF/EPRI meeting in Washington, D.C., in October 1989. (Courtesy Lawrence Livermore National Laboratory)

done. . . . We proposed, and all agreed, that the skeptics and strong believers should work on the same experiments."

Ironically, at the same time the DOE cold fusion panel was finalizing its harsh, negative report. It met on October 30 to produce the final dark score, but before that meeting, cochairman Huizenga was already telling the press that he didn't expect any changes from the July interim report. He told Jacobsen-Wells of the *Deseret News* in referring to EPRI–NSF meeting, "That was a very minor group of people who had been getting positive results for some time." Even within NSF there was dissent. Marcel Brandon, director of the NSF physics division sent an E-mail message to many NSF colleagues: "It seems unfortunate that an NSF office [the engineering division] is now appearing to encourage such discredited work. . . ." By contrast, cochairman Norman Ramsey of Harvard dearly wanted to see the results of the EPRI–NSF meeting. Paul Chu would have obliged, but he said it would be a few months before the report was due to come out. (Unfortunately, the EPRI–NSF report has still not been published.) In guiding the Panel to its conclusions, Huizenga had shown the members purported evidence that Fleischmann and Pons had broken their agreements with BYU. (Question: What did that have to do with science?) One Panel member, H. Guyford Stever, characterized Fleischmann and Pons's behavior as "bad science." Within the year, Stever would be appointed the head of a federal panel charged with plotting the future course of hot fusion.

It was on Halloween that newspapers announced the DOE panel's conclusion: Experiments to date ". . . do not present convincing evidence that useful sources of energy will result from the phenomenon attributed to cold fusion." Meanwhile back in Utah, Fleischmann and Pons, having tried and failed to build a scaled-up version of their minicell during the summer, dropped the effort and continued to pursue a solution to the critical problem of reproducibility. Tough business, without a solid theory to go by. Their way was empirical—tinkering with experimental conditions; they had dozens of cells bubbling away. For his part, Pons felt that most failures were due to not giving an experiment enough time. It took skilled, patient people like Professor Oriani of Minnesota months to get the effect.

The negative ERAB report undoubtedly provided a convenient justification within the DOE bureaucracy for reducing federal funding for cold fusion research. Two million dollars were summarily removed from fiscal year 1990 funding of the Division of Advanced Energy Projects led by Dr. Gajewski. This, in effect, put a damper on cold fusion research, but did it in a way that was not directly traceable to the report. At the EPRI–NSF meeting, Gajewski had made it known that he would support cold fusion projects. But now the "modest support for carefully focussed and cooperative experiments within the present funding system" favored by the ERAB report went up in a cloud of bureaucratic smoke fueled by the negativists.

Back in Utah, there were also new developments. In mid-November, Hugo Rossi resigned from the NCFI, and James Brophy replaced him as interim director. At last NCFI would get a permanent director, or so it was announced in late December. After carefully evaluating the cold fusion question, world-renowned electrochemist Fritz G. Will became convinced that cold fusion was real and decided to take up the directorship starting the following February. Since 1973, Will had worked at the General Electric Development Center in Schenectady on advanced electrochemical processes. The native of Germany, with many patents and publications to his name, was formerly president of the prestigious Electrochemical Society.

If Washington couldn't bear cold fusion, Utah sure would. After about nine months of gestation, on December 6, the State Fusion Energy Advisory Council voted to approve NCFI's work, mocking the federal government's surrender. Though the DOE panel had let cold fusion down, positive results from the DOE's own laboratories, Oak Ridge and Los Alamos, were beginning to surface. Los Alamos was strong on anomalous tritium and neutrons. ORNL had gotten tritium bursts, excess heat, and low-level neutrons. Yet without winking, Huizenga claimed that the DOE panel had considered ORNL and LANL findings. How so? Pons, for one, reacted angrily to the DOE report in the *Deseret News*: "They have made their judgement: they have passed sentence on fusion and they've been proved wrong. They were wrong from day one. The DOE appointed a bunch of negative people to give a negative decision. They will continue to be proven wrong—even by their own laboratories."

✳ Foreign Influences

If cold fusion research in the United States was hampered during 1989 by undue skepticism and a virtual "climate of fear" about being associated with the topic, investigators in other countries seemed to have experienced fewer such hang-ups. One could not have gleaned this from most media accounts, however. Yes, there were reports in major newspapers that various isolated experimenters in Japan, India, or the Soviet Union had claimed this or that result, but the impression was that the same official lack of interest existed elsewhere. This was false in many cases.

Evidence was all around: In late November, Japanese scientists at Nagoya University reported a radically new cold fusion method. Writing in the English-language *Japanese Journal of Applied Physics*, Nobuhiko Wada and Kunihide Nishizawa described high levels of neutron emission—20,000 times background—apparently coming from fusion reactions that began when they applied 20,000 volts between palladium electrodes in a cell of deuterium gas. Another group at Osaka University reported a level of neutrons 2.5 million times background coming from a cold fusion cell. "We ought to kick their butts," was the reaction of Senator Hatch to the news from Japan (*Deseret News*). Fearing that the Japanese had almost certainly already filed patents on these kinds of processes, there was anger in official Utah that they weren't able to obtain as broad a patent coverage as would have been possible with federal support.

India was forging ahead too. In early January 1990, *Nature* reported that Dr. P. K. Iyengar, the director of the Bhabha Atomic Research Center, believed that "U.S. scientists are convinced that cold fusion can take place but are keeping their results secret." He perhaps got that impression because the climate of scientific intimidation and ridicule in the United States was so strong that many researchers of necessity had to conduct their cold fusion work very quietly or underground.

The December 1989 report on cold fusion studies at BARC (April through September 1989) gave straightforward interpretations of experiments underway in India and were a welcome relief to the circumspection so prevalent here. Measured scientific descriptions were mixed with an ebullient description of "the first six months of the new 'cold fusion era.' " More than 50 scientists and engineers, plus many technicians from more than 10 divisions at the Center, were apparently involved in the initial phase of the cold fusion program. The report described work going on in electrolytic cold fusion, deuterium gas loading of metals to produce cold fusion, and theoretical investigations of the new phenomena. The report concluded, "Investigations of cold fusion phenomena carried out at Trombay during April to September 1989, have positively confirmed the occurrence of (d-d) fusion reactions in both electrolytic and gas-loaded palladium and titanium metal lattices at ambient temperatures."

The Indian researchers claimed to have seen their first bursts of cold fusion neutrons generated by electrolysis as far back as April 21, 1989. As researchers elsewhere had found, not all of these early cells became active. Some of the cells that worked had cathodes of palladium, but others employed palladium-silver alloys and even pure titanium. But the Indians had discovered an important general property of the cells that did work: surprisingly low overall ratio of neutrons generated to tritium atoms produced. The ratio was in the range of one-millionth to one-billionth (10^{-6} to 10^{-9}). Just what some theoretical work was indicating and what experimenters in the United States who had gotten tritium were also finding! The report concluded that cold fusion is essentially "aneutronic"—unlike hot fusion, it did not produce copious amounts of energetic neutrons. The Indian researchers also believed that cold fusion in electrolytic cells is a phenomenon that occurs on the surface of a metal electrode, not deep within its structure.

At this peculiar time between belief and disbelief, an accounting of cold fusion experiments offered these reasons to accept that there might be something new under the sun: (1) persistence of unexplained tritium at concentrations from 10,000 to a million times normal background levels as well as numerous reports of tritium at much lower but still unexplained levels; (2) persistence of anomalous heat in many different types of calorimeters, even closed cells with forced D_2-O_2 recombination; (3) the very erratic nature of the phenomena, which added to the impression that there was something *different* to be explained, not that someone's crazy experiment was simply exhibiting a systematic error; (4)

low-level neutron fluxes; (5) neutron bursts in deuterium gas-pressurized palladium and titanium; (6) when faced with numerous seemingly inconsistent facts characterizing a system, it is best to try to find a single, unifying explanation that encompasses them all, ergo a new phenomenon. This dictum is called "Ockham's razor."

Theoretical frameworks were building and experimental evidence for cold fusion was mounting. Huggins had seen a 16-day continuous period of excess heat, "about the temperature of your dishwasher," he said. In the first public revelation of positive cold fusion results from a national laboratory, ORNL reported tritium, heat, and neutrons. The LANL work on tritium production became public in December, after a third party leaked it. Ed Storms and Carol Talcott had seen tritium at 100 times background level.

Paradoxically, while the provocative evidence was mounting for cold fusion, the skeptics were going the other way. A Harvard University physics Nobel laureate, Sheldon Glashow, told reporter David Chandler of the *Boston Globe*, "Many of the people who are involved in [cold fusion], I don't have the highest respect for, and many of the results they're reporting are so absurd as to be rejectable out of hand." To the critics, cold fusion was terminally ill and perhaps already dead.

U.S. Department of Energy, Energy Research Advisory Board Cold Fusion Panel:
John Huizenga, Cochairman, Professor of Physics, University of Rochester
Norman Ramsey, Cochairman, Professor of Physics, Harvard University
Allen J. Bard, Professor of Chemistry, University of Texas
Jacob Bigeleisen, Professor of Chemistry, SUNY, Stony Brook
Howard K. Birnbaum, Professor of Materials Science, University of Illinois
Michel Boudart, Professor of Chemistry, Stanford University
Clayton F. Callis, President, American Chemical Society
Mildred Dresselhaus, Institute Professor, MIT
Larry Faulkner, Head, Chemistry Department, University of Illinois
T. Kenneth Fowler, Professor of Nuclear Engineering, University of California
Richard L. Garwin, IBM Fellow & Science Advisor to the Director of Research, IBM Corporation
Joseph Gavin, Jr., Senior Management Consultant, Grumman Corporation
William Happer, Jr., Professor of Physics, Princeton University
Darleane C. Hoffman, Professor of Chemistry, Lawrence Berkeley Laboratory
Steven E. Koonin, Professor of Theoretical Physics, California Institute of Technology
John Landis, Senior Vice President, Stone and Webster Engineering Corporation
Peter Lipman, U.S. Geological Survey
Barry Miller, Supervisor, Analytical Chemistry Research Department, AT&T Bell Laboratories
David Nelson, Professor of Physics, Harvard University

John P. Schiffer, Associate Division Director, Physics, Argonne National Laboratory
John Schoettler, Independent Petroleum Geologist
Dale Stein, President, Michigan Technology University
Mark S. Wrighton, Head, Department of Chemistry, MIT

11 *Denial and Acceptance*

I think it is very premature at this time to say that we are losing a race in cold fusion when *we have very clearly validated that we are not sure that it's fusion.*

James Watkins, U.S. Secretary of Energy
January 1990

The four stages of reaction to a great invention: 1) It's impossible. 2) It's impractical. 3) It's immoral. 4) All along I said it was a great idea.

High-tech wisdom

If it's fusion, it's certainly fusion of a different color than we've ever heard of before. . . . It's got to be entirely new physics, entirely new physics.

Physicist Richard Petrasso
MIT Plasma Fusion Center, January 1990

✳ Only Fire and the Wheel

MORE BOGUS DEATH CERTIFICATES for cold fusion surfaced in late 1989 and early 1990. Most blatant was a small squib in *Nature* in the January 4 issue, accompanied by a photograph of Stanley Pons and Martin Fleischmann in the sunnier springtime of cold fusion. The anonymously authored article said, "Cold fusion has come and gone in eight months." It stated, "The international scientific community showed it could take on novel scientific ideas, digest them and dispatch them, all with considerable speed and relatively good humor. . . . The scientific event to be proud of in 1989 was not the announcement by two Utah researchers . . . but the worldwide activity that it stimulated." *Nature* attributed the slaying of this "Nonevent of 1989" to the DOE final report, which it said "virtually ended interest in the phenomenon."

It was not the first but the second time that *Nature* had written an obituary for cold fusion. This following on the first gave the impression that some kind of monster of science had to be killed with a stake through its heart, lest it rise once more to cause trouble. In this unpleasant atmosphere, it was the rare skeptic who could present his or her case with grace and sensitivity to the views of the other side. One such critic was Richard Petrasso, battle-hardened in the spring offensive of the cold fusion war in Baltimore, a friend of *Nature's* general scientific perspective on cold fusion, to be sure, but a man of no mean spirit. "Gene, I'm going to make a disbeliever out of you yet," was the ribbing he once gave me after I told him the latest cold fusion gossip that had crossed my desk.

During the Independent Activities Period at MIT, between semesters in January, the PFC sponsors a lecture series about research at the Center. It was the third and final day of the talks in the new year, 1990. Ron Parker two days earlier had given an overview of activities at the Center; there was a talk about the environmental impact of hot fusion reactors—"Is Fusion Mr. Clean?"; and PFC associate director D. Bruce Montgomery spoke about the past, present, and future of fusion.

On a drizzly and misty Wednesday it was standing-room only in the conference room at the PFC as Richard Petrasso was wired with a microphone to give an entertaining and revealing lecture under the rubric, "Cold Fusion: Facts and Fallacies." In the crowd milling around the coffee and donuts, a secretary who knew of my accepting attitude suggested to me that I might have more of an open mind about cold fusion than "others around here." It seemed like déjà vu, and I was anticipating another discussion about how wrong-minded the "two gentlemen from Utah" were. Parker was toying mischievously with a Utah Jazz basketball team hat that rested on a table at the front of the room. He put it near the overhead projector and seemed eager to have fun; it was still the tail end of the holiday season. We were deep in the denial phase of cold fusion, but there were signs of nervousness and a bit of unease with the persistence of a few disquieting results that refused to go away. Yet it was by-and-large a crowd of committed nonbelievers.

Dr. Petrasso, who had received his Ph.D. in physics from Brandeis University, had worked at American Science and Engineering nearby and once was involved with the Einstein orbiting X-ray observatory and with the Skylab space station's X-ray imaging of solar flares on the Sun. Though Petrasso lives and breathes plasmas, he is not really a fire-breathing dragon, but a very gentlemanly debater ordinarily. He was one of a vanguard of researchers who began to use solid-state X-ray detectors for imaging fusion plasmas. Besides X-ray and gamma-ray diagnostics for plasmas, he is very interested in physics education and is friendly with MIT Professor Emeritus of Physics, Philip Morrison—

a Renaissance man of remarkable talent. Morrison came that day to share his views about the process of science.

PFC research scientist Robert Granetz introduced Petrasso, "I'd like to point out that this talk was originally titled, 'Cold Fusion: Facts, Fallacies, and Fantasies.' Rich thought that was a little too long. I thought it should be truncated to just 'Fantasies.'" Some giggling from the audience, but a bit reserved.

Petrasso began: "Ten months ago when Fleischmann and Pons held their startling press conference in which they announced the achievement of cold fusion—it was an event which everyone was aware of no matter from what walk of life. And certainly, we were keenly aware of it here at the Plasma Fusion Center. I think it's fair to say that if Fleischmann and Pons turn out to be correct, that their claim of cold fusion in a cell will go down with the discovery of fire and the invention of the wheel as one of the great epics in Man's history. . . . I will argue that . . . it may not be fusion, but there are some interesting issues remaining."

Displaying his hand-drawn cartoon slide of a primitive campfire and a crude Stone Age wheel, Petrasso asked rhetorically, "When the historians look back on this period 100,000 years from now and they're talking about the great epic events—certainly fire will be one . . . but will the third epic be Fleischmann and Pons with their cold fusion cell?" As he flashed a photo of Fleischmann and Pons and their cell, the audience obliged with predictable laughter. "Keep the audience under control, Bob. That was part of our agreement—particularly the Director!"

Petrasso was addressing the question that had brought notoriety to the PFC in the cold fusion saga: Were the neutron detection claims in the original Fleischmann-Pons paper valid? He began at a tutorial level for the uninitiated, citing an example of a fusion reaction, one in which a neutron (from a possible cold fusion reaction) scored a direct hit on the nucleus of an ordinary hydrogen atom (a proton) in a water molecule. Fleischmann and Pons had claimed that gamma rays that they thought they had measured from this reaction were evidence of cold fusion.

Petrasso displayed a cleverly drawn and stylized imaginary weighing scale, on which "Truth," "Beauty," and "Justice" were inscribed. In principle, if one could weigh a completely closed cold fusion cell before and after it operated for some time, a decrease in mass would be observed—*if* fusion reactions took place inside. In practice, the infinitesimal amount of mass converted to energy, to the tune of $E=mc^2$, would be far too small for practical measurement. Other effects, such as atoms sloughing off the external surface of the cell would swamp the mass loss due to fusion.

Petrasso's argument proceeded to box the absent Fleischmann and Pons into a corner of "impossible" miracles. "In conventional fusion, you have to put them [the deuterium nuclei] into some kind of accelerator [or hot fusion plasma] to get them together. If they interact, there are only three branches that they can go to. This is it, there's nothing else. The top branch yields a neutron and ^{3}He. The other very probable branch is the proton plus triton coming out. A very intriguing issue now: Some people are claiming to detect tritium, but the incredible thing about the tritium is that it doesn't have any energy [as measured in their experiments]! Here I'm indicating that it has one MeV of energy [the conventional value]. What they're saying now is it comes out without any energy. It is 10,000 to a million times background [in its quantity] but it's coming out without any energy!

"The third reaction yields ^{4}He plus this very energetic gamma ray. Ordinarily this hardly ever happens, but one out of 10 million times—at least in conventional fusion. But, in fact, in the earliest pronouncements of cold fusion—and, in fact, some people are still maintaining today that this is the important branch for cold fusion to occur."

Cold fusion proponents might buy certain "miracles," but not Petrasso. "I'm willing to accept a few things," he said,"but I can't view this as a very feasible thing. In fact, my view of it is that this particular branch can only be referred to as a 'four miracle theory of d-d cold fusion.' Why? First of all, it has to happen cold. I can accept that maybe there is some new physics here, but the second thing is that it also goes to this helium-4 branch here—the one that hardly ever happens. The third thing about it is that the gamma energy, instead of coming firing out at you, which has 24 MeV of energy—somehow it mysteriously couples into the lattice in ways that we don't understand. But really what gets me now is that after these three miracles occur, the helium-4 disappears! I refer to this as 'immaculate disappearance.' I'm willing to accept three miracles, but the fourth one is just too much for me to accept." Petrasso was referring to the curious lack of evidence so far for helium-4 in palladium rods or in the gases emerging in electrochemical cells. The absence of helium-4 seemed to him to be the death knell of the once-promising helium-4 branch combined with some new physics of energy transfer to the metal lattice.*

Petrasso admitted that his argument governed only the possible d-d route of cold fusion. He was aware that a number of other possible "conventional" fusion reactions might be occurring—ones involving

---

*As this book went to press, a remarkable paper was about to be published in the *Journal of Electroanalytical Chemistry,* which purported to show a "correlation between the generation of excess power and the production of ^{4}He in the absence of outside contamination." (See Bibliography.)

lithium, for example. However, one could keep the attention of a mixed audience only so long. As far as the d-d reaction went, he was convinced: "It's just not credible from my point of view."

He stepped back in time to dissect the now-familiar evidence that the Utah duo had put forth. He reviewed the arguments against the reality of their neutron data, basically the story of the PFC research that had led to the dramatic presentation in Baltimore, the surrounding media hoopla, an article in *Nature*, and altering the course of the cold fusion investigation (Chapter 8).

"Is cold fusion dead? Are the claims all discredited? I think that it would be fair to say that they're not. There was a recent meeting [the NSF gathering in Washington] at which some advocates presented some new data. . . . The issues that they talked about and which are still sticking are the question of excess of heat, in which now about five different groups have claimed to see about one watt of excess heat coming out of their cells."

The MIT researchers had long since put the heat question temporarily aside, for want of time and funding to do more work; perhaps also the conviction that the apparent absence of apparent nuclear products said it all. But Petrasso and others at MIT continued to be very interested in the tritium question. He told the colloquium, "Some people at Texas A&M, a group down at Oak Ridge [National Laboratory], and some people in India have measured tritium at the level of 10,000 to 100,000 times background. But the curious thing about it is, it's not energetic. Again, if it were the standard reaction, it has to come out with energy of about one MeV, and it's not doing so. So if it's fusion, it's certainly fusion of a different color than we've ever heard of before." Even if the tritium were not seen directly, if it were energetic, it would be expected to interact with deuterium to produce high-energy neutrons—at 14 MeV.

Another open question was the "sink issue." Referring to the Texas A&M work, in particular, Petrasso said, "Sometimes they have what they call a transient tritium event where the levels of tritium come up and as they continue to operate the cell, the levels decrease as a function of time. Now if you stop the operation, the level of the tritium stays fixed, but if you run it, it can actually decrease as a function of time. Personally, I find that to be as remarkable to hear about as the source."

Petrasso had been thinking along the lines of proving somehow that the tritium was caused by the well-known phenomenon of "enrichment" of hydrogen isotopes during electrolysis. In an open cell, as more solution was added to make up for electrolyzed heavy water, naturally occurring tritium could concentrate in solution, though it was hard to see how such spectacular levels as had been found at A&M and in India could come about that way. He also believed that because the tritium was coming out in a nonenergetic form—sedately appearing in the liquid and gas, not as fast-moving tritons that could be picked up by a radiation-measuring instrument—that this was suggestive of some inherent contamination of the palladium or titanium. He admitted that it was "a little bit hard to understand." Petrasso had honestly failed to mention the by now well-established observation that tritium was not found in cold fusion control experiments using light water. This intrigued proponents but was the kind of evidence that skeptics generally treated as second-order questions, not to be taken too seriously before more basic ones could be answered.

Properly left unstated in Rich's talk, however, were rumors about deliberate contamination of cells at Texas A&M—suspicions that had been circulating for months. The issue would finally become public in June, with the appearance of Gary Taubes's extremely controversial exposé in *Science* magazine (Chapter 14). The tritium question continues to be one of the most fascinating pieces of evidence for cold fusion. If tritium is generated in cold fusion experiments, and is not natural contamination, it would be proof positive of cold fusion by electrochemistry. Despite critical negative assessments that some have made about tritium, large numbers of researchers pursuing cold fusion continue to be convinced that it is *not* an artifact. That is my view, too, after having listened to many of the technical arguments by experts on both sides.

Petrasso ended his refrain and came back to his original question: "How will historians view us 100,000 years from now? It could be that we'll have these three. . . ." His colorful slide showed the wheel, the campfire, and Fleischmann and Pons with their cold fusion cell. "But I would say that my bets are that it will be these two. . . ." And in a flash he removed from cosmic history the defenseless Fleischmann and Pons on their plastic overlay sheet, leaving only the Stone Age wheel and the campfire.

✳ Open Questions

No one can escape the many unanswered and unanswerable questions about cold fusion, even when surrounded by friendly natives. Someone asked Petrasso about the excess heat measurements, a subject with which he was less familiar. Part of his tentative response revealed one of the most critical aspects of the cold fusion debate—the relative inability of experts from different disciplines to appreciate or have confidence in one another's results. Nuclear people typically knew very little about the subtleties of calorimetry. Though some of them were trying to come up to speed, many skeptics simply put it aside as too finicky and error-prone—in my view a big mistake.

Petrasso left the audience with the general impression that skeptical nuclear people were likely to hold: "One of my suspicions is that some of those excess power measurements are going to be shown eventually to be in error due to very subtle mistakes. That claim might be surprising, because you would think that we could do energy conservation to very high accuracy. But my guess is that some of them will be shown to have very subtle mistakes in them, but they're not an obvious kind of mistake." Ron Parker evidenced the same kind of disciplinary bias when he observed, "Lots of the time the excess power is less than the power that you need for the electrolysis, so that if the electrolysis stopped for some reason—electrolysis was not taking place—you would think that you were producing excess power. I think that some of the erroneous reports are based on that type of error. There are dozens of pitfalls in this power business."

The nuclear people had virtually washed their hands of this messy heat-measuring matter. Unfortunately, this tended to prevent them from being challenged by one of the most tantalizing clues to a possible new phenomenon. Petrasso, for example, did not find the recent claimed excess heat measurements at Oak Ridge National Laboratory to be compelling, even though it had been done in closed cells—by his own understanding generally a more satisfactory approach. He said, "Looking at the error bars [the extent to which the curves could be assumed to be in error] I don't find it very convincing. I think that on further examination they won't hold up under scrutiny." For him, the nuclear measurements were really the end-all and be-all.

On more familiar matters, nuclear people could be bolder, though just as skeptical. Asked about his belief in Steven Jones's neutron measurements, Petrasso said with a short, polite laugh, "Yes, Steve Jones has done some very interesting work and I haven't any comment—other than to say that Steve Jones's claims were of course far more modest than those of Fleischmann and Pons, and I think more credible. . . . He's only claiming one and a half times above background. Now the flux of neutrons is really very erratic and that's a very hard measurement. He's saying we're seeing maybe three sigma [above background], but I don't find it very convincing, frankly. He may have a case, but I think we've got to look very hard. It's not like the tritium where we're talking about 100,000 times background. I mean, even I can't make a mistake on that, but 1.5 times background? If you know the fluxes from cosmic rays and what have you, they vary as a function of time. It's erratic. So he may be right, but it's going to take a lot more work on his part to convince us."

One mischievous person asked Petrasso what Peter Hagelstein thought about cold fusion. Hagelstein was not able to attend the seminar, though he was quite aware of the technical objections to cold fusion that prevailed across campus in the hot fusion complex. Petrasso replied, "I don't know, I invited Peter here so he could perhaps talk about it, but he is actually invoking now some weak interactions—I mean it's esoteric stuff; very frankly it's hard for me to follow and understand. It's based upon a coherent laser concept. Peter presented his work at the [December] meeting in San Francisco. He's trying to address the tritium question. It's true that he had a reaction which—mind you, he's talking about the weak interaction, and typically that reaction is down by 25 orders of magnitude from the strong [nuclear] interaction. He [his theory] can get tritium without any energy, because he's generating a neutrino. And the neutrino is taking away all of the energy, so you get a triton that has absolutely no energy—it's just sitting there. But I think you probably need to talk to Peter about it. Again, it's based upon some sort of coherent coupling of the lattice to these processes. I don't understand the theory, I don't understand the basis of it, so I think you have to talk with him."

Then an even "naughtier" question from the gallery, "Rich, isn't there a problem that if the neutrino carries away the energy you don't get any excess heat?" Petrasso, a bit eagerly and to audience laughter that he didn't provoke: "Oh, absolutely, absolutely—exactly, exactly—but he has other reactions." This provoked even greater mirth from the hot fusioneers. "O.K., he has other reactions besides that one," Petrasso said, trying desperately to be straight about it. "That's the one that generates the triton. But you're absolutely right. In fact, you're always stuck. Every time you say, 'Well, gee, I got this particle, now we're going to try and do energy or power balance,' and you can't make it work." Ergo, end of the cold fusion story, was the distinct impression that he left.

What about Hagelstein's "phonons"? Petrasso replied, smiling broadly, "Well, there's that too, but I'm not going into that one. Peter has this coupling where he has phonons and photons and so forth—I just can't understand it at this point—I haven't spent enough time."

A high point of the day's program: Philip Morrison gave a captivating talk about the 18th turning to 19th-century discovery that stones actually do fall from the sky; alleged meteor falls were not merely "fragments of the imagination." The subject was to have formed another chapter in his *The Ring of Truth* book, but it was not included, though it had much to say about scientific methodology, and what qualities confer that intangible "ring of truth."

Venturing onto thin ice, I asked Morrison whether he detected even the slightest glimmer of the "ring of truth" in cold fusion? His rapid-fire response capped the day: "Let me put it this way. At various times I have had various degrees of hopefulness for it. I did not say from the beginning that it was all stupid. I didn't think it was right, but I could see that we couldn't disprove it by just making those arguments. I did not from the very beginning believe that they were observing these [ordinary] fusion reactions in the palladium, because the radiation that they claimed was nothing like adequate for that—nothing—down by many, many orders of magnitude. And therefore there had to be something new.

"Well, something new has to be excluded by either understanding what makes it work, or by getting it to happen in many places. And the people who got it to happen—and there were quite a few of them all over the world—never presented a very strong case for how they did it. Theirs was a museum case like the University of Paris [an allusion to his meteor talk]—it was a lot of mixture. The more people that did it more carefully—as for example here as Richard [Petrasso] has told you, I'm sure, the more you saw that there was not this nuclear phenomenon that was claimed. Even the small one claimed was wrong.

"Finally, a very large effort made all over the world—particularly strongly at Harwell in London—really was able to reproduce most of the effects—showing where they came from, what in the perceptual apparatus was wrong. And the calorimetric curves published by the English group in *Nature* in late November, show this amazing thing where the heat developed goes up and goes up like that—a beautiful curve that continues to go up as you run. In the Fleischmann-Pons experiment and in their replication of that experiment with their replication of the Utah calorimetry—they improve the calorimetry and they shield it and they make all precautions to keep heat from leaking out, gases from coming out and so on, and then it goes up like this, this, this. And it tapers off and it goes as flat as a ruler-drawn line over 300 points to the end of the page.

"So you say, well that's it. It may have not been. That still does not *absolutely* disprove these people, but it turns the complete burden around; I think that's what we can say. Because, unless they can show their calorimeter really is working in spite of the fact that when you make one like it it doesn't work, and in spite of the fact that when you make a good one it really does work and shows no extra flow of energy. Unless they can overcome that barrier, they haven't got it. But of course it's a complicated system. You can't be sure that there's not some mysterious impurity in their palladium that did it—once. I would say I cannot be 100 percent sure of it, but I would say that it's a very reasonable bet and don't invest any money in palladium!"

Well put, absolutely razor sharp in its statement and more than fair in its opening of the cold fusion door just the barest crack. Could anyone fault this assessment? To this, one would have to say, with some hesitation, "Yes, he could be wrong." In fact, he was most likely wrong. Morrison's possible misstep, if he erred here at all, was to violate one of his own cardinal rules and make an inadvertent "appeal to authority." The authority in this case were the negative treatments in *Nature.* In particular, the Harwell work has major problems with its calorimetry and is even worse on its conclusions about tritium. A host of intriguing positive results was simply not getting through, even to the always open-minded Philip Morrison.

The Harwell work on cold fusion is likely to be fraught with problems, as a number of physicists and chemists have indicated to me. One problem can be seen right away: Almost half of the Harwell cells were run close to or below what has become empirically evident (for those with positive results) as a threshold for expecting cell activity—a current density of around 60 milliamps per square centimeter of palladium cathode surface. Furthermore, there are a variety of reasons to suspect that the procedure used to fit the Harwell calorimetry data gives an *underestimate* of power, rather than a "conservative estimate" as the paper claims. One of the Harwell calorimeters has poorer thermal design than units that have given positive results. Among other problems, it may have much too large a volume of heavy water for proper thermal equilibrium—the old stirring problem again, but in this case an experiment with a negative result is brought into question. All in all, said one physicist, "The Harwell work was a travesty."

As for the Harwell claims of not having observed nuclear effects, there are also problems. Even though Harwell claims that its background noise level is 10 times lower than Steve Jones's for neutron counting, Harwell adds up its neutron counts over so long a time period that any possible neutron signal may have sunk below Harwell's vaunted clean background level. Harwell used heavy water containing a tritium concentration some 10,000 times greater than that used by researchers who believe they have seen tritium produced. It would have been very difficult for Harwell to discern any tritium generated in their experiments.

So the question remains, on whom will the burden of proof be in the matter of cold fusion—on skeptic or proponent?

12 Approach to an Answer

When a distinguished but elderly scientist states that something is possible, he is almost certainly right. When he states that something is impossible, he is very probably wrong.

Arthur C. Clarke
Profiles of the Future, 1963

Always the beautiful answer
who asks a more beautiful question

e.e. cummings

Megajoules per cubic centimeter? If they were all diamond chemical bonds, that wouldn't be enough to explain it chemically!

Professor in Cambridge, Massachusetts
(Name withheld)

✳ It's Fire!

WITH THE DAWN OF THE '90s, whether or not the skeptics were aware, possible solutions to the cold fusion mystery were beginning to emerge. Evidence of tritium production in electrochemical cells was getting to be more convincing, despite lingering doubts about enrichment or contamination as alternative explanations. Teams in at least three major U.S. national laboratories (Los Alamos, Oak Ridge, and Brookhaven) were doing cold fusion experiments that were producing tritium. So were the Indians, so were the Japanese, so were more than two dozen other laboratories.

If the substance had not been linked with the "preposterous" allegation of fusion at room temperature, the evidence for the production of tritium would have been considered extremely solid. If tritium was being produced from other atoms, then cold fusion simply had to be happening. You simply cannot make tritium by ordinary chemistry, or at least by what used to be *called* chemistry! Chemistry, we were led to believe, deals only with the rearrangement of electron configurations surrounding atomic nuclei as these atoms combine into various molecules. But now it seemed possible to carry out what were ostensibly chemical experiments—such as with electrochemical cells—that could create rearrangements within nuclei.

It was also clear at this time that precious few neutrons were being produced in any of the numerous cold fusion experiments that had been attempted. If the new phenomenon was fusion then it was, indeed, of the neutronless variety, the kind long sought by some of the fringe groups in the hot fusion community—the aneutronic fusion people (see Chapter 18). In many experiments the ratio of neutrons appearing to the amount of tritium being produced was on the order of only about one neutron or less for every 10 million to a billion tritium atoms coming into existence, that is, 10^{-7} to 10^{-9}. In hot fusion experiments involving deuterium reactions, neutrons and tritons appeared as final products in roughly equal numbers.

Finally—and of foremost importance for potential applications—excess heat was popping up everywhere. Research groups had refined the techniques of calorimetry to answer the many objections that the skeptics had put forth. They were still getting excess heat and more reliably so. But even with calorimetry, being a precise and delicate art, it was still possible to continue taking pot shots at the reported excess energy production, seemingly ad infinitum. Yet even this area of skepticism was weakening.

So we have the evidence of tritium, some neutrons, and many indications of excess heat. All the hallmarks of possible cold fusion. In no way was this new process consistent with the usual expectations of the hot fusion community, hence their extreme skepticism. But how to explain the phenomenon? Many people had tried, and some were coming up with very intriguing ideas.

Hagelstein, for one, was struggling to put the finishing touches on what one might call a *complete* theory of cold fusion, one that would account not only qualitatively for what was happening, but that could also come very close *quantitatively* in connecting fusion reactions with nuclear by-products and the excess heat. His hoped for complete theory would also help to explain why some research teams were getting the effect and others weren't, and even why those who were getting the effect could not reproduce it consistently. To convince critics, any theory would have to meet the most demanding scrutiny of the physics community.

Hagelstein's cautious enthusiasm was infectious. In private he was saying things about the phenomenon like, "Its usefulness is basic. I mean, it's fire!" And, "I probably bought into this before anyone else. My position on this since it began is that it's the greatest 'toy' around." As Hagelstein phrased it, he was "convinced of all the 'miracles' " that were necessary to make the phenomenon true, just as Richard Petrasso was equally certain that the 'miracles' were all mistakes. Out of context, these remarks could easily be mistaken for gross self-delusion, but it was clear that they were backed by much reasoning about what was going on. Hagelstein was well aware that many good laboratories were reporting negative results in cold fusion experiments, a fact that he characterized as "a little bit on the frightening side." Yet still he held fast to his views.

Hagelstein saw the heat effect as the biggest—not necessarily the *best*—piece of evidence that cold fusion was occurring, implying at the reported multiwatt power levels that 10^{13} to 10^{14} nuclear reactions of some sort were occurring each second. In his view, the original arguments against heat were: (1) *Physicists* were not obtaining excess heat, (2) Open instead of closed calorimeters were being used, (3) The electrical input power measurements were no good, and (4) The electrochemical cells were not stirred properly, thus leading to false measurements. But improved experiments had overcome those objections. Robert Huggins was now consistently getting heat to show up every time he did an electrochemical cold fusion experiment. Huggins reported a spectacular run with an 8-watt net power production that lasted 275 hours. Other people were seeing heat come and go, though the situation was getting better—the effect was becoming more reliable. Still, the heat was not easy to come to grips with. Hagelstein jokingly described himself as possibly "the only physicist on the planet to say that heat is not chemical."

The tritium was another matter. Bockris and others at Texas A&M came out with the earliest good reports on apparent tritium production, the unexpected exponential rise in tritium content in cells. But now others had gotten tritium results, which appeared not to be the result of contamination. Among them were Kevin Wolf and his colleagues at Texas A&M and the Scott group at Oak Ridge National Laboratory, who had reported their results at the "Cold Fusion Session" of the December meeting of the American Society of Mechanical Engineers (ASME) in San Francisco. Then there were the excellent triple-checked results coming from Los Alamos—lots of tritium being produced there.

Moreover, tritium was beginning to be reported in radically different kinds of cold fusion tests. Another group at LANL led by Tom Claytor had set up a deuterium gas cell in which a bizarre layer-cake of palladium powder alternating with slices of silicon was pressed together. When a field of several thousand volts was set up across this assembly and current passed through all the layers, significant amounts of tritium were found in the apparatus. A careful analysis of the constituents before the voltage was applied showed that tritium was present initially in insignificant quantities. But after current was applied, tritium appeared in the deuterium gas chamber at levels hundreds of times above those initially. This setup, by the way, resembled an earlier arrangement that Japanese researchers had used (see Chapter 10). There seemed to be no problem in generating tens of *microcuries* of tritium. (A microcurie is a measure of the quantity of a radioactive isotope in terms of the number of disintegrations per second. A microcurie of a radioactive substance experiences 37,000 distintegrations per second.) The "hot" Los Alamos test had created about 170 microcuries of tritium.

Neutrons were most interesting by *not* being there. Nor was there good evidence *yet* that tritons or deuterons—the nuclei of tritium and deuterium atoms—were emerging from whatever reactions were taking place with the kinds of MeV energies normally associated with conventional fusion reactions. Hagelstein's startling conclusion: "Tritium and deuterium products may be born stationary (with extremely low energy)." In his view, numerous particles moving with great energy would destroy the "coherence" of the laserlike fusion process that he had postulated. Here was a possible answer to some of Petrasso's misgivings about the lack of energy in the tritons.

The neutron picture was confusing, indeed. The Scott group at ORNL was observing a correlation between heat and neutron emission in its electrochemical cell experiments, but Kevin Wolf and others were not seeing such correlations. Neutrons were also being seen in the gas-phase tests at Los Alamos and in India. No heat was being observed in these gas-phase experiments, except there was an indication that the Indians were beginning to see heat plus neutrons. The Italian group at Frascati had so far not been able to reproduce its neutron bursts, but people at Los Alamos had seen significant evidence of neutron bursts in gas-cell tests. In a bit of international intrigue, Howard Menlove went to Beijing and helped researchers there see the same kinds of neutron bursts that his team at Los Alamos had observed.

All in all, the evidence for this plethora of unusual effects was "very strong," the words Hagelstein used to characterize it. Of course, his views were diametrically opposed to those of researchers like Petrasso. I told Richard Petrasso in December 1989 that I had "crossed the line" and was now prepared to believe that cold fusion was so clearly a real new phenomenon, that the burden of proof must surely rest now with the skeptics—it was their duty to try to debunk cold fusion. Good-naturedly chastising me for having been on different "sides of the line" so many times, Richard said that in his opinion there was but a two percent chance—"five percent tops"—that cold fusion was real. He believed that tritium would ultimately be explained away as contamination or some kind of enrichment process, and that the calorimetry was simply not good enough so far to prove power generation.

✳ Conquering the Coulomb Barrier

A welter of slips of paper festooned the door of Peter Hagelstein's small office, where large thoughts roamed—not always in the glare of the fluorescent bulbs' plasma lighting, but occasionally in the glow of a computer screen in a darkened room. It is a room of organized chaos—a hallmark of an active mind too preoccupied with the mysteries of nature to care much for straightening the teetering piles of technical preprints that covered the floor. A childhood incident illuminates this intensity: Seven-year-old Hagelstein clipped one of the two parallel wires going to a household lamp, put both frayed ends of the now open circuit in a tub of water, then sprinkled in some salt to make the water carry current. The light went on. In late 1989, a light had gone on in his mind, when he discovered a possible way around the coulomb barrier. It began to make cold fusion believable.

It was all well and good to have copious experimental support for cold fusion, but why should anyone believe that the coulomb barrier could mysteriously crumble inside palladium rods? The biggest hang-up to a cold fusion theory, of course, was finding a way to overcome that mountain of electrical repulsion. Time and again, calculation showed that two positively charged nuclei had a negligible chance of fusing at room temperature, no matter what theoretical tricks were played with electron shielding and the like. Conventional nuclear reactions occurring between two such nuclei—binary reactions—simply had to be dismissed in seeking an answer.

Also, if one believed that the tritium was real, it certainly could not be produced in the conventional plasma physics reaction of two deuterium nuclei fusing to produce a one MeV tritium nucleus and in another branch, a high-energy (2.45 MeV) neutron. Since numbers of such high-energy neutrons were not being seen, the tritium had to be "born sitting still," Hagelstein reasoned. One had to swallow a remarkable miracle, indeed, to believe in an energy-producing reaction with the products remaining essentially stationary by conventional nuclear physics standards. Remember, energetic products such as neutrons were what hot fusion people required to make practical use of fusion!

That was not all. Hagelstein realized that cold fusion reactions could not be allowed to overturn the well-understood and largely verified models of fusion reactions occurring at high temperature in stars, nor could cold fusion ignore or violate the known nuclear stability of ordinary heavy water or deuterium gas.

Hagelstein's calculations and those of Kevin Wolf suggested that the maximum kinetic energy with which tritons in cold fusion could be born would be 15 to 25 keV—hundreds of times lower than the energy of tritons that hot fusion researchers were accustomed to. Moreover, the rarity of neutrons in cold fusion—less than 10^{-7} of tritium nuclei produced—led Hagelstein to this remarkable conclusion: "The relative lack of neutron emission can be used to rule out essentially all *known* nuclear fusion reactions which evolve tritium, even if some mechanism were found to overcome the coulomb barrier."

In the spring of 1989, when Hagelstein had begun thinking about coherent fusion reactions, he was still coming up against that stubborn barrier. The coherent reactions between two deuterons that he could imagine were simply too unlikely. The reaction rate was just too low to explain cold fusion experiments. Then in the fall of 1989 had come a leap of imagination—one might say right over the coulomb barrier. If the charged particle coulomb barrier was blocking the way, why not do coherent fusion with *neutral* particles. Not the mystical neutral *meshuganon* particle that theorists had joked about, but perhaps neutrons themselves. And where to get free neutrons? Why, simply steal them from nuclei.

Hagelstein imagined this scenario: If a proton in a nucleus could "capture" an electron from the inner reaches of the electron cloud surrounding it, the two could combine to form a neutron. If the proton in a deuteron, for example, could through electron-capture become a neutron with a fleeting existence—a "virtual neutron"—then there would be two neutrons to play with. Another deuterium nucleus picking up such a virtual neutron would become tritium. A proton (from the small amount of ordinary hydrogen of ordinary water in almost pure heavy water) picking up a virtual neutron would become deuterium and release heat to the palladium lattice. Thus would the coulomb barrier be breached, by a back door approach, so to speak.

Easier said than done, however. To create virtual neutrons requires making use of a very touchy particle interaction that is responsible for some kinds of ordinary radioactive decay, what is called the weak force or interaction—one of the four fundamental forces (including also gravity, electromagnetism, and the strong nuclear force) that physicists believe rule the universe. Hagelstein reasoned that the weak interaction would ordinarily be too flimsy to run the reported levels of cold fusion, but he saw a way around the problem. If the resulting nuclear energy could be taken up by the palladium lattice through the coherent or laser-like process that he had earlier envisioned, the weak interaction might be "strengthened" enough to work its magic and permit cold fusion.

The capture of an electron by a proton through the weak interaction necessarily involves the emission of an attendant neutrino—the zero mass or extremely low mass particle that, once created, travels at virtually light speed and interacts hardly at all with matter. If collective or coherent effects in the palladium lattice could work just right, they might permit what Hagelstein characterized as a "Dicke superradiance of neutrinos" that would mandate that the final state of the reaction products were largely stationary—no high-energy deuterons or tritons should emerge. But this was just what was required to explain the extreme dearth of high-energy particles that experimenters were finding.

So in Hagelstein's cold fusion scenario, the fusion fuel was not simply deuterium, but deuterons, protons, and electrons, all of which were present in heavy water. The role of palladium was that of a catalyst, because of what in his theory he referred to as the metal lattice's "nonlinear response." Besides the heat transferred to the lattice in the coherent fusion process, the reaction products would be stationary deuterons and numerous but essentially unobservable neutrinos. Nuclear energy from deuterium could thus be converted to heat without producing helium-4, which no one was having success finding in spent palladium rods.

Hagelstein had put forth these ideas publicly when he delivered his paper, "Coherent Fusion Theory" at the Cold Fusion Session of the American Society of Mechanical Engineers (ASME) Winter Annual Meeting in San Francisco on December 12, 1989. Much as in a conventional laser a large population of atoms in an excited state is required to decay into lower energy atomic states, in Hagelstein's coherent fusion, a large number of these fusion reactions with virtual neutrons has to occur to produce an observable effect.

A fusion reaction that can occur through the emission of a high-energy gamma ray might, in principle he thought, also proceed through the successive emission of a large number of low-energy photons. His general idea of coherent fusion in a metallic lattice was "very close," he said, "to semiclassical models used in laser physics."

Before Hagelstein had gone to the ASME meeting, the *Boston Globe* did write something about him, but it was not about the substance of his theory. Many reporters had essentially given up reporting on cold fusion science, choosing to report instead the political maneuvers associated with the scientific battles. The *Globe* story was an exposé, on page one no less, of Hagelstein's supposed tenure "fight" at MIT. The headline was "Fusion Defender in Tenure Fight." This tempest in a teapot was reported in *Science*, but details of Hagelstein's theory were only noted in passing. MIT ultimately granted him tenure in the spring of 1990.

✳ Comprehending the Mystery

Peter Hagelstein had much respect for Martin Fleischmann, because he had learned of Fleischmann's discovery of the "surface-enhanced Raman effect." This was an esoteric effect that people in the laser physics community had come to know well. Indeed, the laser community was well aware of Fleischmann's discovery of the effect. In Hagelstein's words, "Fleischmann got up one morning and visualized it. It is now the biggest nonlinearity [an effect that doesn't change in a uniform way with changing conditions] in the laser business."

Hagelstein saw Fleischmann and Pons as electrochemists who were "professional master tinkerers." "Parts and pieces had been lying around and it took world class tinkerers to get it," Hagelstein said. We know that Fleischmann and Pons had evidently spent five years tinkering with electrochemical cells to come up with what they saw as proof of cold fusion. The precise conditions that would bring the effect on might be very delicate, indeed, and it was these conditions that seemed to intrigue Hagelstein the most. He laughed that by virtue of his theoretical understanding and links with experiments worldwide, he had become "the world expert on what would make cold fusion cells *not* work."

Hagelstein's coherent fusion theory at this time required as a byproduct the appearance of vast numbers of neutrinos coming out, but he was well aware that, neutrinos being the extremely difficult-to-detect entities that they are, it was unlikely that any experiments soon would be able to see them. After all, Supernova 1987A 160,000 light years away had produced, three years ago billions of neutrinos per square centimeter in our vicinity, and yet the most sensitive detectors in the world had managed to pick up only several dozen of them. Neutrinos can penetrate distances of light years into lead without being absorbed.

Hagelstein also believed that cold fusion required a "nonlinear material under stress," such as might be afforded by palladium in an electrochemical cell. Within the palladium rod, various phases of the metal would conspire to produce these stresses. But there might be other ways to produce coherent fusion, as was becoming evident in the gas-cell work at Los Alamos.

In the deepest corners of the minds of the skeptics, they could perhaps imagine cold fusion turning out to be real after all, an interesting curiosity perhaps, as muon-catalyzed fusion had been for so many years. Yet hardly a soul in that community would have imagined the prospect of significant applications for cold fusion within several years. But Peter Hagelstein did. To him, it was indeed like fire, a phenomenon of such astounding potential that he could conceive of numerous applications, from household heaters, through chemical and agricultural applications, and outward bound to new forms of space propulsion systems. He had already incorporated some of these ideas in his thinking, starting in the earliest days of the cold fusion revolution.

In January 1990, he felt that the field would break open within the next few years if not much sooner, leading perhaps to hundreds of watts of power generation in cold fusion apparatus, not just the few tenths of a watt that was being reported in most experiments with positive results. [There were serious indications that Professor Schoessow at the University of Florida at Gainesville was already getting these power levels and was soliciting interested investors. Within one year, Fleischmann and Pons themselves would privately acknowledge to trusted associates that they had achieved much higher levels too.] Hagelstein was guided by his theoretical intuition, which told him that high power levels would be possible, but also by the various high power level anomalous events some had reported, such as the Italian investigators D. Gozzi and others and Fleischmann and Pons themselves.

A severe funding crunch was squeezing the life out of even the hardiest cold fusion research "bootleggers"—people who would borrow time on contracts designed for other matters. Federal research funding for cold fusion had all but dried up, thanks in no small measure to the November 1989 DOE report. Initially there had been seed money given out by DOE. Ryszard Gajewski of DOE's Advanced Energy Projects Division had managed to parcel out on the order of $1 million, but DOE had severely cut his budget so he could no longer fund cold fusion studies, at least not until the beginning of the next fiscal year (October 1990). Gajewski eventually left DOE for reasons unrelated to cold fusion research.

California-based EPRI was funding cold fusion work to the tune of about $1 million annually. India seemed to have going the biggest effort of all. Equivalent in prestige to the National Science Foundation in the United States, the more applications-oriented MITI organization in Japan was also providing good support. It was an extreme irony, given all the talk these days about the faltering U.S. competitiveness position vis-à-vis Japan, but the U.S. National Science Foundation had a formal policy at the moment of not funding cold fusion. And they were the ones open-minded enough to have held the seminal cold fusion conference in Washington, D.C., the past November 1989. The Department of Defense was funding one effort but was an even more problematic organization to deal with on this highly charged issue. Then there was the National Cold Fusion Institute in Utah, which had money from the $5-million infusion the state of Utah had given it, but NCFI's ground rule was that work had to be done there.

Dr. Mark Stull, an astrophysicist from New Hampshire who had become an attorney practicing law in Boston, met with Peter Hagelstein in early 1990. Following this meeting, proposals to seek funding for definitive cold fusion experiments would soon emerge, involving a closely knit group of researchers in the MIT community. Some even had ties to hot fusion work, such as Dr. Stanley Luckhardt of the Plasma Fusion Center and independent scientist Dr. Vladimir Krapchev. Though Stull was very experienced in the world of high-tech research and investment, and though Hagelstein was very persuasive that something should be done, the outside sources couldn't quite bring themselves in 1990 to take the plunge and finance cold fusion.

Many cold fusion researchers had wish lists of experiments that they wanted to carry out, but they were hampered by the funding crunch. A chunk of money on the order of $100,000 would have helped almost anyone. Small change, of course, by the standards of almost any kind of major federal research effort, "noise level" as engineers are fond of saying. But it wasn't noise level for cold fusion research in early 1990. It was the level of support for small projects that even skeptics would have approved, if their April 1989 testimony before Congress was to be believed. Many researchers like Hagelstein were simply acting selflessly as catalysts, trying to stimulate ideas within the experimentalist community. But there were often no funds for their own proposed experiments.

Hagelstein was being buffeted by a host of suggestions from within his circle of friends at MIT. His engineer colleagues were urging him to begin experiments right in his own lab. Professor Richard Adler, for one, had been suggesting that an experimental effort should happen within MIT. (Tragically, Dr. Adler would soon die in an accident.) Hagelstein felt that "something was going to break" within the next six months to a year, making the possible rewards of cold fusion development abundantly clear to all who would listen, not just to already converted believers.

He was most intent at this time on proving that protons really were being consumed in the heat-producing reaction. The single proton nuclei of ordinary hydrogen atoms were responsible in his theory for the heat production in electrochemical cells. This would not be an easy task, because it took only a small quantity of ordinary hydrogen in the heavy water to supply the protons to react with the hypothetical virtual neutrons and yield significant energy. The protons would be preferentially absorbed into the palladium lattice ahead of the deuterons. Moreover, once in the palladium, protons would fuse with neutrons some 5,000 times more effectively than deuterium. In a typical cell, 10 to 20 percent of the total number of single protons might be consumed in a month's time of cell operation, an effect that might be seen with a sensitive measurement.

There was also the possibility that lithium from the electrolyte, specifically ^{6}Li, was being consumed and converted to ^{7}Li by fusing with neutrons. Some experiments had reportedly found that already. But Hagelstein reasoned that if ^{6}Li were the dominant fusing product, one might expect to see energetic electrons (beta emission) and ^{4}He by-products, but there was little evidence of this. The nuclear system buzzing on the surface of the palladium rod had to be complex. Hagelstein thought that neutrons could be reacting with all available nuclei, including H, D, Li, O, and Pd. Yes, even the oxygen in the water and the palladium atoms themselves were fair game.

There were ample reasons to be taking cold fusion very seriously, even though confirmation that it was real was not at hand. The first objective of any scientific effort had to be a consistent explanation of the energy balance and nuclear by-products in a working cold fusion cell. Given that, "The dominoes would then fall," Hagelstein believed. Until then, cold fusion would suffer widespread ridicule in the scientific community—"state-supported ridicule," he joked.

The priorities were clear: (1) prove that there was indeed a new effect; (2) achieve consistent reproducibility; and (3) discover the mechanism of the phenomenon(a). Engineering applications would have to wait for this basic level of physical understanding. But to some extent, Hagelstein's cold fusion theory was then evolving independently of experiments through (by his own count) dozens of versions. Each was based on a general set of principles about coherent fusion reactions, but the specific reactions and methods of computing the behavior of the metal lattice would radically change. Ultimately, the exotic virtual neutrons and neutrinos would be abandoned in favor of what he felt was a more workable, yet more provocative, scheme. Hagelstein would find a way to use the energy of the lattice itself to rip neutrons directly from deuterons and fuse these neutrons with other atoms, including palladium. The pieces of this puzzle were even then falling into place.

13 The Turning Point

It is no longer possible to lightly dismiss the reality of cold fusion.

Julian Schwinger, Nobel laureate in physics, 1965
Salt Lake City, March 29, 1990

Although there are still skeptics here—and good luck to you, to believe that all the people are observing systematic errors, there comes a point where one's credulity is stretched too far.

Martin Fleischmann
Salt Lake City, March 30, 1990

You can't kill cold fusion by an edict in a newspaper.

Nate Hoffmann
Salt Lake City, March 31, 1990

✳ The End of *Nature*?

THE FIRST ANNIVERSARY of the Fleischmann-Pons announcement was fast approaching. Despite the expectations of the media and the scientific community at large, cold fusion stubbornly refused to fade away a year after its advent. Many times a crossroads had been reached on the way to confirming or rejecting the possible new phenomenon, only later to turn into a blind alley.

Vitiating this pattern, a single event can be said to have been a turning point in the saga, the remarkable First Annual Conference on Cold Fusion held in Salt Lake City, March 28–31, 1990. The meeting convened in the elegant University Park Hotel, only a few hundred feet from the National Cold Fusion Institute on a mountainside-plateau that marked an ancient high-water mark for the now much-diminished Great Salt Lake. Crystal clear air, spectacular vistas of the snow-capped Wasatch Mountains, and the glistening and fabled capital city below was the setting. The NCFI, the research organization that the state of Utah had generously endowed with $5 million seed money, made the conference happen.

The turning point did not come easily. Even before the first words of the conference were uttered, some cold fusion skeptics seemed to take leave of the last modicum of scientific civility to drive a stake into the heart of what they perceived to be the cold fusion "monster." Robert L. Park, Executive Director of the Office of Public Affairs of the APS, who was not at the conference, characterized the meeting from a distance to enquiring reporters as "a seance of true-believers." Science reporter Robert Bazell of NBC television news opined, ". . . though some will say the matter is not quite settled, it is a safe bet that cold fusion will soon bubble off into oblivion." He equated the quest for cold fusion to "Elvis sightings." Not surprising, because his thinly disguised skepticism a year earlier caused him to delay a full week after the cold fusion announcement before doing a story.

The conference began amidst a furor about several editorials that appeared in the March 29 issue of *Nature*. Even before that, the content of the opinion pieces were well known to many conference attendees through the hard-working fax machines. Titled "Farewell (Not Fond) to Cold Fusion," one attacked the "cold fusion fuss," saying, ". . . it has licensed magic in the particular sense that reports of remarkable phenomena—it could next be unicorns again—claim equal credence even when they fly in the face of *expectation* [italics mine]." If the history of science reveals anything, it is that science does not grow and encompass new knowledge by *expectation*!

The editorial chided cold fusion theorists, saying that the episode ". . . has shown up the frailty of the collective confidence in theoretical science; why else should so many serious people have been bamboozled for so long." The editorial was particularly harsh on Fleischmann and Pons, blaming them for not adequately disclosing their work, this despite the many, many laboratories that were now producing useful results without the benefit of whatever concealed proprietary secrets might still lurk within the NCFI or the Chemistry Building. As it transpired, Fleischmann and Pons made a very comprehensive disclosure of their thermal measurements at the meeting (not yet their revised nuclear measurements) and followed this with even more extensive publication in two journals of the excess heat results.

In the same issue of *Nature* appeared yet another technical article with negative findings by physicists Michael Salamon and Haven Bergeson of the University of Utah and their colleagues, who reported on their search for neutron, gamma-ray, and charged-particle emissions that they had obtained from the Fleischmann-Pons laboratory. With the permission of Pons, these were gathered nearly a year earlier over a five-week period. But whatever their quality or interpretation, at this late date they missed the central issue of what was happening in cold fusion, which certainly was not radiation fluxes alone. Many found it difficult to believe that the timing of this paper's release was pure coincidence. *Nature* required a semblance of substance as well as polemics to support its position.

The main editorial suggested again that Fleischmann and Pons had misled themselves by not sharing knowledge of "what they considered a great discovery" more widely. The piece came up with a fabulous red herring, saying of the cold fusion scene: ". . . it is a shabby example for the young; who can now hope to go about the world telling the tale that science is a collective enterprise in which all shoulders are bent to the same wheel of winning understanding from a common literature without fearing the shout 'What about cold fusion?' " The editorial went on to make a forecast completely without foundation: "What has irretrievably foundered is the notion that cold fusion has great economic potential. The time has come to acknowledge that. It would be a cruel deception of a largely amused public not to admit that simple truth." Could it be that *Nature* was hinting that cold fusion could be real after all, but that it was just not practical?

The editorial provoked outrage from many at the conference, anger that seemed the more justified because *Nature* had sent no official reporter to the gathering, while at least 40 journalists from other news organizations had come. It was quite possible that various undercover reporters (skeptics) would carry out the embarrassing task of scientific reporting to the conspicuously absent editors of *Nature.*

In a parallel commentary in the same issue titled, "The Embarrassment of Cold Fusion," *Nature* associate editor Dr. David Lindley said of Fleischmann and Pons's work, "What was reprehensible a year ago has become absurd." He stated, "*All* [author's italics] cold fusion theories put forward so far can be demolished one way or another, but it takes some effort." (Cold fusion theoretical papers now numbered in the dozens.) He ended his remarks with words never before seen in a science magazine, "Would a measure of unrestrained mockery, even a little unqualified vituperation have speeded cold fusion's demise?" *Nature* had contributed its fair share of vituperation and innuendo, witness the titling of their March 22, 1990, issue's news update on routine matters of federal cold fusion funding: "Cold Fusion—The Party Continues . . ."

Fleischmann and Pons shot back with reprinted comments of their own at the conference, words that had appeared in the *Deseret News.* They accused *Nature* of having a policy of printing only negative papers on cold fusion and suggested that the editors should ". . . cultivate the impartial publication of scientific papers. . . ." In a blistering rebuttal they wrote: "In its extreme form, following the herd in editorial opinion is a manifestation of cultural fascism: the expression of convictions based on inadequately understood theories and facts. Scientific conformism is known as 'handle cranking' or 'me-too science.' Committee reports (which are editorials) are specialized ways of inducing scientific conformism. Electronic mail and fax machines are specialized ways of inducing scientific hysteria. . . . If Lindley doesn't have the time to come now to Utah to gather information firsthand, then why doesn't he at least have the sense to use that well-known shortcut of establishing the scientific credentials of the believers and non-believers, namely the Citation Index*?"

✳ Journey to Salt Lake

It was no secret that most people who had come to Salt Lake City were sympathetic to the idea of cold fusion. They were "believers," as they had disparagingly come to be known. About 230 scientists, engineers, basement experimenters, entrepreneurs, and interested citizens had come to witness more than 40 presentations and panel discussions on cold fusion experiments and theory. Many were not convinced that cold fusion was a real new physical phenomenon, but for them the evidence was provocative enough to make the trip worthwhile.

On a Wednesday evening Delta airlines flight, there were at least four others heading for the meeting, three of them from MIT, including my archskeptic friend, Richard Petrasso. After on-again, off-again plans to attend the conference, Petrasso had made a last moment decision to go. But it was clear that he really did want to find out what new developments might have occurred, as well as to become a prominent de facto spokesman for the critics. And he did that with relish.

Petrasso was still finding it hard to come to grips with the reports of tritium generation. He didn't believe them at all and thought they would eventually be attributable to inadvertent contamination. The lack of experiment repeatability and the quality of reported pre- and post-experiment analysis of composition concerned him. He had listened to reports from other skeptics that some of the scientists doing the tritium work had perhaps "gotten out of their depth" and were not doing the exacting work necessary to pin the tritium question down. But he also must have known that at least at the Bhabha Atomic Research Center in India, there were seasoned tritium measurement experts who were reporting significant levels of tritium in cold fusion experiments.

---

*A computerized data base that can quickly determine how many times a scientist's papers have been cited by other researchers in their papers.

Not surprising, interest in cold fusion among the general population of the Salt Lake City area had diminished, mirroring the declining media attention. "A year later" it was the rare souvenir shop that sold cold fusion T-shirts, mugs, bumper stickers, or other such memorabilia.

Fleischmann and Pons were there as speakers and participants; they distributed their paper on an advanced method of measuring excess heat in electrochemical cells, which later appeared in the July issue of *Fusion Technology*. Severe critics were present too, albeit noticeably few: John Huizenga, who cochaired the DOE review panel that reported negatively on cold fusion the previous November, Douglas Morrison of CERN, Richard Petrasso, and MIT Professor Ronald Ballinger.

Though most papers dealt with experimental results, a handful of theorists besides Hagelstein presented ideas to explain the physical mechanism of cold fusion, such as Professor Robert T. Bush, Drs. S. R. and T. A. Chubb of the Naval Research Laboratory, Professor Giuliano Preparata, and Professor Julian Schwinger.

By and large, the theorists had concluded that if heat-producing cold fusion were real, some hitherto unknown collective or cooperative phenomenon among atoms and nuclear particles had to be at work to distribute energy throughout bulk matter. Either this occurs deep within the deuterium-infused palladium atomic lattice or in or near the surface of the electrodes. The energetic particles that emerge in conventional fusion reactions could not account for the heat in cold fusion, all agreed. Professor Schwinger said at the meeting, "It is clear that cold fusion and hot fusion are qualitatively different phenomena," and then he went out on a limb to assert, "It is no longer possible to lightly dismiss the reality of cold fusion."

✳ Real Heat

Electrochemist Fritz Will, recently arrived from General Electric to become the first permanent director of the NCFI, opened the conference Thursday morning and went to the heart of the controversy: "What were originally believed to be simple experiments that could be readily reproduced in other laboratories turned out to be complex phenomena that defied confirmation in many laboratories and which cannot be explained on the basis of classical nuclear physics. However, persistent and careful work by recognized experts in the fields of electrochemistry, nuclear measurements, and materials science has now led to confirmation of the Fleischmann and Pons results in many laboratories in the United States, Japan, India, Italy, Russia, and several other countries." He initially avoided the term fusion, though the implication could hardly be missed.

Will continued in his precise, German-accented English: "The multitude of results obtained by so many different groups can no longer be explained away as experimental artifacts. The reality of these effects is further underscored by the absence of such effects in carefully executed control experiments, employing hydrogen instead of deuterium or platinum instead of palladium."

He inserted a caveat, which could also be taken as as warning to skeptics: "While key observations relating to cold fusion have been confirmed by many competent groups, it is also true that the phenomena cannot be reproduced on demand, and that an understanding of the underlying mechanisms is not at hand. The phenomena involve surface chemistry and the behavior of metal loaded with deuterium. Appreciating the complexities and well-known irreproducibilities involved in each of these cases individually, many scientists are not surprised that one year of research and development have not been sufficient to unravel the complexities of cold fusion which combines both cases [bulk and surface effects]. . . . We know that the reliable results obtained by a minority must not be regarded as wrong only because a majority of others has failed to confirm these results within one year."

Expectations were high as Stanley Pons stepped to the lectern to deliver the first talk. Unlike the jacket-and-tie formality favored by most there, Stanley Pons was a study in understated attire. If he was overstating his science, as the critics suggested, he was not doing so with flashy clothing. Wearing no tie and sporting tennis shoes and an open-collar shirt at such a momentous gathering betokened his sense of assurance about their results. It had always been so the past year, despite the depressing assaults on their credibility.

In soft, serious tones he mentioned that the paper he was about to deliver had already been accepted for July 1990 publication in *Fusion Technology*. This was said with a note of triumph, since his accusers had made much of the uncertainty the previous March about whether or where the first Fleischmann-Pons paper would be published. Pons pointed to a tall stack of copies of the draft paper, which were available after his talk and were eagerly gobbled up. The paper dealt with the "Calorimetry of the Palladium Deuterium System," but Pons said that two other papers covering nuclear products and related matters had been submitted to another journal.

Pons described a creative mathematical technique that they had applied to their open-cell calorimeter to estimate some of the quantities or parameters that affected its operation—such things as "heat transfer coefficients" and the like, numbers that told how much heat was flowing through the wall for a given difference in temperature from the inside to the outside. He laboriously went through some of the mathematical details, which were likely not easy to readily verify even by many in the audience versed in these matters. It was a technically masterful concept, which was aimed at reducing uncertainties in the heat measurements as well as making possible the *economical* running of numerous cells simultaneously. He claimed that the technique could assure a heat measurement error level of only 0.1 percent, or in this case a mere 10 milliwatts (whichever was greater in a particular run). He assured the audience that, if anything, the technique would give a conservative, systematic *underestimate* of the amount of excess heat being produced.

With this method, over the last year they had carried out some 2,000 calibrations on 200 individual cells. During some of the excess heat bursts that were encountered, which lasted upwards of several weeks, they measured large excess power levels amounting to 75 to 112 watts per cubic centimeter (W/cc) of palladium rod. This was on the order of twice the heat generated per equal volume of fuel rod in some water-cooled nuclear reactors, said Pons.* Some of these bursts would last for weeks or even months. The evidence was there for all to see. The only reasonable open question was what unidentified effect was causing them and how should they be interpreted. In some instances the system would approach boiling temperature. "It is inconceivable that that can be due to any known chemical reaction," he asserted with quiet assurance. In the conference proceedings published many months later, he and Fleischmann wrote: ". . . it is inconceivable that chemical or non-nuclear physical energy could be stored in the system at these levels and then be released over prolonged periods of time." Over a typical experimental run lasting three months, hundreds of megajoules (MJ) of energy were being released per cubic centimeter of electrode—enough energy to lift hundreds of tons, tens of feet up (if the energy were all converted to useful work).

Richard Petrasso boldly stood up and asked the first question, probing into the most glaring omission in Pons's discussion: "Do you have any nuclear measurements, because those are very important?" Before responding to Petrasso, Pons looked down as if to avoid facing his opponent. Injudiciously passing off the question, Pons said that Martin Fleischmann would address it later. Then he retorted, "I think they are important too."

Ronald Wilson from GE, supposedly a collaborator with Fleischmann and Pons, offered a mathematical objection to the duo's "nonlinear regression" technique. This achieved high currency in the next day's edition of the *New York Times*, but Wilson belatedly withdrew his criticism by meeting's end after *his* technical misunderstanding had

*Indeed, the power density in a conventional boiling water nuclear reactor is about 50 watts per cubic centimeter.

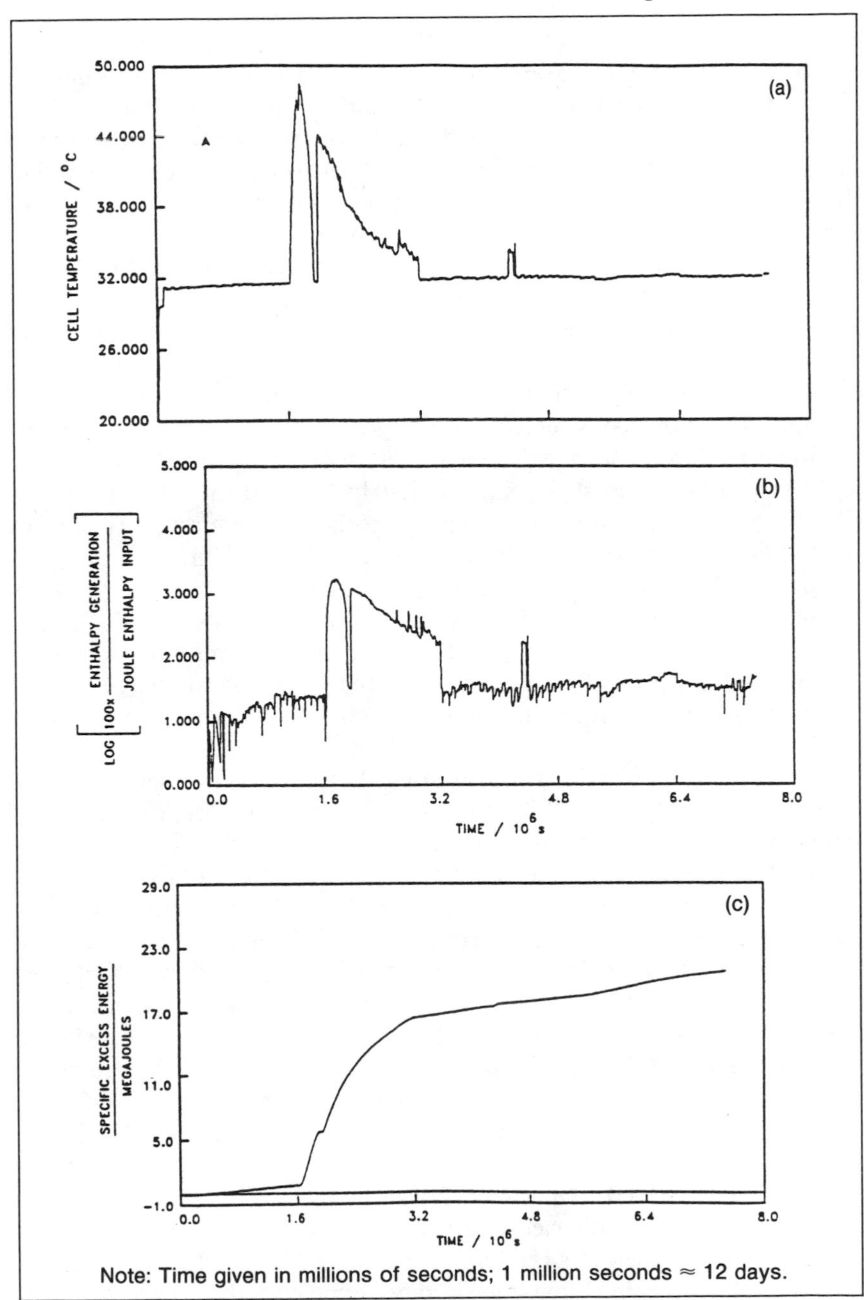

Note: Time given in millions of seconds; 1 million seconds ≈ 12 days.

At the First Annual Conference on Cold Fusion, Fleischmann and Pons presented new results that appear to show excess power that, when added up over time (c), translates to impressive total energy production—tens of megajoules. An unexplained heat burst, which lasted over two weeks, is also clearly evident in (a) and (b). (Courtesy Drs. Martin Fleischmann and B. Stanley Pons)

been pointed out. Steve Jones injected melodrama by posing a question while displaying a color transparency made from a July 1989 issue of the *Deseret News*, which purported that a boiling water device had been achieved: "Do you still feel that a near-term practical device is possible?" With neither emotion nor faltering Pons said, "All our present research is aimed at it." Inconsistently, Jones favored explaining some of Earth's internal heat by cold fusion, but he could not suffer talk of more intense heat being generated in cold fusion jars.

The morning was devoted largely to heat measurements, the formal term being calorimetry. What better follow-up to the open-cell work of Fleischmann and Pons than to hear about heat measurements done in completely closed cells, a technique instantly recognizable in the air-tightness of its approach. Michael McKubre of SRI International, whose cold fusion group is still being funded by the nearby Electric Power Research Institute, discussed the high-precision measurements that his team had made with deuterium-palladium electrochemical cells pressurized with deuterium gas to 60 times atmospheric pressure. They intended to infuse as much deuterium as possible into the Pd lattice. SRI had begun its work soon after the Fleischmann-Pons announcement and was attempting to discover whether the phenomenon was a surface or bulk effect, arriving at the tentative conclusion that it was in the bulk.

The team emphasized determining the ratio of number of deuterium to palladium atoms (D/Pd) in the electrodes during the course of their experiments with two different kinds of calorimeters. This they did by continuously measuring the electrical resistance of the palladium electrodes and knowing how it varies with the D/Pd loading. Getting D/Pd as high as one could—over 1.0 if possible—had become the empirical figure of merit in many of the cold fusion efforts. McKubre reported bursts of heat lasting several hours or tens of hours, producing excess energy amounting to several hundred thousand joules. The bursts were irrefutable. Often excess power would die away and then reappear within a single day. In the printed proceedings the group would state: ". . . we consider it unlikely that [the] excess energies . . . can be accounted for by chemical processes."

McKubre added a new and puzzling twist—one that the Indian group at BARC had reported originally. About a week after one of their palladium electrodes was removed from its cell, they placed it between layers of photographic film. After 12 days' exposure, the developed black and white film showed intense white "hot spots" of fogging, with nebulous interconnecting filaments. It was very cosmic, looking much like an astronomical photograph of a cluster of stars and surrounding nebulosity. Using an ancient image, McKubre described the wondrous thing as looking like the Shroud of Turin. Indeed, skeptics probably

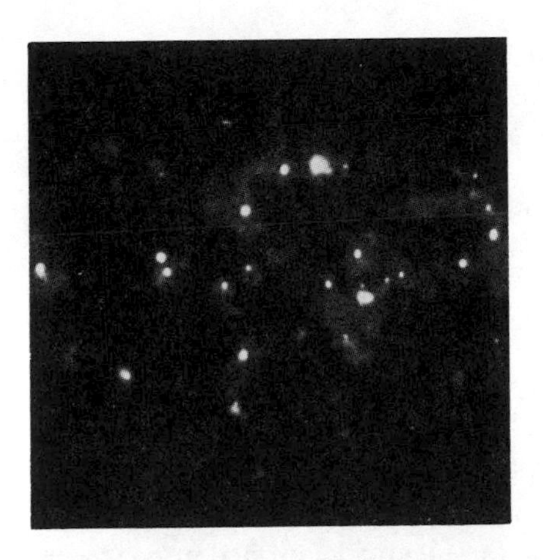

The research group of Dr. Michael McKubre at SRI International, made this impressive autoradiograph by placing a palladium electrode from an excess power-producing cell on a piece of covered photographic film and leaving it there for 12 days. The fogging of the film appears to show evidence of penetrating ionizing radiation coming from sources within the electrode. (Courtesy SRI, International and the Electric Power Research Institute)

thought it had as much validity as that medieval Italian relic! The implication of film-fogging was that some penetrating form of radiation had left the Pd electrode to trigger the molecules in the film. This would be known technically as an autoradiograph, a much-used approach in many fields of science—especially these days in molecular biology to identify molecules that have been tagged with radioactive tritium. The group had tried the test on a blank (unused) electrode and obtained no fogging. The origin of the photo-fogging, possibly some kind of low-level ionizing radiation, was simply a mystery.

In the question and answer period, the nouveau Shroud was attacked by a few as a possible artifact brought about by chemicals leaching out of the electrode during the exposure, a possibility that the SRI group has vehemently denied. After all, why would there be localized hot spots in the exposure? Though McKubre was clearly a "convinced" proponent of cold fusion, he ended diplomatically with the statement, "I am not going to leave you with conclusions, I'm going to leave you with a dilemma."

Oliver J. Murphy spoke for John Appleby's group at Texas A&M, which was also working on calorimetry and exacting analysis of Pd electrodes for their surface atomic structure. Their specialty was *microcalorimetry*—high-precision closed electrochemical cells that had very small diameter electrodes—a difference that might have given their work an advantage. Nearly 80 percent of their experimental runs yielded apparent excess power. "We have been receiving some criticism, some of it bordering on the amusing, having to do with Maxwell's demons or Murphy's laws," Murphy joked in his rapid-fire Irish-accented speech. But he and his colleagues had little doubt that what they were dealing with was truly a new effect. "To us it's very real," he said.

The group reported an intriguing trick that appeared to work consistently. The use of lithium-hydroxide (LiOH) as the conducting salt in the electrolyte caused the heating to drop off, but it would just as surely come back when lithium-deuteroxide (LiOD) was readmitted to the cell. Deuterium (D), whatever its form, seemed to be the common denominator in all of the positive experiments. The group also found that sodium-bearing electrolyte, sodium-hydroxide (NaOH), killed the heating effect, as did lowering the electric current density too much.

The chemical and isotope analysis of electrode surfaces, performed with various sophisticated instrumentation, revealed a complex surface suffused with dendrites—microscopic fingerlike growths of material that reached out from the electrode into the solution. "You are going to have a very unique surface when you do these experiments," Murphy remarked. Moreover, many regions of the electrode surface appeared to have quite disparate compositions. Palladium, lithium, and deuterium atoms made up a complex alloy of material at the surface, with evidence of various kinds of crack formation in the electrode. Even an uninitiated observer could begin to grasp why these supposedly "plain and simple" electrodes could perform erratically and give inconsistent results—including null results—in different laboratories.

Unlike SRI, the Appleby group concluded that whatever was causing the heat seemed to be coming from a surface layer of reactions on the electrodes. The lithium ion, Li, seemed to them to be critical in obtaining excess power. They concluded that the energies being manifested were ". . . much too large to be explained by continuous chemical processes taking place in a thin superficial skin of palladium." Combined with evidence for tritium formation that others were getting at A&M and elsewhere, the group argued that its results favored the "existence of solid-state nuclear processes." But they had found no obvious nuclear products, such as helium, in their autopsied electrodes, nor, as it turned out, had anyone else at the conference.

Other positive heat results rolled in. D. P. Hutchinson, who led one of the groups working at Oak Ridge National Laboratory, reported that the group had stopped its work the previous December for lack of funding, but before ending had totaled up 17 megajoules of excess energy in one of its cells. Martha Schreiber, speaking for the Huggins group at Stanford University, described their closed-cell calorimeter in which they obtained no excess energy in light water control experiments, but in one case obtained in a heavy water cell running for 10 days some 2.5 MJ of excess energy per cubic centimeter of palladium. A search for helium-4 in the spent electrodes evidenced nothing above natural background.

The bottom line on all of these experiments seemed to be that the excess heats were real. In view of their impressive magnitude, they had the distinct smell of a nuclear process. But proving what the hidden nuclear reaction products were was much tougher than anyone had bargained for. The process was a victim of its own success. It simply does not take all that many fusion reactions (compared to the total number of palladium or deuterium atoms in a system) to get these levels of energy release.* Was it the lithium reacting with deuterium to form helium, or the lower abundance ^{6}Li isotope fusion "burning" with deuterium to create ^{7}Li plus hydrogen? Possibilities, but no evidence.

At a rate approaching $2 million per year, EPRI was funding a number of these projects: McKubre's effort at SRI, Huggins's calorimetry at Stanford, some work at Texas A&M, and various consultant and support service work. Manager Joseph Santucci spoke of the EPRI initiative and was "very comfortable saying that the excess heat is not due to chemistry," that "the source of the heat and the tritium may be different." In his view, "It's very puzzling and we have to get to the heart of it." In his summary he had mentioned "tantalizing" evidence of X rays coming off the spent electrodes—by implication connected with the puzzling SRI autoradiograph. X-ray expert Petrasso took offense at such an offhand remark about potentially significant confirming evidence: "I don't think you can be *alluding* to X rays—this is a bit irresponsible."

The words from India did not disappoint those who anticipated challenging news from the mysterious East. P. K. Iyengar, director of the BARC in Bombay, where more than 1,000 scientists work on various nuclear energy projects, told how acceptance of cold fusion had sprung naturally from his culture: "We in India believed in this because of our traditions. . . . Our experiments further strengthen our belief that cold fusion is occurring." Experiments had begun in earnest at BARC in April, with positive results coming in fast and furious almost from the beginning—in great contrast to the halting Western experience. This led some skeptics to believe that the vast holdings of heavy water (used as a neutron moderator in their nuclear reactors) with its attendant tritium impurities might be a source of inadvertent contamination. Bhabha claimed an amazing overall 70 percent success rate in measuring tritium production and bursts of neutrons coming from a wide variety of palladium-titanium-deuterium systems—gas cells and liquid cells. In the Indian paper submitted for the proceedings, the scientists concluded: "Experiments carried out by a number of totally independent groups

---

*Since one MeV—about the amount of energy released in a typical fusion reaction—is equivalent to 1.6×10^{-19} joules or 1.6×10^{-13} megajoules (MJ), it only takes 10^{13} reactions to exceed a megajoule of energy output. But a cubic centimeter of palladium has 7×10^{22} atoms, so if the lattice has a 1/1 ratio of D to Pd atoms, only a minute fraction of all lattice sites containing deuterium will have to experience fusions each second.

employing diverse experimental setups have unambiguously confirmed the production of neutrons and tritium both in electrolytically loaded and gas loaded Pd/Ti lattices."

The BARC experiments, as outlined by Iyengar's animated colleague M. Srinivasan of the Neutron Physics Division, were extraordinary in both the number of variations attempted and the relatively consistent finding that tritium atom generation was ahead of neutron production on the order of a million to a billionfold. On the very first day of their experiments (April 21, 1989), bursts of neutrons and tritium had come from eight out of eleven cells. Half of the electrolysis experiments were "doubly successful" in giving forth *both* neutrons and tritium. Tritium increases were often seen right after neutron bursts. To critics who claimed that the copious tritium could be coming from the large amounts of tritium-laden heavy water in Bhabha's fission reactors, Srinivasan replied, "Just because we have a heavy water reactor doesn't mean that the whole world is coated with tritium." These were experienced measurers of tritium and neutrons, whose results were not fairly dismissed by cultural prejudices about science in the developing world.

To cap these provocative results with visual proof that something very unusual and probably nuclear had been found, Srinivasan displayed an autoradiograph made from palladium-silver foil that had been infused with deuterium gas. It resembled the one from SRI. It was presumptive evidence of perhaps ionizing radiation from tritium decay. Srinivasan showed other intriguing autoradiographs obtained from a variety of related deuterated metals experiments. If the pictures weren't lying, India had snagged on film the footprints of the cold fusion Genie.

✳ Sharp Theories

Fiery Italian physicist Giuliano Preparata had been one of the first to propose mechanisms to explain cold fusion; he and his colleagues published a paper in *Il Nuovo Cimento* in May 1989. "The humble work of a theoretician trying to make sense out of something that 'does not make sense' " was how he described his work to the conference. Even before news about cold fusion broke, Preparata had been thinking about *coherent* oscillations of particles in solids. Those ideas were conceived to explain cold fusion about the same time that Peter Hagelstein on the other side of the Atlantic was coming up with *his* coherent fusion theory.

Other than using the terminology of coherence, the two theorists were really suggesting quite different cold fusion mechanisms. Despite the lack of evidence for a helium-4 reaction product in numerous heat-producing experiments, Preparata still believed that helium was emerging undetected from the palladium into the gas phase. While Hagelstein spoke of proton-burning to get heat (from the trace amounts of ordinary hydrogen in the heavy water), Preparata stuck with d-d fusion, but hoped to find a mechanism that suppressed the tritium or helium-3 branches in favor of the ordinarily rare helium-4 outcome.

In May 1989, Preparata had concluded that "cold nuclear fusion is a real physical process." Why deuterium fusion reactions should work so differently in the solid state than in vacuum was a mystery that he and his colleagues hoped to solve by "a major shift in our perception of the ways in which condensed matter organizes itself." Their paradigm was a "plasma" of charged particles within a lattice that were oscillating collectively around equilibrium positions.

Preparata had examined the spectrum of positive and negative results in cold fusion experiments—including the "lukewarm" fusion reported by the Brookhaven National Laboratory chemists (see Chapter 10) and tried to explain these various "lines," as he called them, within a single framework. In Utah, the general direction he reported was the "plasma" of electrons inside the solid lattice carrying energy away from the deuterium fusion reactions, and in doing so suppressing the two usual outcomes of d-d fusion—the helium-3 and tritium branches.

Preparata was offended to the core that *Nature* had so vehemently ridiculed cold fusion. "If you read *Nature*, you only get one line," he shouted to much applause. He offered the audience a parallel with molecular biology: "The fact that we don't understand cancer doesn't mean that what they [molecular biologists] are doing is bullshit!" The fact that you cannot reproduce tritium doesn't mean it's not there. . . . You have to do science without arrogance and with patience."

John Bockris, who chaired the Thursday afternoon session at which many of the theory papers were presented, announced Peter Hagelstein: "A well-known name is approaching." Martin Fleischmann relished seeing this ambassador of goodwill from MIT, which had gotten the reputation as a bastion of cold fusion skepticism. It was so ironic, Fleischmann could not squelch his smile as Hagelstein began. Though 17 minutes were inadequate to do justice to his written paper, he delivered his theory's salient features: coherent fusion, the formation of his "virtual neutrons," heat evolution from proton-burning, and tritium production through deuterium capture of a neutron. The reaction product responsible for the heat would be neatly buried in the overwhelming background of original deuterium in the heavy water.

Other theorists followed. Scott and Talbot Chubb—an unusual energetic nephew and uncle pairing—dwelt on the "Quantum Mechanics of 'Cold' and 'Not-so-Cold' Fusion." They too had the idea that coherent phenomena were at work in the lattice, both in cold fusion accomplished by electrochemistry and in the lukewarm cluster impact fusion of the Brookhaven accelerator team. Later in the conference Yeong E. Kim from the Purdue Physics Department tried to resurrect the idea of screening with the negative charges of electrons as a mechanism to overcome the coulomb barrier between positively charged nuclei. His twist: combining the screening with special circumstances for the velocities of deuterons that would greatly enhance fusion rates. "Electron screening is the savior of nuclear physics," said Kim.

Robert T. Bush put forth his "transmission resonance model" to explain everything from heat-producing cold fusion to cluster impact fusion. He ended with an enthusiastic pitch for his theory for which he had become famous from Santa Fe onward, "The model gives an excellent fit to the data, strengthening the likelihood that it is correct, and boosting the credibility of the effect itself."

The sheer volume of all these analytical tours de force—mathematical chicken scratchings across many pages—testified to the intensity of enthusiasm to derive a viable explanation for cold fusion. Without deep faith that something new under the sun was really at work, these prodigies would not have summoned the will to persevere. They were far out on their limbs and pushing the limits of physics hard.

For the level of sheer electric anticipation, it would have been difficult to match the atmosphere surrounding physics Nobel laureate Julian Schwinger's talk on Thursday evening. Almost from "day one" he had believed in cold fusion. "Apart from a brief period of apostacy, when I echoed the conventional wisdom that atomic and nuclear energy scales are much too disparate," he said, "I have retained my belief in the importance of the lattice." Quoting Niels Bohr, who had remarked that "an expert in a subject is one who has already made all possible mistakes," Schwinger said, "I stand before you as one who, in the field of cold fusion, is rapidly attaining expert status."

Schwinger suggested that heat-producing cold fusion was due to a proton-deuteron reaction producing helium-3 (^3He) and transferring to the palladium lattice the energy of what would have been (in plasma physics) a horrendous 5.5 MeV gamma ray. He proposed to get his protons for this kind of proton-burning from the trace amount of ordinary hydrogen in nearly pure heavy water. Not to be confused: Hagelstein was burning his protons with "virtual neutrons" and getting deuterium, Schwinger was burning them with deuterium and getting helium-3.

Schwinger said that the palladium lattice "acts to suppress the coulomb repulsion between a proton and a deuteron," an effect that he proceeded to calculate. The "asymmetry" between the p-d reaction and the d-d reaction enhances the former, Schwinger maintained. Moreover, he saw the beginning of an understanding of why the phenomenon may be so finicky experimentally. One of the parameters that he derived, if altered by a mere 25 percent would change the fusion rate by a factor of 10 million, "a degree of sensitivity that borders on chaos," he noted.

He also explained how tritium could simultaneously be generated within the lattice at a rate that roughly spanned the range of experimental reports.

Schwinger's theory went directly against the grain of hot fusion dogma. "The correct treatment of cold fusion will be free of the collision-dominated mentality of the hot fusioneers. . . . It is clear that cold fusion and hot fusion are qualitatively different phenomena," he emphasized. "What you have heard, of course, is not the end of an investigation, it's the beginning."

Why, I asked Schwinger on the side after his talk, was a Nobel laureate physicist striking out in a direction that had become so disreputable in the general physics community? His simple answer: "My nature is that when everybody is going one way, I like to go the other way."

✳ Tritium, Neutrons, and More

On Friday, the conference considered nuclear products in greater depth. John Bockris presented the Texas A&M tritium findings in electrochemical cells, which were among the most spectacular of all cold fusion data that had been obtained—results that were being darkly questioned in back rooms by the skeptics. Journalist Gary Taubes hovered about the press room, waiting to spring his "tritium as *deliberate* contamination" theory within the next few months (see Chapter 13). The A&M team had observed tritium concentrations of 10^4 to over 10^7 disintegrations per minute per milliliter of electrolyte—approaching a million times background. This corresponded to tens of billions of tritium atoms being created each second for every square centimeter of electrode surface. (Results from some tests at BARC in India had produced as many as 10^{12} tritium atoms.)

Bockris and his colleagues believed that there was much evidence that cold fusion was a surface effect, that it was not occurring in the bulk palladium. They theorized that high electric fields near the tips of dendrites might be accelerating and smashing deuterons together at the surface, making them fuse to produce tritium.

Bockris acknowledged that the A&M track record of repeatability was not as good as he would have liked. Fifteen out of 53 electrodes had produced tritium; thinner electrodes (less than a millimeter diameter) had more like a 70 percent success rate; only five out of 28 electrodes had produced excess heat. It was possible, Bockris said, that in "failed" cells tritium could have been missed and "sparged out" [mixed with other gas and removed] in the gas flow of the open cells, because tritium ordinarily came in bursts—perhaps not always detected by the irregular sampling schedule. In one cell, on two occasions there appeared to be a correlation between two heat rises and tritium bursts. Adding to the strangeness of the phenomena was that A&M required weeks before the initiation of tritium evolution in a cell, whereas at BARC tritium typically came up very quickly within a day of electrolysis start.

Ed Storms and Carol Talcott of Los Alamos National Laboratory had been working on the tritium problem in closed electrochemical cells for many months. Though their success rate was only about 10 percent at this time, and they could not pin down the parameters that would induce a cell to turn on, they stood solidly behind their tritium. They were talking about 150 cells and 5,000 tritium measurements, an experimental tour de force with many kinds of checks for contamination in the environment. It was an Edison-like trial and error approach in treating electrodes with gunk and grime to track down what would boost them into activity. Said Storms to critics, "We can put aside the question as to whether it's real."

Charles Scott brought more news about tritium from Oak Ridge National Laboratory. His group had gotten not only extended periods of excess heat in closed cells, but tritium and neutrons as well. "In every test greater than 200 hours we get excess power," Scott said. "Excess power does occur—unequivocally—and can be extended for hundreds of hours." One run had already topped 2,000 hours. The Los Alamos and Oak Ridge tritium was high enough over background to convince the researchers that it was not contamination, but it was admittedly far below the levels that Texas A&M and India had gotten.

In a panel discussion to sum up what had gone before, Fleischmann said, "I think that it's now pretty clear that there is steady state heat production from Pd electrodes." Electrochemist Charles Yeager from Case Western supported Fleischmann: "This will be noted as a decisive turning point in the history of the affair. . . . What we are seeing cannot be explained by trivial mathematical error. . . . Let's hope that we can take the effect up by many orders of magnitude."

With all the provocative evidence—the heat and the tritium, one would have thought the sternest skeptic would wither. Not so, as CERN's Douglas Morrison revealed: "I started out as a believer [in April 1989]. It's fair to say I'm now a skeptic." Mike McKubre shot back at Morrison's suggestion that theory could not account for what was being reported: "Morrison misunderstands the purpose of theory." At which point Yeager recalled the seminal Michelson-Morely experiment of 1887 that found the speed of light to be constant despite the motion of Earth through the much-discussed invisible "aether"—contrary to expectation. (Incidentally, Morely was a chemist!) Yeager had the wisdom of the long view in science: "The thing will work its way out, but it will take time and be very expensive."

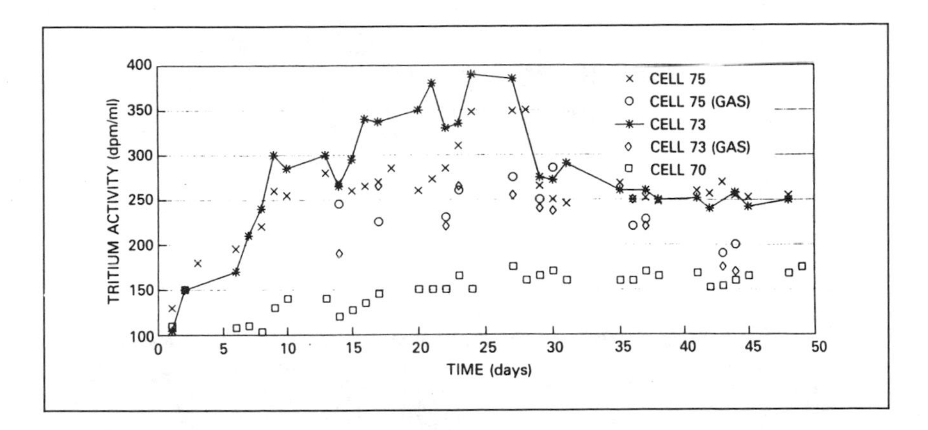

Tritium production reported in electrochemical cells by Drs. Edmund Storms and Carol Talcott of the Los Alamos National Laboratory. (Courtesy Drs. Edmund Storms and Carol Talcott)

Richard Petrasso would not give in: "I'm more than willing to grant heat in the lattice, but I'm concerned about finding reaction end products." Fleischmann responded: "Absolutely so! In order to do this you must devise a very clean experiment. It is very expensive to do this and is not consistent with our general direction. You must use high resolution SIMS [an instrument technique] into metal. You must have an unbiased view as to what might be there. It is a 'Catch-22.' If there is no belief that the effect is there, then there will not be the money to do the experiment." The ultimate indignity to the hot fusioneers was the reminder from Huggins: "People talk about breakeven in these large [hot fusion] experiments. *We* are now far beyond breakeven."

Last but not least were the neutrons, a Friday afternoon and evening dessert after the meat and potatoes of heat and tritium. Francesco Scaramuzzi, from Frascati, came bearing not only neutron bursts as he had at Santa Fe, but now tritium—from titanium samples in high-pressure deuterium gas. They were doing the routine of cooling to liquid nitrogen temperature followed by warm-up. Menlove of Los Alamos had copied their approach, and now Frascati was imitating Menlove's detector system and getting good results. No bursts of neutrons came from control experiments with hydrogen gas.

Since Santa Fe Menlove had gone back to his drawing board and now had an even more superlative neutron-measuring system with cross-checks galore for background interference—security level upon security level. There were small bursts of two to ten neutrons, with larger events up to 300. Burst duration was typically 100 microseconds and occurred when the sample was between −30°C to +40° C. It was tes-

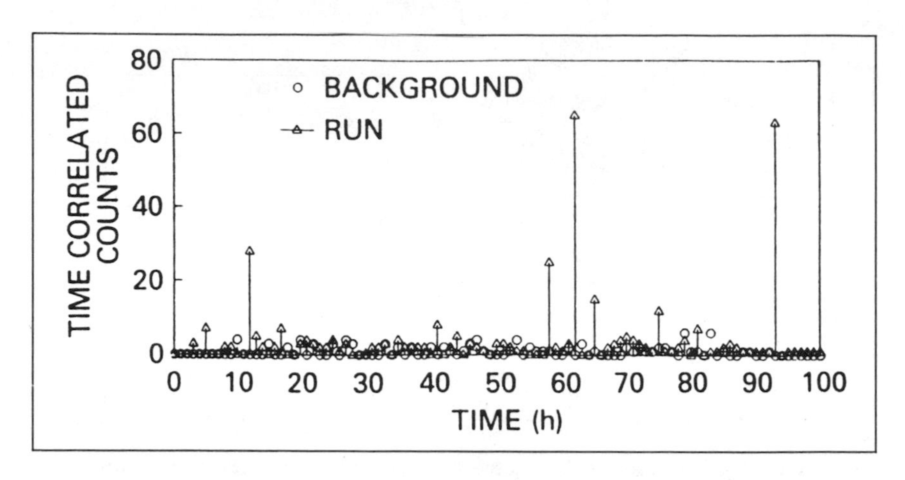

Bursts of neutrons from a deuterium gas-pressurized cell containing titanium shavings. (Courtesy Dr. Howard Menlove, Los Alamos National Laboratory)

tament to the position of *Nature* that his work had not appeared in its pages.* Here was essentially proof positive of fusion occurring where it was never supposed to.

Steve Jones, the original neutron man, came back to the platform. He was still overly fond of saying that the neutrons he was certain he was measuring were some 13 orders of magnitude below the conventional fusion requirement for excess heat. Now working with DOE and EPRI funding, he reported plans for an experiment to be conducted deep in a mine at Park City. A detector was to measure both neutron and proton emissions simultaneously in a collaboration between BYU and the University of Utah physics departments. "For which we hope to get the Nobel Peace Prize," quipped Jones.

✳ All's Well That Ends Well

Weary from a deluge of information from two solid days of talks, the conferees were refreshed by crisp summaries of what the past year had wrought. Physicist David Worledge of EPRI, who was courageously shepherding the funding of many cold fusion research efforts, tried his hand as tour guide. An observer aboard the overflight would have no-

*Menlove told me that by the end of 1989, four out of the five reviewers of his paper for *Nature* had approved it, but still the magazine would not publish the work. *Nature* wanted more data, so Menlove set about collecting it. When he got back to *Nature* with it, the journal wanted him to begin the review process anew which he rejected with frustration.

Dr. Julian Schwinger, Professor Emeritus of Physics at UCLA and Nobel laureate, explained his theory of cold fusion at an evening session of the First Annual Conference on Cold Fusion, in March 1990. (Courtesy University of California at Los Angeles)

Dr. David Worledge, a physicist at the Electric Power Research Institute, Palo Alto, California, helped to promote and fund investigations of nuclear and thermal effects in cold fusion experiments at a number of research centers across the United States. (Courtesy Electric Power Research Institute)

ticed certain peaks and depressing swamplands. The negatives seemed overwhelming, but there was reason to hope in the positive results. Tritium was a gigantic mountain—with signal-to-background ratios that ranged from a hundred to a million. The tritium results looked strong to Worledge, but he could not bring himself to say that it was being *generated.* Whatever the case, reproducibility was poor. Another peak: Many labs had seen neutrons, although at low count rates. "I don't trust neutron counts done by electrochemists or heat measurements done by nuclear physicists," joked Worledge, with more than a little serious intent.

Calorimetry was getting better, and it was fair to say, said Worledge, that the recombination of O_2 and D_2 bugaboo to explain away the excess heat had been ruled out. Closed cells had taken care of that. Power levels clearly over 20 watts per cubic centimeter of palladium were certainly real. As to the more important question about the power added

Dr. Fritz G. Will, Director of the National Cold Fusion Institute, Salt Lake City, Utah. (Courtesy University of Utah)

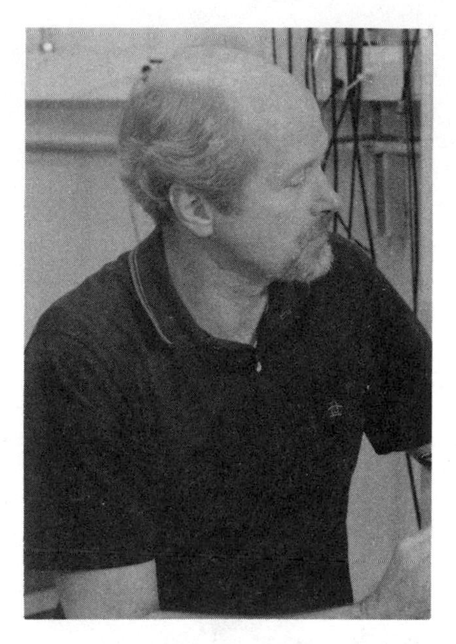

Dr. Howard Menlove of the Los Alamos National Laboratory, whose detection of bursts of neutrons from deuterium gas-pressurized cells holding titanium metal chips are among the most convincing evidence for cold fusion. (Courtesy Los Alamos National Laboratory)

up over time—*integrated* power or energy—there were at least four groups that had measured over a megajoule of energy per cubic centimeter (MJ/cc). The most that one could expect from chemical reactions—even if all atoms were consumed—would be 0.05 MJ/cc. Impressive and unexplainable. Perhaps too much the physicist, Worledge couldn't make the transition to strong belief in a nuclear explanation for the heat: "The experiments look good to me, but I don't know what to make of them." The bottom line for Worledge: "Cold fusion is not proved yea or nay, but it would be ludicrous not to go on." He favored a general direction of increasing reproducibility and doing different kinds of experiments. Lack of reproducibility, he reminded the assembled, is the serious waster of time that is slowing the whole field and inhibiting the global scientific community from becoming more actively involved.

On the way to the meeting's dizzying conclusion on Saturday, a free-form panel discussion evidenced bursts of certainty and bewilderment:

Nate Hoffman: "I'm exactly in the middle between skeptic and believer."
Menlove: "We need to change our vocabulary on reproducibility. . . . Earthquakes are believable, even though they are not reproducible. Tremors go on all the time. If we sharpen our techniques to see the tremors, we'll be on our way."
Wilford Hansen: "Can we say that tritium has been proved, excess heat has been proved?
Bockris: "I'll stick my neck out and say yes in both cases."
Storms: "I firmly believe that it has been produced."
Iyengar: "It is impossible to disprove that tritium has been produced."
Hoffman: "If you ask are there nuclear reactions occurring, I'd say a 50 percent chance."
Steve Kellogg: "I feel like Alice in Wonderland being asked to accept so many miracles before breakfast."
Preparata: "There is no greater injustice than to compare things which are incomparable. . . . I think you should have the courage to tell the world that cold fusion has been proved to exist. . . . There are too many serious people, too many serious experiments."

Martin Fleischmann ended the conference with a historical review of the electrochemical mysteries and wonder about deuterium crammed into a palladium lattice that led up to his and Stanley Pons's startling discovery. His synopsis could be considered in two ways: Either it was the product of a brilliant mind that had opened a breathtaking new window onto a previously unimagined realm of nuclear phenomena, or it was the demented sequel of a "mad idea" run completely amok. In light of the extraordinary observations that the conference had heard, the former seemed by far the more appropriate fit. Correct in all its particulars or not, this was scientific imagination of the highest order, informed by streams of interdisciplinary research from electrochemistry, quantum mechanics of the solid state, and nuclear physics. "We would not have started this investigation if we had accepted the view that nuclear reactions in host lattices could not be affected by coherent processes," said Fleischmann. He then acknowledged his appreciation for those who had stood on his shoulders to look further: theorists of coherent phenomena in the lattice Julian Schwinger, Giuliano Preparata, Peter Hagelstein, the Chubbs, and Robert Bush.

In the conference proceedings published a half-year later, Fleischmann ended his summation with a defiant challenge to those who were cavalierly dismissing cold fusion without having weighed the evidence carefully: "It is hardly possible that the repeated observation of such a wide range of disparate phenomena can be explained away by the operation of a whole set of *systematic* errors nor that we have been attending a seance of true believers."

John Huizenga, who had chaired the DOE panel that had issued such a striking negative report the previous November, left the conference unmoved in his skepticism: "I don't see anything new here that wasn't available to us when we prepared that report. There is no reason to change one statement in it. There's a large discrepancy between the heat and the particles, and until they can explain it, it's not fusion." He would never tire of this incantation, which he made to the press over and over again. Richard Petrasso held a similar though more informed view: "The reports here are very interesting, but there is nothing here to change my mind. They haven't addressed the fundamental issue: What are the end products? You have to have particles commensurate with the amount of heat produced, and they are not there." Petrasso still had the eyes and ears of a doubter, but at least he could appreciate the sincerity with which the proponents were posing their case.

Yet despite this skepticism, in fact perhaps *because* of it, many conference participants felt that there had been a turning point that week in the cold fusion story. Most cold fusion critics simply did not seem to be playing fair. They weren't all as severe as Robert Park or David Lindley. But even the reasonably dispassionate critics seemed to have a blind spot. They would relentlessly decry the lack of clear evidence for fusion products consistent with the heat. Yes, they were right—there was "no clear evidence." But it had been made abundantly clear to them that there could be end products like deuterium or helium-3 (or even the still elusive helium-4) that could account for the heat. These products were far from easy to assess. Furthermore, the evidence for some nuclear end products associated with the processes seemed overwhelming—the tritium and the neutrons. A storm of controversy would soon arise over the tritium but it would be very far from knocking it out. The tritium issue aside, many neutron experiments looked to be on absolutely solid ground.

So as we left the conference, I talked with archskeptic Douglas Morrison of CERN (a "believer" during the first few weeks after the cold fusion announcement) and agreed to take him up on his proffered wager about cold fusion. We formalized the gamble in a written agreement, signed the gentlemanly understanding, and left the site of the turning point. I look forward to a mighty fine bottle of wine in 1993 (if not sooner) and expect that as a one-time cold fusion "believer," Douglas will be very happy to buy it. Cheers!

14 *Still Under Fire*

Reports of tritium production can always be questioned as to contamination. Defense against each suspicion has to be made each time a new suspicion is expressed.

N.J.C. Packham, K.L. Wolf, J.C. Wass, R.C. Kainthla, and J. O'M. Bockris, October 1989

Most screwy ideas turn out to be screwy ideas. . . . [Cold fusion was] preposterous to begin with.

Robert L. Park
July 6, 1990, *Science*

A lot of people undergo personality changes when discussing this topic.

James McBreen of Brookhaven National Laboratory
Salt Lake City, March 1990

✳ An Unseemly Missive

THE MARCH LOVE-FEST in Utah concealed fiery passions that waited to be unleashed by both sides. The appearance of the negative paper in *Nature* by University of Utah physicist Michael Salamon and his colleagues was apparently the last straw for Fleischmann and Pons. Through attorney C. Gary Triggs (a childhood friend of Pons), Fleischmann and Pons dispatched to Salamon a chilling letter within days of the conference (April 3). The faxed missive traveled from the Morgantown, North Carolina, office of Triggs to the Physics Department at the University of Utah. The letter implicitly threatened a law suit against Salamon if he did not voluntarily retract his paper because of its claimed "untenability." Failing that response by Salamon, the letter asked that he permit his coauthors to voluntarily withdraw their names from the paper. Triggs had apparently sent copies of the letter to most of Salamon's coauthors.

The backdrop for this bizarre episode: In the spring of 1989, Pons had allowed Salamon to work in his laboratory for five weeks to attempt to detect nuclear products coming from the cold fusion cells. Salamon sought but did not find gamma rays, neutrons, or energetic tritons. He reasoned that if tritium was being generated inside the palladium it could not be seen directly coming out, because it would have lost its energy inside the electrode. But that branch of the conventional d-d fusion reaction also makes an energetic proton, which would be expected to interact with the palladium nuclei and generate a gamma ray of about 0.5 MeV (in the process of fluorescence). Salamon had found no gamma rays that he could attribute to such a mechanism.

That negative finding aside, there still remained the question of whether his measurements were made with cells that were deemed "active" by Fleischmann and Pons—was he or was he not working with a heat-producing cell. He claimed yes, Pons said no. Whatever the case, Salamon's results went in and out of the public arena until they were finally trotted out on the pages of *Nature* to "celebrate" the one-year anniversary of cold fusion.

Triggs's letter to Salamon cited six complaints: (1) The *Nature* paper was factually inaccurate. (2) Some data may have been "selected" for presentation. (3) The reported experiments failed to show the presence of a particular device set up by Fleischmann and Pons as a calibration standard for gamma rays. (4) The publication date of the *Nature* paper was "engineered for editorial reasons by *Nature*" with or without Salamon's knowledge. (5) The possibility that the experiments were "predesigned" to give negative results. (6) The existence of "serious inconsistencies" between the paper and other data circulated by Salamon. The letter further charged that *Nature* had been apprised of these concerns, but that the magazine evidently did not take them to heart, nor did it pass back to Fleischmann and Pons comments that Salamon may have made in response. Triggs's letter referred to "undue ridicule and negativism created by the publication of this paper."

Regardless of the merit or lack of merit of Fleischmann and Pons's arguments, it is difficult to imagine what benefit they expected from dealing with their antagonist and his colleagues this way. The action certainly did not make for "good press" for cold fusion. Did they really believe Salamon or his associates would capitulate and withdraw their paper? Perhaps the threat of a lawsuit (and the publicity thereof) was a tactic to punctuate or draw attention to their position that Salamon's paper was at best of little or no technical significance, and at worst was an anti-cold fusion missile. Despite the distaste that they had for *Nature's* editorial policy, it would have been wiser for them to let the future course of science work things out. It is hard enough to adjudicate difficult scientific matters in the laboratory, why enter the capricious legal arena?

Physicists immediately circled their wagons for Salamon, a number of them even phoning the American Physical Society to ask about contributing to their colleague's legal defense. Salamon voiced his utter disdain for the legal threat, calling it "antithetical to the spirit of free academic inquiry." This was mild compared to his accusing the University of Utah of indirectly funding the legal threat against him. It was well known that Triggs had been paid over $60,000 for patent work done in connection with cold fusion (on the order of a 10th of the University's total expenditure for such work). As it was later learned, University of Utah funds had not actually paid for this particular assault against Salamon, but to some, this appeared as a conflict. In any case, Triggs immediately denied that Utah funds had been so used. Salamon's prepared statement for the public read, in part, "I am extremely disturbed, in fact disgusted, that the University has apparently been financially supporting such detestable activities . . ." On the advice of his attorney, however, he did not reply directly to the Triggs blast.

Even though Utah had not paid directly for Triggs's missive to Salamon and others, officials there were extremely concerned by the apparent conflict of interest: Triggs on the one hand working for them on the cold fusion patent process, and on the other attacking a faculty member—even if that professor was perceived by some to be unfairly assaulting Fleischmann and Pons in the scientific literature. Cooler heads must have intervened, because within two months (early June) the affair took an almost comic turn: Triggs dispatched a second letter to Salamon apologizing for any "misconceptions" that might have been created by the first message, and avowing no intention to limit Salamon's academic freedom and pledging to resolve the matter "in the court of science through publication."

In early June, when much of this controversy surfaced in the press, another serious matter boiled over at the University of Utah. After a bitterly contentious meeting of the Academic Senate of the 23,500-student university, President Chase Peterson announced that he would retire at the end of the 1990–91 academic year. Earlier, the University had reported that an anonymous donor had given $500,000 to the university's National Cold Fusion Institute. In truth, Peterson had transferred the money from the University's Research Foundation account, a well-intentioned maneuver that he later admitted was a mistake. His effort to "avoid jealousy" by other departments had backfired, he said, and adversely affected the credibility of the NCFI. At about the same time an unrelated controversy arose over the return of a $15-million donation that would have required renaming the medical school and hospital. The brouhaha about the not-so-anonymous gift led 22 faculty members to request both a scientific review and financial audit of activities at the NCFI.

The conjunction of these two blows—the adverse publicity of the threatened lawsuit and the announced departure of cold fusion supporter Peterson amidst controversy would have been enough to cast a pall over Utah's cold fusion effort, but this was not the end of the June troubles. A very ill wind would soon blow in from Texas.

✳ A Shining Star Falls

Since the heady days of the Santa Fe meeting, nuclear chemist Kevin Wolf had been the wonder-man of cold fusion. It was he who had reported what were claimed and considered by many to be extremely solid tritium results—clear evidence of cold fusion in electrochemical cells. The paper that he and his colleagues had published in *Journal of Electroanalytical Chemistry* (October 10, 1989) noted ". . . the observation of tritium produced in eleven D_2O electrolysis cells at levels 10^2 to 10^5 times above that expected from the normal isotopic enrichment of electrolysis. Particular attention has been paid to possible sources of contamination."

But in early June, Jerry Bishop of *The Wall Street Journal* reported that Wolf was now saying that at least some of his palladium rods probably had had tritium contamination from the beginning. Wolf said that he had belatedly discovered that his palladium, obtained from metals processor Hoover and Strong, Inc. of Richmond, Virginia (via the Texas Coin Exchange), was suspect. The only two of his dozens of cells that ever produced tritium turned out to have come from that very source. The idea arose that this might also explain why the research group of his Texas A&M colleague, John Bockris, was obtaining many tritium-producing cells. Two-thirds of Bockris's cells used palladium from Hoover and Strong. Bockris, though clearly shaken by these findings initially, continued to maintain that a nuclear reaction producing the tritium was still a strong possibility. In an interview with William Broad of the *New York Times*, Wolf said, "Our results are consistent with contamination. . . . And it's non-trivial. We can offer no support for tritium being produced by cold fusion." (June 8, 1990.) Wolf, though no longer believing his own tritium results nor the more positive results of the Bockris group (although he has taken no action to formally withdraw his name from the original joint publication on tritium), was not giving up on cold fusion. He still believed that his and others' neutron measurements were solid. Dozens of electrochemical cells were still bubbling away in his laboratory.

This did not stop some once sober and circumspect members of the science press from proclaiming the end of cold fusion. Joining the chorus led by *Nature, Science* magazine now entered the picture, with reporter Robert Pool writing in the June 15, 1990, issue, "Now Wolf appears to have knocked perhaps the last prop out from under the shaky claims of cold fusion." Pool wrote as if oblivious to the reports of tritium that had been made by more than a dozen other groups. True, possible tritium contamination found by one group had, by inference, cast a pall over all others. But putting a cloud over a generic result is a far different matter than proving everyone else had tritium contamination. Wolf himself was far more cautious and said, "I'm not trying to explain the world's results, just our results." (*Wall Street Journal*, June 7, 1990.) Wolf's backtracking might be taken as part of the reason for his not attending the cold fusion conference in Salt Lake City in late March, at which he had been scheduled to chair a session. (The public explanation was that he was ill.)

Wolf's pronouncements prompted others to certify cold fusion's death. Physicist Kevin G. Lynn at Brookhaven National Laboratory was quoted by the *New York Times*, "I think it's over. A lot of this has been bad science, with a few sincere people making an honest effort to understand what was going on. Now these people have results plagued with systematic errors. It's the nail in the coffin."

The Wolf backtracking was not the revelation of carelessness—not even ardent skeptics accused Wolf of that. There were two possible explanations: One involved a new understanding of how difficult it might be to assure that palladium is not contaminated by tritium. Wolf discovered his alleged tritium contamination even in palladium rods that had been heated in a vacuum before use. It was a great paradox how positively charged tritium that was supposedly so tightly locked within palladium could emerge from a negatively charged electrode. Wolf's October 1989 paper in the *Journal of Electroanalytical Chemistry* had mentioned that tritium would tend to be driven "*into* the cathode rather than evolve tritium from within it."

A second possibility seems much more likely: Wolf had been working for over a year to find out whether the tritium was due to contamination, but his procedures might have been flawed. One of the best ways to measure tritium in palladium rods, he thought, was to dissolve the entire rod in an acid solution that completely broke down the metal structure. Bye-bye palladium rod. The tritium thus liberated could be measured with the usual sort of nuclear counter—one that measures pulses of light caused by tritium atoms disintegrating. But one must be extremely careful; it is possible to fool the detector with chemical coloring effects from impurities in the dissolved rod entering into the measured liquid solution. This has been demonstrated by other researchers who found no contamination whatsoever in more than 150 different palladium samples–including those from Hoover and Strong.* Wolf had dissolved some 125 different palladium pieces to find tritium. All but three pieces showed no tritium; of the three that did seem to evidence it, two were virgin materials and one was a rod that had been used in a cell with light water (where tritium production was not anticipated). So this "spot contamination" problem, if it is true at all, seems to affect only about 2 to 3 percent of palladium samples. More likely Kevin Wolf's experiments were simply inappropriate. NCFI director Fritz Will says, "We duplicated his technique and found it to be open to all kinds of artifacts. As he applies it, it is unsuitable to come up with any good results."

It is possible, but barely so, that there is some unknown mechanism whereby tritium can persist in localized "pockets" or spots and not be driven out by heating. It would be very interesting, in itself, to know what could cause this! At the very least, cold fusion experiments have helped to sharpen research tools with studies like these.

✳ Enter a "Fraud Buster"

Recipe for a superb cocktail of cold fusion-confusion: Mix one part of a possible finding of previously unknown tritium contamination with 10 parts of innuendo and circumstantial evidence about *deliberate* tritium doctoring at the same university but in another laboratory.

Reports that an exposé about *possible* fraud at Texas A&M would appear somewhere had been circulating for months. The article finally appeared in the June 15, 1990, issue of *Science.* First it was speculated that *Nature* would carry it, then the more popular *Discover* magazine where the established investigative science reporter, Gary Taubes, had ties. It was bad news for him that *Nature* had exercised admirable caution and turned it down. *Nature*, being no friend of cold fusion, would certainly have embraced the piece, if the editors had been convinced of its merits.

"Cold Fusion Conundrum at Texas A&M" as it appeared in *Science* was written by Taubes, perhaps edited and a bit softened by staff members of *Science.* Taubes provided the alleged *motive, opportunity,* and possible *method* for unspecified graduate students or others to have tampered with the tritium results obtained by the Bockris group, but it offered no "smoking gun." The subheading beneath the article's title blared, "The administration's laissez faire response to worries about

---

*K. Cedzynska, National Cold Fusion Institute, "Tritium Analysis of Palladium Samples," paper given at the BYU conference on Anomalous Nuclear Effects in Deuterium/Solid Systems, October 22–24, 1990.

possible fraud raises questions about the proper balance between academic freedom and the need to guarantee the integrity of research."

The magazine put out a press release on June 12 in the form of a media advisory designed to promote the six-page detailed fraud-busting story. The news release's provocative lead paragraphs quoted Taubes's article, ". . . suspicions were raised almost from the first that the tritium in the A&M cells was put there by human hands." It followed with, "Was this a fusion reaction, was it inadvertent contamination, or was it something more insidious?" The media advisory offered this quote from Taubes: "In an atmosphere of increasing public scrutiny of the scientific process by legislators like John Dingell (D-MI), the scientific community must have ready answers for such questions . . . they take on added importance in this case, because of its high profile and the tens of millions of dollars and thousands of scientific man-hours spent chasing after the chimera of cheap, plentiful energy from 'fusion in a jar.' "

The article did not offer a single piece of direct evidence of fraud, only the circumstantial evidence that at such and such times and places there would have been the opportunity for "human hands" to inject electrolyte samples with tritium-containing water (tritiated water). The article liberally mixed in the names of graduate students and researchers who were conveniently placed at these times and locations, being careful not to point the finger at a specific culprit. In so doing, it spread the blame around, potentially causing even more damage to a number of persons than an accusation against a single one might have done.

One of the major proponents of the possible fraud theory at Texas A&M, Charles Martin, was quoted: "I can't go before a committee and accuse anyone of scientific fraud when all I have is circumstantial evidence." Martin, you may recall, had been in the Texas A&M group that made the first announcement of corroborating Fleischmann and Pons's claims of excess heat, but his group had made an error that voided that particular claim and he never was able to get excess heat in subsequent experiments. Martin and his colleagues' April 10, 1989, claims for excess heat were never publicly withdrawn, even though it was known that an improper electrical connection in the apparatus had produced spurious results. Due to various rivalries and disagreements with John Bockris and his colleagues, Martin had left Texas A&M to take another position.

As an example of how flimsy was this circumstantial evidence for tritium doctoring, Taubes cited at least two occasions on which high levels of tritium were found—each of them near a time when funding officials came to the University. However, he failed to mention the other times of visits by sponsors when high tritium levels had *not* been announced—a statistical misrepresentation. (The figures were run by the statistics department at A&M and they found no strong correlation with the visits.)

The article made an even more fundamental mistake by asserting at the outset, ". . . Bockris's tritium data remain not only the single most extraordinary 'cold fusion' effect, but also the only compelling evidence in support of the original cold fusion claims." This false proposition—elevating the Bockris results to keystone significance—is used to exaggerate the importance of possible fraud at Texas A&M. In fact, there is much other evidence to support cold fusion that is as compelling as the tritium: bursts of heat, integrated excess energy, low-level neutron emissions, evidence that the phenomena occur in related systems (for example, gaseous deuterated metals), and blank cell controls without deuterium.

Thus, even if tritium fraud were ultimately to be pinned on Texas A&M by the discovery of a smoking gun (something that seems increasingly unlikely ever to happen), this would do nothing to knock down the other kinds of anomalous effects that are being found in cold fusion experiments. Also, many other experimenters were reporting tritium too, albeit now temporarily under a pall because of a remotely possible preexisting contamination of palladium, the issue that Kevin Wolf had advanced in the media, not in scientific channels. Finally, *all* tritium reports could eventually be proved to be inadvertent contamination and *still* cold fusion could be a new nuclear process because of theories that sanction this.

In lieu of firm evidence, Taubes's article attempted to build its case on a very shaky premise: that claims for the reality of cold fusion are inherently not credible because they appear to "violate the known laws of physics." There is much evidence of the latter perspective sprinkled throughout, for example, "If the tritium had been created in the cell by any known nuclear reaction, from a few hundred thousand to a few million neutrons per second should have accompanied its creation." Or, ". . . but no gamma rays were seen, which indicated that no nuclear process had taken place." The result is to make the report of any tritium at all—particularly high levels—seem apparent evidence for adulteration. Taubes surmised that tritiated water might have been the material used to spike the cells.

Anyone seriously investigating cold fusion phenomena had long since realized that a nuclear explanation for any of the anomalies, particularly the excess heat occurring in multiple kinds of calorimeters, could not be conventional nuclear reactions. Standard plasma physics branching ratios for tritium, ^{3}He, ^{4}He, and neutrons simply did not pass muster. A serious researcher also had to contend with what is now obvious: that subtle materials properties of electrodes and/or other cell components were causing effects to appear and disappear erratically. It was therefore extremely hazardous to allege possible fraud on the basis of the comings and goings of tritium, however odd they might seem. The Taubes approach to the cold fusion enigma was not to entertain the possibility that there could be some utterly new physics waiting to be resolved.

Texas A&M University and its Dean of Science, John Fackler, stood proudly behind its scientists. "The University's researchers will continue to aggressively pursue this [cold fusion] question and to study the conflicting data coming from the several laboratories at Texas A&M, elsewhere in the U.S., and internationally," the University statement read. "Our specialized capabilities in nuclear science, electrochemistry and thermodynamics make it possible for us to approach this question from several important perspectives. We are proud that Texas A&M researchers have had the freedom to undertake an unfettered exploration of this controversial phenomenon and to question each other. Such unhampered investigation has been, and will continue to be, instrumental in delineating what is, and is not, reproducible in 'cold fusion.' "

Physicist David Worledge of EPRI, who oversees the funding of cold fusion research at Texas A&M and elsewhere and has adopted a moderate "it's not yet proved" position on the phenomenon, also countered the Taubes allegations: "The extraordinary spectre of intentional contamination should be essentially ruled out by the facts that (1) the results occur in different organizations, (2) security measures are in effect at all three laboratories, (3) at least one of the Texas cells was inaccessible beneath shielding and detectors, (4) in at least one instance, tritium was increasing in samples taken over three days, and (5) Storms's data [Los Alamos] show evidence for many small tritium bursts in some cells."* And as Julian Schwinger has written, "Intermittency is the ultimate rebuttal to charges of fraud in tritium production."

The attempt at fraud-busting by Gary Taubes was over a year in the making. Almost from the beginning, as his conversation with me in 1989 revealed, Taubes did not believe that there was likely to be anything to it, and he proceeded on a direct path to prove that cold fusion was an insubstantial "chimera," as he called it. He simply did not believe that heat-generating Fleischmann-Pons cold fusion was possible, as evidenced by his negative attitude during his interview with me at MIT in 1989 on one of his information-gathering field trips. I was very frank with him.

Taubes was very aggressive in his pursuit of what he thought was true—that fraud at Texas A&M played a major role in keeping cold fusion alive. He had traveled to the University, collected over 50 hours

---

*Proceedings of The First Annual Conference on Cold Fusion, March 28–31, 1990, National Cold Fusion Institute.

of taped interviews, and had zeroed in on one or more possible tritium-adulterers. He had become deeply involved in an investigation of their personal lives in an effort to come up with a motive; in one case, he even traveled to England in search of details on the suspect's family life. Taubes was focusing on personal factors that may have led a certain graduate student into deliberate tritium tampering. This is the hallmark of skepticism run amok that has pervaded the entire cold fusion episode. Instead of the search for scientific truth, at every turn we find people desperately trying to shoot cold fusion down with suspicion and challenges about motives.

Furthermore, there is no mention in the June 15, 1990, *Science* of the many other results that strongly indicate something very peculiar going on in electrochemical cells. Reports of excess energy generation, heat bursts, and neutron emissions are not even touched. *Science* magazine, in fact, did not report on any of the findings presented at the March Salt Lake City conference. *Science* even refused to publish perhaps the best evidence against deliberate spiking at Texas A&M, the work of Edmund Storms of LANL, which shows that a cell *deliberately* spiked with tritiated water gives rise to a very slowly and smoothly declining tritium concentration—radically different than the ragged, rapid decline characteristic of the unadulterated cells. A publication that formerly treated the cold fusion affair impartially had now crossed the line. Cold fusion, far from being home-free, was still under heavy fire.

The following October, the report of a three-member internal "Cold Fusion Review Panel" at Texas A&M stated: "In brief, we have found no evidence which would lead to a conclusion that some of the cells were spiked with tritium. . . . None of the [circumstantial evidence] provides a convincing argument that spiking occurred. Although it has been suggested that tritium findings of the Bockris group were correlated with funding agency visits, conversations with the funding agency (EPRI) refute this. The group was already well-funded by EPRI and other granting agencies. In addition, EPRI was constantly in contact with the research group and the visits per se had no particular significance in the funding decisions.

"While it is not possible for us to categorically exclude spiking as a possibility," the report added, "it is our opinion that possibility is much less probable than that of inadvertent contamination. This may provide a natural explanation for anomalies and lack of reproducibility, if the process of cold fusion were found not to be responsible." Given that no one, not even Taubes, has ever seriously suggested that "inadvertent contamination," that is, the equivalent of stupidly and accidentally pouring tritiated water into a cell, might have happened, the report all but ruled out the tampering suspicions. In an oblique assessment of Taubes's approach, the panel wrote, "The tactics used by an experienced investigative reporter, G. Taubes, in a clearly adversarial role were much more intimidating and exerted much more psychological pressure than can or should be expected from an outside committee."

✳ Return to the Beginning

The cold fusion story returned full circle to March 23, 1989, when in July 1990, Fleischmann, Pons, and Hawkins and their colleagues published their definitive statement on producing and measuring excess heat in electrochemical cells (July 25, 1990). If the experiments and analysis reported in their highly detailed 56-page technical paper had been performed prior to their 1989 announcement, and if this convincing account had been available, the cold fusion story might have evolved far differently. Fleischmann and Pons would have been taken more seriously; they would have had a very solid foundation on which to base their claims. Now it is difficult to get the most important skeptics to react seriously to that report.

Roger Parsons, editor of the *Journal of Electroanalytical Chemistry*, introduced the new paper, saying: "The preliminary note by Fleischmann and Pons published in this Journal in April of last year, has probably generated more controversy, and even hysteria, than any other paper we have published. . . . They have now produced a detailed description and analysis of their calorimetric experiments which support the work described in their preliminary note. We consider that this publication will go some way to bring the discussion of this problem back to a proper scientific level. . . . We hope that with the publication of this paper, the discussion of cold fusion will enter a more rational phase." Unfortunately, this hope was not realized.

The paper reiterates that they have achieved excess power generation up to 100 watts/cc of palladium electrode. With increasing electric current density (amps per square centimeter of electrode), the excess power rises. They again noted bursts of power exceeding 40 times that input to the cell and total energy emanating from their rods in the range 5 to 50 MJ/cc—a 100 to a 1,000 times greater than any conceivable chemical energy release. These rods were 0.1, 0.2, and 0.4 centimeter diameter. In control (blank) experiments replacing the palladium cathode with one of platinum, or in tests using light water, the excess power was essentially zero even while applying increasing electric current density. Just what the critics had so loudly demanded as "proof" of fusion, but which now is met with stony silence.

Their final conclusion: "There can be little doubt that one must invoke nuclear processes to account for the magnitudes of the enthalpy [heat] releases, although the nature of these processes is an open question at this time." Though in this paper the authors did not report new tritium or neutron results to replace their earlier measurements retracted in the spring of 1989, they said these would be published elsewhere. They remarked, however, "It is hardly tenable that the substantial number of confirmations of the calorimetric data using a variety of techniques can be explained by a collection of different systematic errors nor that tritium generation can be accounted for by any but nuclear processes."

Despite increasingly detailed accounts of cold fusion experiments like the latter paper, critics keep chanting a tattered phrase, "Show me results!" Typical was Gary Taubes's response to letters to *Science* from David Worledge of EPRI and John Bockris and Duwayne M. Anderson of Texas A&M, which were critical of his June 1990 tritium "exposé." (*Science,* August 1990.) Taubes wrote, "What is needed is the reporting of data and experiments that can speak for themselves, and a year and a half after the 'discovery' of cold fusion those data and experiments are still talked about but not seen." A more ostrichlike statement would be hard to construct.

15 Whither Cold Fusion?

To predict the future we need logic; but we also need faith and imagination which can sometimes defy logic itself.

Arthur C. Clarke
Profiles of the Future, 1963

. . . It can be taken for granted that before 1980, ships, aircraft, locomotives, and even automobiles will be atomically fueled.

David Sarnoff, past head of RCA, 1955

Even if [cold fusion] were only fifty-percent probable, every laboratory in the world should be working on it.

Ernest Yeager to John Bockris, 1990

✳ Here Today, Here Tomorrow

EVIDENCE THAT THE COLD FUSION STORY will be here a long time grows rather than diminishes. Recently compiled reports of positive evidence for cold fusion have come from 90 or more research groups in at least 10 nations and at five federal laboratories in the United States (pages 246–248). New dimensions of the puzzle keep popping up like dandelions on a manicured lawn. Like the sturdy product in a television ad, cold fusion is "here today, here tomorrow." Of this even most scientists are unaware. As John Bockris told a meeting of chemists in April 1990, "The general opinion of scientists around the country is that [cold fusion] is all a joke, and that it's terribly funny that people should do any work on it, because it was a gigantic mistake, which was made by two fine fellows in Utah. It's all finished now and we can look back at those times and laugh."

The *Japanese Journal of Applied Physics* featured an unusual article (April 1990) by E. Yamaguchi and T. Nishioka at the NTT Basic Re-

Groups Reporting Cold Fusion Evidence

Investigators	Institution	Heat	Trit-ium	Neu-trons	γ-rays	^{3}He	Report Type*
Adams/Criddle	U Ottawa, Canada	X	X				4
Alquasmi	U Kiel, W. Germany		X	X			4
Alikin	Perm State U, USSR		X				3
Appleby	Ctr. El. Chem. Energy, Texas A&M	X					2
Arata	Kinki U, Japan			X			1
	Belorussian State U, USSR	X					3
Bertin	U. Bologna, Italy			X			
Blanco	Oak Ridge NL			X			
Bockris	Texas A&M U	X	X				1,2
Bose	BARC, India		X	X			1
Cai	Chinese Acad Sci, China	X		X			2
Celani	Frascati Res Ctr			X	X		1
Claytor	Los Alamos NL		X	X			3
Cherepin	Metal Phys Inst, Kiev, Russia		X			X	3
Dash	Portland State U	X					4
De Maria	U Rome, Italy	X		X			4
Din, D.Z.	Nuclear Energy Inst, Shichuan & Beijing			X			2
Eng. Group	NCFI/U of Utah	X	X				6
Faler & Vegors	Idaho State U			X			5,6
Fukada	Kyushu U, Japan			X			4
Gao, G.T.	Eng Phys Inst, China			X			
Gou, Q.Q.	Science & Tech Inst, Chendu	X				X	2
Gozzi	U Rome, Italy	X	X	X			1
Huggins	Stanford U	X					2
Hutchinson	Oak Ridge NL	X					2
Ikegami	NFSI, Japan		X?	X			5,6
Ikezawa	Chubu U, Japan			X?			3
	Barc, India		X	X			1
Jones	Brigham Young U			X			1
Jordon	Case Western U			X			
Jorne	U Rochester			X			4
Krishan	Indira Gandhi Ctr, India			X			

(*continued*)

search Laboratories, which has a reputation roughly like that of the AT&T Bell Laboratories in the United States. Yamaguchi and Nishioka detected a gigantic burst of over a million neutrons per second (sustained for two to three seconds) from a thin plate of palladium coated with special oxide and gold films on either side. Employing a pressurized gas cell, they had initially infused the three-centimeter square plate with deuterium. Apparently what prompted the unusual outpouring of neutrons was lowering the gas pressure around the millimeter-thin palladium. Soon two more huge neutron bursts appeared, each a few minutes

Groups Reporting Cold Fusion Evidence, *Continued*

Investigators	Institution	Heat	Tritium	Neutrons	γ-rays	^3He	Report Type*
Krishnan	BARC, India		X	X			1
Kuzmin	Moscow State U	X	X	X			3
(no name)	Karpov Inst, Russia	X					3
Landau	Case Western U			X			4
Liebert	U Hawaii	X					4
Maeda	KURRI, Japan			X?			4
Mathews	Indira Gandhi Ctr, India		X	X			
McBreen	Brookhaven NL	X	X				4
McKubre	SRI International	X	X				2
Menlove	Los Alamos NL			X			2
Miles	Naval Weapons Ctr	X				(^4He)	1,2
Milikan	UC Santa Barbara	X		X			4
Mizuno	Hokkaido U, Japan			X			1
	U Mexico, Mexico		X				
Nayar	BARC, India		X	X			
Niimura	Tohoku U			X			4
Noninski	Sofia, Bulgaria						1
Ohta	U Tokyo, Japan	X					4
Okamoto	TIT, Japan			X			4
Oriani	U Minnesota	X					4
Ozawa	Hitachi, Japan			X			4
Oyama	TAT U, Japan	X?					4
Pons & Fleischmann	NCFI/U of Utah	X	X				1
(no names)	Qinhua U, Beijing, China		X	X	X		2
Radhakrishnan	BARC, India		X	X			1
Raghaven	AT&T			X			
Raj	BARC, India					X	1
Rout	BARC, India			X			1
Saini & Raye	BARC, India	X	X				6,7
Sakamoto	Tokai U, Japan			X			4
Sanchez	U Madrid, Spain		X	X			1
Santhanam	Tata Inst, India	X					
Scaramuzzi	Frascati, Italy		X	X			1
Schoessow	U Florida	X	X				6
Scott	Oak Ridge NL	X		X	X		2
Seminoz-henko	All-Union Inst, Monocrystals, Russia		X			X	3
Shyam	BARC, India			X			
Srinivasan	BARC, India		X	X			1

(*continued*)

after beginning depressurization. On further trials with the same plate, the investigators were unable to reproduce the effect, but the scientists conclusively associated the huge neutron emission with the explosive release of deuterium gas from one surface of the plate.

Coincident with the first neutron blast, the flat plate buckled like a potato chip and the gold coating on one surface changed color—consistent with the gold film alloying with the underlying palladium at a

Groups Reporting Cold Fusion Evidence, *Continued*

Investigators	Institution	Heat	Trit-ium	Neu-trons	γ-rays	³He	Report Type*
Storms & Talcott	Los Alamos NL	X	X				3
Szpak	Naval Systems, San Diego	X	X		X		6
Tachikawa	JAERI, Japan			X?			4
Takagi	TIT, Japan			X			4
Takahashi	Osaka U, Japan		X?	X			1
Taniguchi	OPRRT, Japan			X			1
Tian, Z.W.	Xiamen U, China			X			2
Venkateswaran	BARC, India		X	X			
Wada	Nagoya U, Japan			X			1
Wadsworth & Guruswamy	NCFI/U of Utah	X					2
Wakabayashi	PRC, Japan		X?				4
Wan	National Ching-Hwa U, Taiwan	X	X				5
Wang, D.L.	Nuclear Energy Inst, Shichuan, China			X			2
Wang, G.G.	Nanjing U, Nanjing, China	X		X			2
Werth	Englehard Industries	X					
Wolf	Texas A&M U			X			2
Xiong, R.H.	SW Nuclear Phys Inst, Shichuan, China			X			2
Yeager & Adzic	Case Western U	X	X				2
Zelenskiy	Kharkov Inst, Russia		X	X	X		3
Zhou, H.Y.	Beijing Normal U, China		X	X			2

TOTAL NUMBER OF GROUPS: 92
NUMBER OF COUNTRIES: 10 (U.S., Japan, India, Italy, USSR, Canada, W. Germany, China, Bulgaria, Spain)

Dr. F.G. Will, Director of The National Cold Fusion Institute
Courtesy of The National Cold Fusion Institute
September 12, 1990
* Key: 1 = Refereed Journal Publication
2 = Conference Proceedings
3 = Nonrefereed Report
4 = Conference Presentation
5 = Newspaper Article
6 = Personal Communication
7 = Submitted to Journal

temperature probably exceeding 1064°C. The temperature of the plate's steel sample holder rose 50°C. The heat may have had nothing directly to do with the fusion reactions that evidently gave rise to the neutrons, but it was a most unusual new phenomenon in a field that seems to grow more and more curious. "Cold Nuclear Fusion Induced by Controlled Out-Diffusion of Deuterium in Palladium" was what Yamaguchi and Nishioka called their approach.

In July came *really hot* cold fusion! The cold fusioneers met in Honolulu, Hawaii, at a special session of the World Hydrogen Energy

Conference that was sponsored by the Hawaii Natural Energy Institute. Associate Professor Bruce Liebert, B.Y. Liaw, and their colleagues at the University of Hawaii announced finding substantial excess heat with a high-temperature cold fusion cell that contained a molten solution of materials. They were working at temperatures from 350 to 500°C and claiming excess power bursts of up to 15 times the input power that lasted for many hours.

Again the common denominator was deuterium. Without deuterium somewhere in a system, virtually no one was reporting any kind of anomalous effects. Both palladium and titanium cathodes were tried and found to yield substantial excess power in the molten system. In one case, electrical power of 1.68 watts coursing through the cell yielded 25.4 watts of excess thermal output. The Liebert-Liaw group used a molten lithium chloride-potassium chloride mixture (LCl-KCl) that was saturated with lithium deuteride (LiD)—the latter being equivalent to the electricity-conducting material in low-temperature heavy water cold fusion cells.

Though the title of their paper was understated, "Elevated Temperature Excess Heat Production Using Molten-Salt Electrochemical Techniques," a sentence in its abstract was not: "If this effect can be reproduced at will, and duplicated at other facilities, not only would these results provide virtually irrefutable evidence for the excess heat effects attributed to *cold fusion*, they would also mean that practical use of such electrochemically-assisted processes for power generation is much closer to reality than previously anticipated." Liebert told Nick Tate of the *Boston Herald*, "We're not claiming that this is necessarily fusion. But it's some nuclear process that is occurring at the solid state, and the significance is that it's a non-chemical (reaction). Our results tend to support the cold fusion proponents a lot more than other experiments have." (July 23, 1990.)

The group stated in its final paper, which appeared in the conference proceedings, "When all reactions that are known to occur in these systems are considered, no rationale can be obtained that would justify attributing a thermochemical reaction to the excess power generation. Thus, these results suggest that this effect is nonchemical . . . the origin of the excess heat generation can only be attributed to a nuclear process or, maybe, several processes, which are unknown as yet." The group's preliminary work in the same device with LiH, rather than LiD, has so far given no indication of excess heat—the kind of all-important control experiment for which critics had clamored. Shown this kind of evidence now, the critics remain dutifully silent, hoping that this annoying evidence for cold fusion will simply disappear. It won't.

If real, these molten system results are very serious power levels. They are hardly trifling, subtle effects at the limits of detection. The

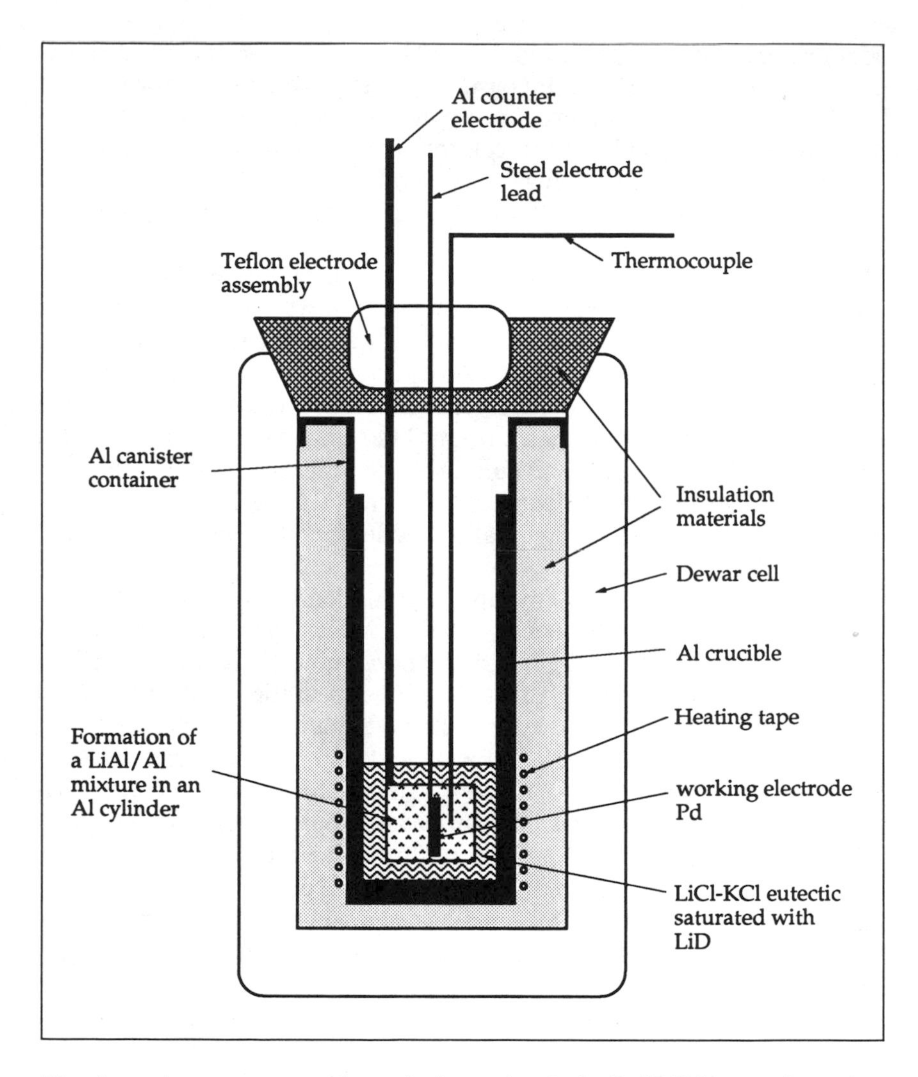

The elevated-temperature molten salt electrochemical cell of B.Y. Liaw et al.—a schematic diagram. The chamber is of aluminum metal (Al) and rests in an insulated Dewar vessel. (Courtesy B.Y. Liaw, The University of Hawaii)

excess power generation in one case is reported to be over 600 watts per cubic centimeter of palladium—much more than any known heavy water cell has so far evidenced. The total energy observed coming from the system for tens of hours was about 120 megajoules per cubic centimeter (MJ/cc) or 34 kilowatt-hours per cubic centimeter Pd. Such remarkable results verge on proof that something very new and powerful is at hand. But that is not all. . . .

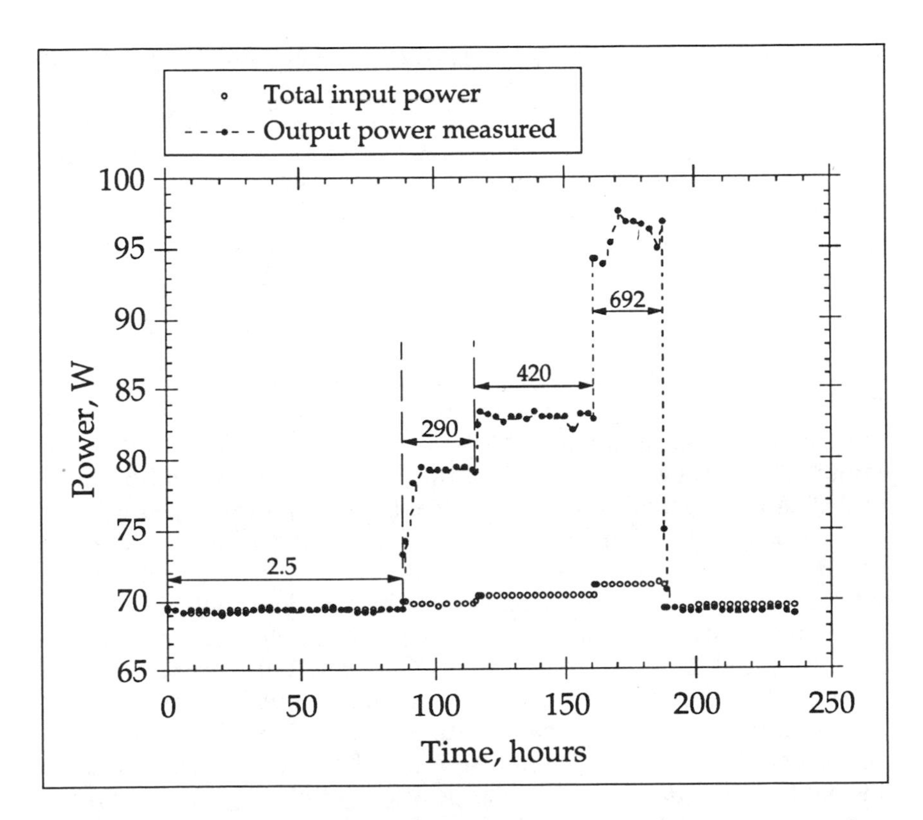

Comparison of the input and output power measured during high-current excursions in the molten salt electrochemical cell of B.Y. Liaw et al. Numbers in the figure are the current densities in milliamps per square centimeter at which deuterium was charged into the palladium electrode. (Courtesy B.Y. Liaw, The University of Hawaii)

✳ Proof Positive?

An irony of the cold fusion saga: The March 1990 First Annual Conference on Cold Fusion in the Salt Lake City bailiwick of Fleischmann and Pons was a turning point in cold fusion history. But at the meeting in October 1990 in neighboring Provo—BYU-Jones territory—researchers discussed results that some might call final *proof* that cold fusion is real. Evidence that tritium is being generated in cold fusion experiments became so compelling as to be essentially impossible to deny any longer. The fiction of widespread inadvertent contamination, despite every precaution and all evidence against it, and the even more tortuously conceived story of willful adulteration at a single university sank beneath the waters.

The international review meeting, "Anomalous Nuclear Effects in Deuterium/Solid Systems," chaired by Steve Jones and Nate Hoffmann met to discuss nuclear effects issues in cold fusion studies. About 150 researchers from Argentina, China, Europe, India, Japan, Korea, the United States, and the Soviet Union gathered at Brigham Young University to hear more than 60 papers dealing exclusively with the exotic anomalous nuclear effects seen in cold fusion, not the thermal effects. Quite a threshold has been reached when scientists convene a meeting on a *subdiscipline* of cold fusion phenomena. The conference was sponsored by EPRI, DOE, and BYU. And—wonder-of-wonders—the conference proceedings were to be published by the American Institute of Physics!

A highlight of the meeting: physicist Ed Cecil of the Colorado School of Mines discussed evidence that energetic tritons—the nuclei of tritium—come out at about 5 MeV from a thin deuterium-loaded titanium foil through which an electric current is passed and which is cycled though extremes in temperature (from liquid nitrogen temperature, −196°C, to room temperature). Cecil had discussed earlier phases of his work at the Santa Fe conference in May 1989. For the present work he used a so-called silicon surface barrier detector, known to be highly resistant to neutron or gamma-ray background interference. Tritons appeared to come from localized regions in the titanium foil at consistently repeatable emission rates of many hundreds per minute. This is thousands of times above natural background. The results imply that energy is being liberated in fusion reactions at tiny sites within the metal. This represents an apparent power level of about a kilowatt per cubic centimeter in those localized regions! Cecil tried his experiment with ordinary hydrogen instead of deuterium and found—with a confidence greater than 98 percent—that high-energy "events" were associated only with the deuterium-treated titanium. In all, he registered 24 bursts of particles, with 12 of his 26 foils evidencing the phenomenon.

The Princeton University-trained physicist told a reporter, "That's the amazing thing about this; conventional wisdom says it shouldn't be fusing at all. But something is happening in there. . . . There is an awful lot of good evidence that a nuclear reaction is taking place. . . . We're at the 80 to 90 percent confidence level."

Even at the Utah conference in March 1990, George Chambers of the Naval Research Laboratory had reported charged particles with MeV energy coming out of titanium foils that he had bombarded with beams of deuterium ions. The deuterium ions going in were only in the kilovolt (350 eV) energy range. Without a fusion mechanism, you simply cannot get a 5.9 MeV particle out when the input particle is 350 eV! Chambers' conclusion then was that the charged particles coming out "could not be explained by conventional physics." That assessment has even more weight now that others have verified his results. Little doubt that the charged particles are tritons, little doubt that cold fusion is, indeed, real. His further work reported at the BYU conference corroborates Cecil's work. Kevin Wolf, however, who only recently began trying to replicate such experiments, had no positive results to report.

Howard Menlove from LANL said that he was still seeing the neutron bursts that he had discussed at the Cold Fusion Workshop in May 1989 at Santa Fe. M. Srinivasan of the Bhabha Atomic Research Center noted BARC's similar results and commented that the neutron bursts might be coming from just a few of the individual titanium chips, as his group had found. Menlove agreed. Srinivasan displayed a slide made from an "autoradiograph" of a deuterium-gas-activated chip in which apparently a significant quantity of tritium—a millicurie—had formed.

Steve Jones reported extensions of his earlier reported neutron work, such as putting his experiments in a deep mine and doing further checks for various kind of interference. Jones expressed confidence that he was continuing to see neutrons, both from deuterium gas cells and electrochemical cells. Kevin Wolf from Texas A&M reported more positive neutron results, as did F. Scaramuzzi, Dr. Zhu from the Institute for Atomic Energy in Beijing, and Dr. A. Takahashi from Osaka University and the Matsushita Electric Industrial Company in Japan.

Tom Claytor from LANL reported increasingly refined experiments in his deuterium gas cells that employ high voltages to generate tritium. His group, he said, is now able to generate tritium reproducibly. They did not get it whenever ordinary hydrogen was used.

✳ Open Questions

The compelling evidence for nuclear effects in deuterated metal systems gives enormous credibility to a possible nuclear explanation for excess heat measured in similar experiments—despite persistent attempts by Steven Jones and others to dissociate the two effects. After all, if the heat comes in such profusion that it seems to require a nuclear explanation, and if at the same time there are indisputable nuclear effects detected, even if these do not specifically explain the excess heat *quantitatively*, this seems strong presumptive evidence that the basis for the excess power may be nuclear.

Still the question remains: What is causing the excess heat if not ordinary chemistry or some bizarre, unknown form of mechanical energy storage and release? If the process be nuclear, what is the "fuel" and what are the reaction products—the "ash" of the hidden nuclear "fire." An answer could come in a two-step process: (1) First, experiments to identify the fuel and end product(s) irrespective of the complex mechanism by which the presumed fusion is brought about and (2) elucidating the physical mechanism for the reactions, for example, the specific type of lattice-deuteron behavior that promotes the reaction.

Concerning the heat, if we do not even know what we are burning, we are really nowhere in developing practical applications. If we knew the reaction and we also understood its mechanism, then we could begin to stabilize and control the process and perhaps scale it up. Maybe its intensity could be run up orders of magnitude with different materials or different geometries—alternate parameters of some kind, such as temperature. The Hawaii molten salt work provides an inkling of this.

Hagelstein, for one, now believes that the 40-plus version of his coherent fusion theory offers a mechanism and class of nuclear reactions that qualitatively, and in many aspects quantitatively, accounts for the host of possible cold fusion phenomena: rates of heat production, neutron emissions, low-energy tritium, various high-energy charged particles, and erratic behavior linked to metallurgical properties of the electrodes. Dispensing with many earlier difficulties, neutrons can be removed directly from nuclei such as deuterium and can be "donated" directly to other nuclei such as palladium. The formulation takes energy from the lattice in such a way as to make this stripping and donation of neutrons feasible. With confidence unusual even for Hagelstein, he says of the new mechanism and reaction, "I'm *certain* that this is the right one, that this is what is going on." Yet others were waiting to publish *their* supposedly solid theories, too.

Other theoreticians claim to be nearing their version of "the answer." Julian Schwinger published in late 1990 in the Proceedings of the National Academy of Sciences further theoretical insights. And Robert T. Bush with his Transmission Resonance Model for cold fusion claims to have made a major breakthrough in correlating experimental conditions and observed results.

What experiments are needed to get answers to questions posed by theories? Foremost is the need to find consistently reproducible experiments yielding heat (or nuclear products, for that matter). Sorry to say, work in this direction may continue to require a trial and error, empirical approach. It won't be easy and it's hard to say how long it will take. As Julian Schwinger observed, "The short, bloody history of cold fusion indicates that 'similarly prepared' is not a trivial condition."* Yet there are very good indications that the reproducibility problem is being put to rest. Reliable reports filtering out of the McKubre group at SRI in late 1990 were that they can now switch on and switch off the excess power pretty much *at will* by carefully controlling and monitoring electrochemical conditions.

If protons are being burned from the trace amount of light water in an experiment, how to detect their loss when the quantities disap-

---

*Julian Schwinger, "Cold Fusion: A Hypothesis," *Zeitschrift für Naturforschung*, Vol. 45, No. 5, May 1990: 756.

pearing are so incredibly tiny? Looking for proton loss had been a requirement of both Schwinger's and Hagelstein's theories. After this, how can we account for other possible, perhaps "auxiliary," reaction end products such as tritium, ^{3}He, and neutrons? These need to be nailed down quantitatively.

Hagelstein has suggested looking for changes in cell performance with different trace concentrations of ordinary hydrogen in the heavy water. Schwinger has suggested the same, but also proposes looking at changes in the thermal balance of presumably "dead" light water cells following small additions of heavy water. Both experimental and theoretical indications are that power levels will rise when experiments are performed at higher temperature. There is no shortage of ideas for experiments, but there is definitely a dearth of funds to carry them out. With more money, the science might have been further along already, because much time and effort has been wasted scrounging and bootlegging.

✳ Funding

A good way to begin to improve the funding situation would be to reverse the disastrous conclusions of the November 1989 ERAB report that dismissed cold fusion. The position of the panel then was untenable; it is now even more so. Reversal of the ERAB report and opening federal research funds for focused experiments is essential. This would merely be following through with what even the cold fusion critics proposed at the April 1989 Congressional hearing on cold fusion. Whether $10, $20, or $50 million is needed, who knows? Among several hundred U.S. researchers, $20 million would probably go far right now. There should be small individual grants to "basement tinkerers" too, of whom there are many already. For example, Tom Droege, a superb engineer who has built state-of-the-art instrumentation for the particle physicists at Fermilab, now works in his basement in Batavia, Illinois, as he perfects an extraordinary calorimeter. With great regularity as they monitor cold fusion cells, Tom and his brother John see excess power generation and many other interesting phenomena that neither they nor anyone else seem to be able to explain with known chemical processes. Tom claims to have spent 70 hours a week on his cold fusion experiments since the Utah announcement.

A fair amount of cooperation and close coordination between laboratories working on cold fusion already exists. Establishing mechanisms to enhance that kind of interchange and bring in all kinds of contributors to this supremely interdisciplinary field would be important, as would sharing of nuclear and chemical analysis facilities held in common.

It also would be beneficial if publications that have criticized cold fusion were to come around at least to a more balanced view of the research—if not to outright acceptance. Recanting previous hostile views may not be easy, but time will heal the wounds that are inevitable in the earthquakelike shift of a scientific paradigm.

Private efforts to capitalize on cold fusion research cannot be neglected. That is where the cold fusion industry will ultimately reside, if there is ever to be one. Already there are a number of venture capital groups with scientifically savvy officials who have been traveling from lab to lab in search of the best research opportunities. Many universities have filed, in toto, dozens of patents on cold fusion devices, and these intellectual properties may be ripe for licensing. Some of the earliest cold fusion ventures started in Utah. The Fusion Information Center, Inc. was formed almost immediately after the announcement. Some companies registered names in Utah by early May 1989, such as: Cold Fusion, Inc.; Electro Fusion, Inc.; Fusion Technologies, Inc.; Fusion Consultants, Inc.; Deuterium Energy Products, Inc., and Fusion Resources, Inc. Even now, there are quiet cold fusion ventures springing up around the country, and they are not all in Utah. One former inertial confinement fusion physicist in the Midwest has discretely launched a cold fusion venture—now that he is convinced the phenomenon is real.

Will the funding crunch permit Utah's National Cold Fusion Institute to survive past the summer of 1991, thus following through on the work begun there in 1989? The question hung heavy in the air as NCFI was enveloped in yet another controversy in the fall of 1990. On November 8, Utah's Fusion Energy Advisory Council met in closed session to hear a presentation by Stanley Pons to an outside panel of four scientists. A group of 22 disgruntled University of Utah faculty members earlier had demanded this review to determine whether continuing NCFI's work was warranted.

The four scientists, chosen to be mutually acceptable to the council and NCFI officials, including Fleischmann and Pons, were: Stanley Bruckenstein, Professor of Chemistry at the State University of New York at Buffalo; Loren G. Hepler, Professor of Chemistry and Chemical Engineering at the University of Alberta, Canada; materials scientist Dale F. Stein, president of Michigan Technological University; and Robert Kemp Adair, Professor of Physics at Yale University. Though Dale Stein had been a member of the DOE panel that made a very negative assessment of cold fusion in November 1989, he pledged to be open-minded in this new review.

At issue was whether and how the $1.3 million remaining from the $5 million cold fusion funding approved by the legislature should be spent, and perhaps whether new funding should be sought. The original plan was to have had considerable outside funding flowing into NCFI by this time, but in view of cold fusion's stormy history, only a paltry $110,000 had been raised from outside, while $3.7 million in state funding had been spent. This dismayed Institute director Fritz Will, because he had hoped to have considerable outside resources—enough to make NCFI completely independent of state funding by 1992.

Complicating matters, the media had a field day with the alleged "disappearance" of Stanley Pons—his absence at the first intended review session by the Council on October 25. Adding to the mystery, according to many media reports, was that the Pons Salt Lake City home was on the market, its phone disconnected, and the Pons children had been taken out of school. An October 25 Associated Press story said, "As if political pressure, worldwide skepticism and demands for review are not headache enough for the directors of Utah's cold fusion research, now they cannot find their top scientists. Not only do they not know the whereabouts of electrochemists Martin Fleischmann and B. Stanley Pons, who started the controversy over cold fusion, they cannot say if the pair will return." A headline in a *New York Times* story by William Broad read, "Utah to Start Search for Cold Fusion Scientist." "Miffed at the disappearance of a top cold fusion scientist, the State of Utah yesterday formulated a plan to track down the enigmatic researcher and hold him accountable for his work," wrote Broad.

It turned out that Pons and his family were in Europe. Pons had formally requested a one-year sabbatical from his University responsibilities, but had departed before getting approval and was responding to University inquiries only through his attorney, Gary Triggs. Fleischmann was in England awaiting surgery to correct a problem with a facial nerve. Neither scientist showed up for the scheduled October 25 hearing and vigorously disputed Fritz Will's contention that they had been properly notified of the hearing. Will expressed his extreme displeasure at having to communicate with Pons through his lawyer and asserted that Fleischmann and Pons were "unwilling to cooperate with the Institute or the state committee in participating in any useful review of their work." (*New York Times*, October 26, 1990.) To the press, Will characterized Pons's behavior as "almost self-destructive."

It seemed that civil relations had broken down between Fritz Will and University of Utah Dean of Science Hugo Rossi on one side and Fleischmann and Pons on the other. The strains of organizational strife and assaults on the research from within and without the University had taken their toll. The immediate cause of the acrimonious break between Will and Fleischmann and Pons was Will's seemingly unavoidable siding with the faculty group's request for the external review of NCFI's work. Fleischmann and Pons regarded this as a "grave mistake," Will told me. Further bitter internal recriminations followed, along with legal maneuverings, and a complete breakdown in communications. Fleischmann, for his part, regarded NCFI as a "leaky sieve"—unsuitable for the proprietary work that he and Pons wanted to pursue. It had always been true that Fleischmann and Pons had reluctantly agreed to work within the confines of a formal, publicly funded research institute. They preferred instead to pursue their work independently.

The absurd media tempest about the "disappearing scientist" ended with Pons returning for the November 7 meeting, at which he reportedly was unusually self-assured and relaxed, revealing details about a ninth patent application that had been filed.* He told the panel that he wanted to return to work in Utah later. Fusion Energy Advisory Council members were encouraged at the meeting and the final report by the four outside scientists. Given the poor public image of NCFI, the individual reviewers' reports were remarkably supportive. Even skeptical physicist Adair had nice things to say about the research. Sprinkled in the reports were sentiments such as: "The scientific competence and objectivity of NCFI personnel is impressive"; "Dr. Pons was fully cooperative and open during his discussions with the Committee"; "They have provided strong evidence (not proof) in favor of the production of excess heat"; "Although it is possible that their claims of excess heat production are mistaken, it is neither accurate nor fair to say that they have done bad calorimetry"; and "The technical programs appear to be well designed and conducted with vigor and intelligence." After the review, Council chairman Raymond L. Hixson said, "The science is sound, and we felt it has been all along."

It seems that Pons is now working on cold fusion with Fleischmann both in Utah and abroad. Fleischmann works primarily in England, while Pons is speculated to be working in France at a laboratory or laboratories that may have Japanese sponsorship. This would not be so surprising, because the Japanese have shown a much greater continuing interest in cold fusion than have scientists in the United States. A recent survey of 1,600 Japanese researchers, engineers, CEOs, and other businesspeople revealed that about half as many believe that cold fusion is a key technology to replace oil as believe that nuclear power could play that role.† The group chose cold fusion as one of the "20 Most Important Technologies or Products that Japan Should Undertake in the 21st Century."

Cold fusion research in the United States, however, is deeply troubled. Barring a "minor miracle," said Fritz Will, the NCFI will shut down operations on June 30, 1991. The Institute has been working since

---

*The patent application reportedly concerned a method for achieving a cold fusion power density up to 1,000 watts per cubic centimeter of electrode.

†*NKSS Journal* (*Japan Economy and Industrial Newsletter*) September 3, 1990.

late 1990 with a much-reduced staff, ironically at the time of some of the most exciting results confirming tritium production and neutron bursts in new kinds of gas-discharge experiments. Director Will could not realistically go back to the Utah legislature for more funding, he said, because the original purpose of the state support was "seed money." In the present climate, he suggests, it has not been possible to convince potential sponsors that excess power has been proved beyond doubt, hence no external funding is forthcoming.

The very cautious Will remains convinced, with others, that the nuclear effects in cold fusion are new scientific phenomena that have been proved "in all likelihood." He is much less confident about the excess heat results and a possible connection with the nuclear effects. "One has to work harder and harder in order to find whether or not there is a relationship," he says. On the other hand, he completely discards the idea that the excess power has a *chemical* origin. He thinks that there could conceivably be some kind of mechanical energy storage and release mechanism at work in microcracks within the palladium. However, he agrees that if large energies—megajoules per cubic centimeter—can be verified not to be an experimental artifact, then the only explanation would be a new nuclear process. Regrettably, the dramatic heat bursts obtained in October 1989, have apparently not reappeared.

The swan song for NCFI will likely come during the Second Annual Meeting on Cold Fusion, to be held June 29–July 4, 1991 in Como, Italy. A deeply disappointed Will said, "It hurts me to think that in a few months we will no longer be able to pursue research—some approaches we have been taking are absolutely novel, which others to the best of our knowledge haven't tried. The obligation of a scientist is, if there is a new phenomenon that might exist, to pursue that until one understands what the nature of this new phenomenon is. It is a pity to see that the sentiments of people have been flying so high among the nuclear physics community that they have given cold fusion such a black eye that it is not even possible for the sponsoring agencies to be able to look at new proposals in the area of cold fusion and apply the same arguments to it as for proposals in other research fields.

"In other words, if one hands in an excellent proposal on new batteries or fuel cells, the chances are much higher of getting money than for an equally excellent proposal on anything that has to do with cold fusion, and *that* is a pity. That is where the bias that exists in the United States and in some other nations reminds me of the Dark Ages, in that one does not give science a chance and scientists a chance to pursue something that looks new and interesting. Even those—and I belong to those—that want to go to the bottom of these phenomena, will find it increasingly difficult to get sponsorship."

As for the original pioneers Fleischmann and Pons, Will told me he was "absolutely sure" that their actions and outlook have been colored by the unrelenting attacks. "These two people have been haunted by many people, by some of the media, and the negative attitudes displayed by much of the—let's say the nuclear physics community. That stimulated in them a strong sense of persecution and being handled very unfairly—no question."

✳ Applied Cold Fusion?

If the excess heat observed in cold fusion cells is what it could quite conceivably be—a nuclear-driven aneutronic process—then it would be possible to apply that heat to a host of practical devices. These could range from small heating units in the kilowatt range for individual dwellings to central-station electric power plants. There are so many uncertainties, however, that it makes little sense to try to project in detail what will happen with applied cold fusion.

The phenomena are at the moment quite erratic and would have to be made more stable and controllable before they become useful. No one yet knows how this could be done, though presumably if we understood the physical basis for the comings and going of the heat, something could be done to give it staying power. There is also the thermodynamic efficiency issue: Higher temperature thermal processes are fundamentally more efficient in producing useful work than ones at lower temperature. If one is not interested in motive power or electricity and merely wants to heat a building, then the efficiency issue diminishes. If you have temperatures near the boiling point of water, as appears to be so at least sporadically already, the hot water is then usable for environmental heating (providing no dangerous power excursions or significant releases of tritium occur).

It might be very economical to have such a home heater, providing that palladium does not turn out to be an essential ingredient of the electrodes. If titanium or some other metal alloy sufficed, then a really cheap home heating unit would be possible—conceivably in the several thousand dollar range. For all practical purposes, the fuel might well be "free." Heavy water runs from $.50 to $1.00 per gram at the moment. While not knowing exactly the heating mechanism or fuel consumption per kilowatt hour, it is likely to be some tens of thousands of kilowatt hours per gram of heavy water. For example: Fusing all the pairs of deuterium atoms in several grams of heavy water would yield over 30,000 kilowatt-hours of energy—enough for a few years of home heating. Whatever the fusion reaction—proton-neutron fusion to deuterium, proton-deuteron fusion to ^{3}He—the order of magnitude of heating from a mass of deuterium should be comparable.

Of course it would not be that simple, because a considerably larger mass of heavy water might have to be maintained at some high purity level for the heater to work. A gallon of heavy water costs only about $1,000, however. A home heater might typically have to be recharged just a few times in its useful life. Good-bye monthly fuel deliveries! Also problematic would be the occasional need to maintain or replace electrodes, which swell, crack, and change surface composition.

Generating central station electricity might entail more difficulties; it hinges perhaps on increasing the temperature at which cold fusion reactions occur. A well-known consequence of the laws of thermodynamics is the increase in efficiency with a rise in the difference between a process's operating temperature and the ambient temperature at which waste heat is rejected to the environment. The molten salt technology pioneered at the University of Hawaii may thus have a lot to offer, but there are prospects that the temperature of a heavy water cell could be run up to the boiling point and beyond. So it would be unfair to dismiss cold fusion's possible contribution to the electric power grid. Some theorists have suggested that cold fusion processes might be coupled directly to produce electricity, bypassing any thermal conversion.

Plasma physicist Harold Furth, a great skeptic of cold fusion, recently dismissed its possible role even were it to be real. In *Science* he wrote, "Electric plants of the same output ratings tend to have similar overall size—and the prospect of building power stations out of billions of palladium cells, instead of a few tokamaks, would not be a clear winner."* He ignored the possibility that titanium or some other material less costly than pure palladium might work and that each "cell" might produce substantially more than a watt of power. He also conveniently forgot to note that cold fusion, as reported, is aneutronic; it has few neutrons to degrade and make confinement vessel walls radioactive.

Just possibly there is a cold fusion-powered automobile in your future. If so, it might very well be a fusion-electric vehicle, with the fusion reactor continuously generating electricity (through a thermal cycle or directly) that would charge up advanced high-energy density chemical storage batteries. The batteries would power the automobile's electric motor(s).

Of course the real beauty of this technological Utopia would be the negligible effect on the environment—no carbon dioxide to worry about threatening even remotely global warming, no sulfur-dioxide or nitrogen-dioxide to cause acid rain, no oil spills, no mining accidents, nothing to promote lung disease, and the end of any worry whatsoever about

---

*H.P. Furth, "Magnetic Confinement Fusion," *Science*, Vol. 249: 1522–1526.

the safety of fission nuclear power. Not to forget the other joy of the cold fusion wonderland: an infinite, incredibly cheap fuel supply.

✳ Fusion of Any Flavor

Whatever does or does not develop with cold fusion, it would be absolute folly to injure what the hot fusion program has already accomplished. On this issue, there is agreement between the two temperature camps. Martin Fleischmann, in his response on April 26, 1989, to a question from Representative Morrison, testified: "I have been on the record throughout as saying that existing programs should not be affected by the discovery of some new phenomenon. The existing programs are well founded in theory, well founded in terms of the experimental results which have been obtained. Stan Pons and I share the view that we shall need fusion, the generation of fusion power, in the coming centuries—probably already in the next century—and it may well be that devices based on the research which has been carried out so far will prove to be optimal for certain types of application. If our research turns out to be successful, it could be that it turns out to be suitable for the same application or a different range of applications.

"I think it would be a mistake to narrow the options. . . . I think there will come a point in time when it is a question of trying to realize that as a demonstration unit, and, in fact, to put it into commercial practice. At that point, there has to be clearly a decision taken on which to pursue. But I would be very unhappy if the existing lines of research were affected by what we have demonstrated so far."

The director of the Princeton Plasma Physics Laboratory at the time, Harold Furth, testified the same day: ". . . let me echo a statement very eloquently expressed by Dr. Fleischmann. Namely, the overriding fact is that society needs fusion, and the great positive future to me of the recent events is that they have drawn the attention of society very vigorously, far more vigorously than we could have done, I'm sure, to this need for fusion—as the great energy crunch of the next century comes into view, and as we need to prepare to deal with it in some economical and environmentally benign manner. That's a great plus, but the immediate need is verification of the reality of the thing that we've been talking about for the last six hours . . . we should pursue the best road to fusion power, and we should make a good plan. Maybe, in view of what has been happening recently, we will pursue that plan somewhat more vigorously than fusion power has been pursued in recent years."

16 Fusion Confusion and Scientifico-Media Madness

A reporter gets into a very dangerous situation when he begins to decide what the public should know or should not know.

Jerry Bishop, April 24, 1990

Although cold fusion excited our imagination, in the end it was just another corrected mistake.

Steven E. Koonin and Nathan S. Lewis
Los Angeles Times, March 25, 1990

The safety officer wanted to shut us down, but we showed him the *Nature* papers and he said it was O.K.!

M.H. Miles of the Naval Weapons Center
Salt Lake City, March 31, 1990

✳ Giants from the Big Apple

COLD FUSION HAD ALL THE INGREDIENTS of a hot story: a possible revolutionary breakthrough, an amazing geographic coincidence in the discovery, little science battling big science, rivalries between scientists, one branch of science versus another, regional conflicts, possible errors, and more. A single word—doubt—was, however, the commanding issue of the cold fusion controversy. Assessing the validity of the Fleischmann-Pons effect—or Jones's neutrons for that matter—remained the paramount scientific and journalistic enterprise, but an unexpected spin-off benefit came about. The episode put science journalism—and journalism in general—through an extreme test.

On and after March 23, 1989, where did both the average citizen and the individual who is above the norm in scientific literacy and awareness get the most appropriate news about cold fusion? In what form did biases creep in, either on the proponent or skeptical side? How did the various factions in the scientific community use or abuse the mass media—print, television, and radio? What does the cold fusion controversy say about the peculiar relationship between the science community and the news media?

These are not easy questions and will no doubt be the subject of many dissertations in coming years in history of science departments and schools of science journalism. Indeed, less than one year after the Utah announcement, several people had already completed or begun major term papers or thesis investigations on communicating the cold fusion controversy to the public. This is not the place to conduct a *comprehensive* review, but an anecdotal survey of some interesting happenings and results illuminate many of the issues.

Plaudits for best coverage of cold fusion in newspapers must surely go to science writer Jerry Bishop, Deputy News Editor of *The Wall Street Journal*. Indeed, his series of several dozen pieces on cold fusion actually did win an award—from no less than the American Institute of Physics (AIP), which is the parent organization of the American Physical Society. His receipt of that prize is an amazing story in itself. The veteran science writer, who has covered every major science story for *The Wall Street Journal* since 1955, was selected to receive the Institute's 1990 "Science-Writing Award in Physics and Astronomy for Articles, Booklets, or Books Intended for the General Public" (a $3,000 prize) from a competing field of 59 entries. A great honor for a fellow who grew up on the largest ranch in Texas, went to journalism school, and wound up in the heart of New York City. Jerry still speaks in accents of the Southwest and sports a western hat and string tie.

Though I had admired and respected the way Jerry had covered cold fusion—having received much breaking news on the subject myself from the *Journal's* pages, when I first heard about his award I was astonished. How could the premier organization of American physicists, perhaps more than 90 percent composed of skeptics, select Jerry's cold fusion series? Was it an error in the organization's press release? Wasn't the AIP the same group that, in publishing its usually excellent annual review of all the previous year's physics happenings, had astonishingly and deliberately excluded mention of cold fusion in its 1989 year-in-review booklet? To be precise, editor Phillip F. Schewe wrote in the preface: "You may notice that the year's most prominent physics-related story is not covered in any of the chapters in *Physics News in 1989*." He followed with a one-paragraph summary of the whole affair, which said that confirming evidence for both the Fleischmann-Pons and more modest Jones results has been "scant." After bowing to the negative results reported in *Nature* and the "no convincing evidence" conclusion of the DOE panel, Schewe wrote, "Therefore it was decided not to carry a story on this subject in the current edition." On the other hand, Irwin Goodwin, senior associate editor of the AIP's *Physics Today,* told me in 1991 that in his view, "There was a rush to judgment in the affair from the beginning."

There was no mistake. A panel, composed of both respected journalists and apparently a few open-minded physicists, had made the selection. But some top physicists on the board of the AIP boycotted the award ceremony in Washington (April 17, 1990). Dr. Robert Park, at the AIP's Washington Office, whose "seance of true believers" description of a scientific meeting will live in infamy, led the protest against what was so clearly honest and reasonably well-balanced reporting of an episode in which many physicists had become involved.

The boycott was bad enough, but the award ceremony was worse. The essence of what was said: Here is your award; we're not all glad that we're giving it to you; and next time you write about cold fusion be sure to say that there's nothing to it. In presenting the award Dr. Kenneth W. Ford, executive director of The American Institute of Physics, said that it had become "increasingly clear that the Utah claims were without substance" only six weeks after the announcement. Alluding to "some controversy surrounding this award," Ford said, "There are some who regret that he did not use his great reportorial writing skills to make clearer to his readers in the latter part of the year that there is, in fact, no credible scientific evidence for cold fusion at the level claimed by the Utah researchers." He admonished Bishop that in the future he should "make clear to the general reader the fact—as now perceived by the overwhelming majority of the scientific community—that cold fusion as a practical power source is, in fact, an illusion."

A person less charming and philosophical than Jerry Bishop might have taken extreme offense at this nonsense and told off the presenter (though perhaps first waiting for the check to clear!)—using some salty Texas ranch language in passing. But the business-suited reporter, whose hair is always worn in a ponytail dangling beneath the cowboy hat, took the offense with good grace in his usual imperturbable way. Nancy Enright, a science publicist for the American Chemical Society who attended the AIP award ceremony, later told me that "it was one of the weirdest experiences" she had ever had.

The AIP quickly changed its rules to put the journalism award, like its physics prizes, under the advance review by the AIP's board of governors. As could be anticipated, journalists on the judging panel threatened to resign if the new rule were not changed.

Ironically, Jerry's award and the now infamous AIP award ceremony was followed only a week later with the presentation to him of the American Chemical Society's prestigious 1990 Grady-Stack Award for science writing. Chemists' behavior in the cold fusion affair contrasting with that of physicists once again, though this award was not specifically for Jerry's cold fusion series, rather for his long and good work in science writing. At the award ceremony held in Boston, Jerry recounted how he had found out about the Fleischmann-Pons work and became one of the first reporters to write about it.

On Wednesday, March 22, 1989, Jerry received a telephone call from the University of Utah with the message that there would be a press conference the next day announcing a major discovery. Pamela Fogle, spokesperson for the University on the matter, told Jerry that some researchers had achieved controlled fusion for a period of time that was on the order of 100 hours. Jerry told the assembled science writers, "I was sure she had gotten it all wrong."

There the matter might have rested, but Jerry happened to be having lunch that day with his colleague, the affable Ben Patrusky, who is executive director of the Council for the Advancement of Science Writing (CASW). Ben said to Jerry, "Don't you remember that three years ago we had a guy from Utah [Steve Jones] who talked about muon-catalyzed fusion?" The talk would have been at the annual meeting combined with lectures on frontier research that science writers find invaluable. But Jones was at Brigham Young University, not the University of Utah.

Upon returning to his office, Jerry contacted Jones at BYU. Jones said cryptically, "All we can say is that our results don't confirm their [the University of Utah's] results." That was more than enough to pique Jerry's curiosity and make him dispatch reporter Kenneth Wells to the next day's news conference (he probably would have gone himself had he known how big the news was to become.). The rest is history. *The Wall Street Journal* was the first U.S. paper to report the pending Utah announcement in its Thursday, March 23 edition. In fact, that was the way I myself first learned about cold fusion. A colleague at the MIT News Office, Robert DiIorio, called me late on the afternoon of March 23 from a meeting elsewhere on the MIT campus to tell me what he had just read. This gave me time to catch the already fast-moving "cold fusion train" that afternoon.

Jerry waxed philosophical on the problem of covering science stories, particularly those with significant controversial components. One school of thought, which he ascribed to reporters such as editor/writer Philip Boffey, then head of *The New York Times* science section, or Robert Bazell, science reporter for NBC television news, is that the matter has to have "some sort of 'scientific seal of approval' " before being reported. The contrasting philosophy, which Bishop holds, is that the media should be in the business of conveying information that is first and foremost, *useful.* "A reporter gets into a very dangerous situation when he begins to decide what the public *should* know or *should not* know. . . . You cannot decide what the public *should* know, but what they *want* to know," he said. In the case of the cold fusion announcement, "We thought there was an event taking place that people would want to know about."

Could there be any doubt that the public–including physicists, chemists, and other scientists–were intensely curious about the Utah announcement? Did not "*The Wall Street Journal of Physics*" become a very useful information medium, whose reports one could credit or disbelieve as one wished? By his own definition, Jerry's continuous and extensive reporting on cold fusion filled an information need. Some physicists later came to believe that cold fusion was getting too much coverage, and thereby perhaps achieving validity it did not merit. Jerry's philosophy was not to ascertain validity where such divining was impossible, but to let the public decide for itself after being presented a portrait that was as balanced as possible. What could be more fair?

The *New York Times* might be forgiven for its caution and relatively slow uptake in following the cold fusion story. After all, it had much more seemingly urgent local matters to contend with. On March 23, 1989, the day of the Utah announcement, banner headlines spanning the full width of the paper read, "Justices Void New York City's Government; Demand Voter Equality in All Boroughs." Ironically, the subheader read "Confusion Growing" and the reporter's name was Linda Greenhouse. On March 24, however, there was no such compelling reason to relegate the amazing cold fusion story to page 12, where the first *Times* cold fusion story wound up. The diminutive headline, "Nuclear Power Gain Reported, But Experts Express Doubts," ran under a photo of Iran-Contra affair figure Fawn Hall and her mother arriving at court in Washington. Perhaps in a portent of the very measured and cautious coverage the *Times* would give to cold fusion, the paper managed not even to use "fusion" in the banner.

The story by veteran science writer Malcolm W. Browne largely dismissed the Utah report based on lack of information from Utah and on anonymous comments by physicists at Los Alamos National Laboratory. Yet in the *Times* March 25 issue, when more details were clearly available, there was no cold fusion story to quench the thirst of the public for information. The front page on March 25 did herald the Exxon Valdez oil spill in Alaska's Prince William Sound.

But the *Times*, in its usual ability to focus extraordinary resources on a great issue, in due course managed for some time to present a generally balanced view of the cold fusion war. In fact, a year later, when William Broad of the *Times* covered the Utah conference on cold fusion (March 1990), if anything he gave more credit to the cold fusion claims than did Jerry Bishop, properly noting that cold fusion continued to be an unresolved mystery. However, he stumbled in October 1990, with an extremely negative retrospective on the cold fusion affair, "Cold Fusion Still Escapes Usual Checks of Science," which did not even mention the important conference at Brigham Young University on nuclear effects in cold fusion that was held only a week earlier. The *Times* coverage that week devoted itself instead to the supposed unexplained "disappearance" of Stanley Pons.

Many papers in the country took their cues on cold fusion directly from the *Times* or the *Journal*, as well as the newswire services, the Associated Press and United Press International. Papers that were near the sites of the breaking news—the frenzy of attempts around the country to verify the phenomenon—generated their own stories. One could learn much from these local reports, even with their provincial flavor—sometimes more than could be gleaned from the national papers. For obvious reasons, especially rich with cold fusion coverage were the two Salt Lake City papers, the *Deseret News* and the *Salt Lake Tribune*.

✳ The Problem

The basic difficulty in covering any controversial issue in science is twofold. First a reporter needs to understand enough of the underlying scientific issues to establish a basis for evaluating claims and counterclaims. This is ordinarily tough enough, even when the science is not as complex and as subtle as it is in cold fusion. Second, there is the journalistic need to appear *reasonably* impartial in a battle between warring scientific camps. The goal of completely impartial coverage of any controversy, however, is well recognized by science journalists as unattainable—and it perhaps should not even be sought. A reporter brings her or his biases to a story and is swayed in one direction or another as it unfolds. It is the reporter's job to track that trend, all the while keeping a wary eye on the chance that she or he may have misjudged it, or that the viewpoint that appears to be "winning" the majority of the scientific community may not really be the correct one. What is inexcusable in science reporting is not to mention the existence of an opposing view—even if that view appears to have the weakest support in the scientific community.

Early in the cold fusion controversy, most reporters seemed to adhere to these general points, and the coverage was by-and-large balanced. The opposition of many physicists to cold fusion was evident in much reporting, though perhaps there arose too strong a chemists versus physicists storyline. The distinction between hot fusion and cold fusion was made abundantly clear to the public and the potential consequences for humanity if cold fusion proved to be real were neither exaggerated nor ignored. Hot fusion, however, unquestionably did not get enough praise for its majestic efforts and lofty goals. Any time reporters hear about billions of dollars being spent with no practical working device in sight for decades, there is bound to be trouble. The time horizon of many journalists is often as limited as that of politicians. Scientists' plans are often long-range, requiring years to be carried out. There is usually no magic way to accelerate a particular development—*unless* a genuine breakthrough occurs.

Sometimes understandable confusion crept in on the difficult scientific issues in cold fusion. It was all too easy for reporters not familiar with some of the esoteric theories that were being put forth to explain the enigma, simply to ignore them. That certainly happened. This left a "power vacuum," as it were, in which science journalists gave too much weight to the unfair demand from skeptical scientists to have *immediate* evidence of what nuclear products were causing the alleged excess heat.

Some journalists gave far too much weight to the controversy about how the cold fusion discovery was announced and the supposed lack of adequate peer review; the pejorative "science by press conference" charge became a notion that journalists bought too easily. If Fleischmann and Pons had kept to themselves for five-and-a-half years, that should have been enough to convince journalists that the Utah duo were not rank charlatans out to conquer the world. How the news of possible cold fusion managed to be held so closely for such a long period was the more extraordinary question. The peripheral issues of patent rights, business deals, and secrecy were often elevated by implication to bona fide reasons for suspecting shady science. And as for peer review, it was happening right before the journalists' eyes—a welcome invitation to observe the scientific review process in real time. It was not an abdication of responsibility.

Contrary to the railing of many skeptical scientists against the Utah press conference, such interactions between scientists and the public are not uncommon and follow a long tradition. As Professor Marcel C. LaFollette of George Washington University remarked about the cold fusion announcement:* "The history of United States science journalism, however, clearly shows that such interactions are common. From the earliest part of this century, scientists have participated enthusiastically in telling the public about science. They have described their own research, analyzed (and criticized) the work of colleagues, and given free-wheeling interviews to journalists, promoting great results from

---

*Marcel C. LaFollette "Scientists and the Media: In Search of a Healthier Symbiosis," *The Scientist*, July 9, 1990: 13.

small research. Moreover, scientists have done so willingly, despite an atmosphere in science that does not always regard popularization with approval."

Examples abound of what LaFollette wrote about. For example, the hot fusioneers have had their own media blitzes too—apart from the well-known "Zeta" brouhaha in Britain in 1958 (Chapter 5). A report on the rise and fall of one amazing forgotten news event appeared in *Science* in September 1978. The media whirl for hot fusion that occurred on the weekend of August 12–13, 1978, apparently via an unauthorized "leaked" report by an overly eager promoter, led to the banner headline lead story in the Sunday *Washington Post*: "U.S. Makes Major Advance in Nuclear Fusion." As *Science* reporter William Metz described it: "Radio and TV stations throughout the weekend reported the story with the urgency of an international crisis, and by the end of the 2-day media blitz, many citizens apparently got the impression that after years of waiting for proof, fusion had finally been achieved. The message was so strong and so positive that it seemed—for 48 hours at least—that the energy crisis was over, solar energy and nuclear power were no longer needed, and that the future would be assured through fusion." Alas, on Monday the DOE sponsor of the Princeton tokamak fusion research acknowledged that there had been a significant advance in plasma temperature, but in view of other technical problems it should not be considered a major "breakthrough." What cooled off the media fusion fire? You guessed it, a Monday afternoon press conference! Why the news then? You guessed that too; fusion funding was at that time up for internal DOE review.

The basic problem in cold fusion coverage in 1989–90 may have been that so much contemporary science is incremental and plodding in its accomplishments, that people ignore the longer historical perspective in which breakthroughs—paradigm shifts—*do* punctuate normal science.

The most obvious shortcoming of cold fusion reporting was the general media's loss of interest following their initial few months of intensive coverage. Cold fusion, like the man with the dog that could climb a tree, had had its glorious "15 minutes of fame." After *Nature* magazine and the DOE panel had rendered their negative verdicts in the summer of 1989, precious little was heard of cold fusion. Many science journalists simply bought into the *Nature*-DOE panel line and gave up. And why not? So thick had been the disparagement of Fleischmann and Pons and all their followers, that the mud stuck. It became "socially unacceptable" in the science journalism community to give too much weight to any of the cold fusion rumblings that continued to be heard. Few made an effort to ask what those noises might mean. Just as many good scientists had "burned out" in chasing the elusive cold fusion Genie, so had many science journalists. They were sick of the ups and downs, the lack of a clear decision after so many months, and with good reason feared ridicule if they pursued the continuing strange scientific reports.

The power vacuum was filled with the opposition viewpoint of the hot fusioneers. By late 1990 the journalistic "consensus view" had solidified to: "There is probably nothing to cold fusion, but even if cold fusion is real, it probably won't be very useful." An example was the cautiously worded assessment that respected science writer William J. Broad of *The New York Times* included in his October 9, 1990, update on *hot* fusion: "The allure of 'cold' fusion was that it seemed to promise enormous energy from simple devices that worked at room temperature, in contrast to hot fusion machines, which must operate at temperatures above those on the Sun, and are vastly complicated and expensive. But after a year of intense investigation, most experts have dismissed the notion that cold fusion, if it exists, will ever be a significant energy source."

Astronomer and noted author Carl Sagan gave *his* perspective on cold fusion and its coverage in the press when he responded to a question posed at a gathering of science writers at Cornell University in November 1989. "In the case of cold fusion," he said, "we have a contention that you do something with palladium and with some hydrogen isotopes—on a table top, at room temperature—and you can make fusion happen, or at least generate fusion products, or at least make some heat that otherwise can't be generated. That's the contention. And it may or may not have ultimate commercial applications, which is why everybody is interested in it, not because there might be some novel physical process.

"Now how do we decide that?" he continued. "Do we decide it by *polling* the membership of the American Physical Society? No! Polls don't work. They might not be knowledgeable or the minority might be right; it's happened many times in science. Do we write an article saying, 'Well, there is a disagreement, but the prevalent opinion is thus and so?' No. What we do is we say, 'The scientists don't know! They can't figure it out.' Some people say this thing, some people say that thing—too early to say! Let's wait a few years. I guarantee that five years from now, this will be a dead issue. It will either be, there is such a thing or there isn't such a thing. We will not be sitting in some middle ground wondering. The stakes are too high. Either way, the definitive *disproof* of Fleischmann and Pons or the definitive *proof*. The rewards are so great that scientists—competitive, querulous lot—will decide one way or another."

✳ Cold Fusion and Superconductivity at High Temperature

The incredible cold fusion story that began in 1989 followed close on the heels of the high-temperature superconductivity media event in 1987. The last day of 1986 brought a *New York Times* headline announcing the major advance in superconductivity that had come from an IBM laboratory in Switzerland, which was followed by pioneering work at a laboratory in the United States. After 75 years of dealing with superconductivity in metals at temperatures never far above absolute zero, physicists had discovered how to bring about the flow of electric current with zero-resistance in special ceramic materials at substantially higher temperatures. Superconductivity promised to be seen eventually at room temperature.

Many observers have compared the high-temperature superconductivity affair and cold fusion, invariably making the obvious point that, unlike cold fusion, high-temperature superconductivity proved to be very reproducible if the published, peer-reviewed recipes were followed. Even high school students could and did tinker with it. (In 1989–1990, there were also reports of a few school science fair projects on cold fusion!) It seems, however, that no one stopped to analyze the differences in the media handling of these two scientific events, both of which promised revolutionary technological changes. But John Travis in pursuing his bachelor's degree at MIT did just such a careful analysis, drawing a solidly supported conclusion that flies in the face of expectation: "Unlike the press coverage of cold fusion, which has been inappropriately labeled irresponsible by many, the coverage of superconductivity was inadequate. The skepticism and cautious language that should have been used in such an important development was sorely lacking." The superconductivity headlines were replete with claims of a coming revolution in efficient electric power transmission, fantastic new electric motors and generators, compact high-speed computers just around the corner, and an electric car in every garage. As fantastic a development as high-temperature superconductivity is, these applications have not turned out to be for the near term.

Science journalists, who began to realize that their enthusiasm had gotten the best of them in high-temperature superconductivity, were thus doubly or triply cautious in handling cold fusion. John Travis concluded that, by-and-large, coverage of cold fusion was balanced compared to coverage of high-temperature superconductivity, and that some of this balance may have come from their recent experience with the latter story.

Another difference: High-temperature superconductivity was an acceptable, imaginable extension of previous work at low temperature, and it had a ready-made stable of scientists for reporters to talk to. John Travis remarked, "High-temperature superconductivity had established experts, Nobel laureates in fact, that the journalists could question, while cold fusion queries were usually directed to hot fusion people who found the whole idea improbable. Ironically, this difference prompted poor reporting of superconductivity. There was an unhesitating acceptance of scientific opinion." Established scientists in superconductivity gave reporters poor information, mainly because they failed to understand the properties of superconductors that are important in *applications.*

✳ A Dash Through Media Land

If the American Institute of Physics was panning cold fusion–generally giving it the cold shoulder in *Physics Today* and excising it from its annual review of physics, the American Chemical Society could not be said to have "promoted" it. But the society at least gave the affair ample coverage in *Chemical & Engineering News* and handled it with balance. In the second sentence of the introduction to "What's Happening in Chemistry?–1989" is the acknowledgment: "Despite the fact that 'cold fusion' is being hotly debated, it was one of the most widely reported science stories this past year and perhaps of the decade." In fact, the respected Institute for Scientific Information in Philadelphia had queried its spectacular data base and found cold fusion to be the highest ranked research area in 1989 in terms of scientific citations!

The *Boston Globe* in MIT's backyard covered cold fusion intensively at first and then sporadically, as public interest in the subject waned. One *Globe* reporter managed to get a front-page story out of the backroom sniping that was going on within MIT about Peter Hagelstein receiving tenure. This appeared shortly before Hagelstein gave his first public talk on cold fusion at a December 1989 scientific meeting in San Francisco. The following spring he received tenure, but neither the *Globe* nor the *Boston Herald* appeared to take note of either this or the content of his theory. *Globe* columnist Alex Beam wrote a humorous essay on cold fusion that was reprinted in the *Salt Lake Tribune* (May 5, 1989). It ended on a wild note: "Barely five weeks old, cold fusion, the most delectable scientific non-event since Kahoutek's Comet, is fizzling fast. As reports debunking the improbable breakthrough pour in from labs around the country, 'fusion in a jar' is looking like a latter-day Veg-O-Matic, the kitchen appliance that works when you see it on TV, but not when you get it home."

In the early days, *Time, Newsweek,* and *Business Week* all produced flashy cover stories on cold fusion–perhaps the first time that the tiny apparatus of small science had made it onto the cover of weekly news magazines, but thereafter the topic slid into oblivion as far as weeklies

Cartoon by Danziger of *The Christian Science Monitor*, April 24, 1989, mocks a frequently discussed division in the scientific community. (Copyright 1989 TCSPS)

were concerned. The London-based *Economist* did very fine reporting on cold fusion, as did the exotic pronuclear publication *21st Century Science & Technology*, which is operated by followers of quixotic politician of the right Lyndon LaRouche. No doubt to the discomfort of hot fusioneers, his movement has long boosted the need for fusion power—often by pamphleteering at airports. Robert Pool at *Science* magazine gave cold fusion excellent and generally balanced coverage—until the magazine published the Taubes "exposé" about possible tritium spiking at Texas A&M. In November 1990, Pool likened cold fusion to the "Cheshire Cat" of *Alice's Adventures in Wonderland*, boldly suggesting that it had faded away, leaving only the grin—the possibility of low-level neutron emissions. He asserted that tritium reports had a "mundane explanation" and that claims of excess heat production were likely explainable by known physical processes. Pool told me in 1991 that he personally does not think there was really tritium spiking at Texas A&M. Ivan Amato's stories in the weekly outstanding science publication *Science News* were exemplary throughout the cold fusion affair.

Nature has already been discussed. It was good to see the skeptical viewpoints and negative results laid out in this ordinarily indispensable journal, but the magazine's silence on positive results bubbling up all

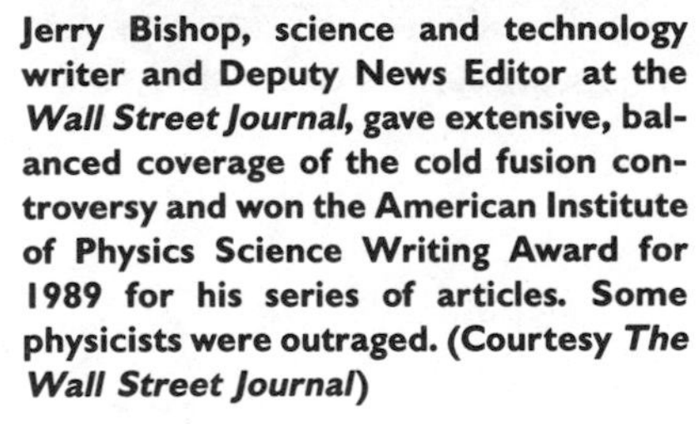

Jerry Bishop, science and technology writer and Deputy News Editor at the *Wall Street Journal*, gave extensive, balanced coverage of the cold fusion controversy and won the American Institute of Physics Science Writing Award for 1989 for his series of articles. Some physicists were outraged. (Courtesy *The Wall Street Journal*)

John Maddox, the strong-willed editor of *Nature* magazine, whose negative appraisal of cold fusion research led to scathing editorials against it. (Photo by Zoë Dominic, London)

around was deafening. Many examples of outstanding scientific work have not received entrance to *Nature* or other prestigious publications, yet they ultimately proved their worth. No less than fission pioneer Enrico Fermi was turned down by *Nature* in 1934. His wife, Laura Fermi, wrote in her memoir of the incident (*Atoms in the Family: My Life with Enrico Fermi*, Laura Fermi, Chicago, University of Chicago Press, 1954): "He felt the need for a vacation from theoretical work, having just completed an abstruse theory on the emission of beta rays from nuclei in natural radioactive processes. This theory was soon considered one of his major works, but at the moment it was causing annoyance and disappointment. The scientific magazine *Nature*, to which Enrico had sent his paper, had turned it down with the statement that it was not quite suited to that magazine. Enrico's 'Tentative Theory of Beta Rays' was consequently published in Italian, in the *Ricerca Scientifica* and in the *Nuovo Cimento*, and soon afterward in German in the

Zeitschrift für Physik, but not in English." All of which shows, "The more things change, the more they stay the same."

History will likely record that a major scientific publication with a long and proud tradition evidently lost its patience with a new and difficult phenomenon. The core thinking of *Nature* editor David Lindley was revealed in a comment that he made to John Travis, which appears in John's thesis: "Once it became clear that [Pons and Fleischmann] were not entirely open and honest in the way that scientists frequently are, they were fair game for muckraking journalists." A *Nature* editorial in November 1990 suggested that for the cold fusion "soap opera" there is "ample opportunity to make fun of these goings-on [the politics surrounding Utah's NCFI], and in the light of the history of cold fusion no reason not to." *Nature* compared cold fusion's "demise" to that of "other failed revolutionary movements"—with "Marxist-Leninists" battling "Leninist-Marxists" no less!

Editor of *Nature*, John Maddox, was perhaps even more revealing when he stated, on the "NOVA" television program, "Confusion in a Jar," which was broadcast in early 1991: "I think it will turn out after two or three years more investigation that this is just spurious and unconnected with anything that you could call nuclear fusion—thermonuclear fusion. I think that, broadly speaking, it's dead, and it will remain dead for a long, long time."

17 *Hard Lessons in Science*

There is another way to truth: by the minute examination of facts. That is the way of the scientist: a hard and noble and thankless way.

John Masefield, 1924

Science is one of the few areas of human life where the majority does *not* rule.

Samuel C.C. Ting
Cowinner of the 1976 Nobel Prize for Physics

Scientists suffer, along with the rest of us, from the ironies that evil sometimes comes from good, that one noble vision may exclude another, and that good scientific ideas occasionally obstruct the introduction of better ones.

Bernard Barber
Science, September 1961

THOUGH YET TO RUN ITS FULL COURSE, the cold fusion controversy has taught reluctant students some tough lessons about the process of science. Once again an intense conflict has brought many cherished institutions and assumptions of science into sharper focus. For many years, historians and philosophers of science will ponder what the cold fusion story has taught us.

✳ Resistance to Paradigm Shifts

A certain level of inertia is critical to science. Science would not be science if it were torn and stretched every which way by weekly claims about contradictions and errors in its framework. Yet science also lives and grows by revolution, as its history exhibits with spectacular examples in almost every decade. Think of relativity or quantum mechanics, consider the discovery of the structure of DNA, the expansion of the universe, and on and on—all revolutions within the 20th century.

Science historian and philosopher of science Thomas S. Kuhn's book, *The Structure of Scientific Revolutions,* is famous for its thesis about the process of paradigm shifts. The cold fusion controversy was a classic example of such a potential shift in the making. If cold fusion is a real new phenomenon, this was a shift in the making; in the event that it is ultimately shown to be a mind-boggling concatenation of mistakes, it will have been a paradigm shift that *tried* to happen.

If the foregoing story hasn't convinced you of this, look at this one example: Early in the controversy, Robert L. Park, Professor of Physics at the University of Maryland and an official in the Washington office of the American Physical Society, published a scathing opinion piece (*The Chronicle of Higher Education*, June 14, 1989) in which he chastised Fleischmann and Pons for what he characterized as incomplete and improper disclosure of their findings. He wrote with great confidence: "The most frustrating aspect of this controversy is that it could have been settled weeks ago. If fusion occurs at the level that the two scientists claim, then helium, the end product of fusion, must be present in the used palladium cathodes." Then he criticized Fleischmann and Pons for allegedly refusing to have their rods analyzed for helium content.

Park was imposing the paradigm of hot fusion reactions on the Fleischmann-Pons experiment to "require" a certain piece of evidence, in this case the presence of helium-4, in order to be convinced that cold fusion was occurring. It turned out, as we have seen, that theorists came up with very interesting mechanisms, unlike those in hot plasma fusion, that could explain heat-producing cold fusion without requiring helium-4 as a reaction end product. In retrospect, Park's confidence in the ease with which the controversy could be settled mere "weeks ago" was entirely misplaced, as even many diehard skeptics would now agree. Almost two years later he seemed no more enlightened when he told me, "You don't need to worry about the heat if there is no helium."*

Resistance to new scientific ideas is nothing new in science. The ample evidence of this that sociologist Bernard Barber compiled in his famous September 1961 *Science* essay is convincing. Two of the lesser-known embarrassments cited: Pasteur's discovery that fermentation was biological in character was resisted by those who clung to established theory that it was a purely *chemical* process; and Lord Kelvin's denial

---

*Weeks later, *M. H.* Miles and others announced their spectacular results showing an association of helium-4 production with proportional levels of excess heat in cold fusion cells. (See Bibliography.)

that helium could be produced from radium (ordinary radioactivity) and his lifelong belief in the indivisibility of atoms.

Max Planck's often-quoted remark about the opposition his radical ideas faced shows a lesson bitterly learned, though his sentiment is clearly not universally true: "This experience gave me also an opportunity to learn a new fact—a remarkable one, in my opinion: A new scientific truth does not triumph by convincing its opponents and making them see the light, but rather because its opponents eventually die, and a new generation grows up that is familiar with it."

✳ The Majority Fails to Rule

In a practical sense, there is no such thing as "scientific truth," though most scientists believe that a final, objective truth exists that transcends our ability to know it. All we have is the ability to endlessly critique and examine assumed facts and theories that have come into the pantheon of "generally accepted scientific truth." That the theories of science are subject to being proved incorrect or incomplete, that is, are falsifiable unlike religious dogma, is what science is all about. There is always the chance that even a well-established concept will be overturned, so it is not reasonable to deny a platform—however small—to rationally behaving scientific critics. Alfred Wegener's theory of continental drift comes to mind—a much-disparaged scientific notion through the first half of this century, which became commonly accepted geophysical wisdom as late as the 1960s.

There is, however, a difference between scientific consensus and polling. Consensus in science emerges naturally, sometimes gradually, sometimes in fits and starts. It cannot be achieved by setting up a panel of supposed experts, whose job it is to pass judgment on the validity of extraordinary claims. If those claims are still under active investigation, they will remain so, despite the publicly supported votes yea or nea on a question, or in ceremonial demonstration "votes" such as occurred at the APS meeting in May 1989.

✳ Dangerous Analogies

One of the worst approaches to dealing with a scientific controversy is to cite battles of the past to illustrate how wrong the present proponents of a new idea may be. Conversely, it is not without hazard to suggest that because many scientific revolutions in the past were successful, that a present *potential* revolution is likely to occur. In the final analysis, each controversy has to be judged on its own merits, but it is wise to keep an eye on past history. In the cold fusion controversy, critics repeatedly brought up failed scientific theories of the past to show how wrong cold fusion was likely to be. This was "guilt by association" of the worst kind and was connected with an even more pathological notion—pathological science. Some of the favorite stigmas with which to skewer cold fusion were: the nonexistent "N rays" that made their appearance around the turn of the century, soon after the discovery of X rays, and the great "polywater" hunt of the 1960s and 70s, which turned up not dangerous or revolutionary polymerized water, which some feared might congeal the world's oceans into a deadly stew and others hoped would revolutionize technology. Polywater was found to be nothing more than contamination of the experiments by ill-defined impurities.*

✳ The Pathology of "Pathological Science"

The concept of "pathological science" began with a talk given by Nobel laureate in chemistry Irving Langmuir at General Electric's Knolls Atomic Power Laboratory in December 1953. At the height of the cold fusion controversy in October 1989, *Physics Today* reprinted his transcribed talk. A few other writers skeptical of cold fusion, such as Douglas Morrison at CERN, also took up the reincarnated psychosocial theory and it soon became an acceptable form of scientific defamation. I believe, however, that Morrison himself personally has acted in a gentlemanly and respectful way to those he was criticizing. He pitied them and perhaps even admired cold fusion followers for their boldness, which is not to say that he did not cause them great harm.

Langmuir recounted a number of interesting scientific misadventures in which he claimed to perceive a recurrent pattern of self-deception followed by various phases of denial. Among these instances were the advent of Blondlot's N rays in 1903, mitogenetic rays in 1923, the "Allison Effect" in 1927, the incorrectly interpreted Davis and Barnes experiment in 1929, and on through extrasensory perception and UFOs. It's not that his analysis of each of these matters was far off the mark, rather it was the drawing of blanket conclusions about the supposed characteristics of "pathological science" that was troublesome. The characteristic symptoms of pathological science as outlined by Langmuir were neatly summarized by the editors of the *Physics Today* piece:

1. The maximum effect that is observed is produced by a causative agent of barely detectable intensity, and the magnitude of the effect is substantially independent of the intensity of the cause.

---

*For an excellent treatment of this incredible episode in science, read *Polywater*, by Felix Franks, MIT Press, Cambridge, MA, 1981.

2. The effect is of a magnitude that remains close to the limit of detectability or, many measurements are necessary, because of the very low statistical significance of the results.
3. There are claims of great accuracy.
4. Fantastic theories contrary to experience are suggested.
5. Criticisms are met by ad hoc excuses thought up on the spur of the moment.
6. The ratio of supporters to critics rises up to somewhere near 50 percent and then falls gradually to oblivion.

It would take a lengthy treatise to relate all the reasons why it might or might not be said that cold fusion fits this pattern. On balance, I think that cold fusion is far from a good fit. For one thing, there is a very solid, stable, and diverse group of "supporters" of cold fusion that continues to do pioneering work, while uncovering greater and greater dimensions of the phenomenon. They aren't simply retreading already covered ground and are not declining in numbers. Nor has anyone claimed any extraordinary accuracy for cold fusion results (Point 3). If anything, there has been the free admission by proponents that measurements are tough, erratic, and not always reproducible. Yes, fantastic theories have been proposed (Point 4) to explain cold fusion—theories "contrary to experience." But wasn't Einstein's special relativity a fantastic theory "contrary to experience"? "Excuses" for the problems still being encountered in cold fusion are not ad hoc (Point 5); they have been carefully considered in a systematic way by experimenters and theorists alike.

Points 1 and 2 do not apply to cold fusion unless one really tries for a forced fit. Even skeptics are beginning to acknowledge that the excess heat is occasionally real, though they are loathe to give it a nuclear explanation. And many skeptics agree that low levels of neutrons *have* been detected in some experiments. Furthermore, looking at Point 2, one sees that it perfectly characterizes neutrino detection experiments (many measurements necessary and low statistical significance of the results)—very much part of the main line of physics experiments and by no means pathological.

Self-deception is, indeed, a problem in science, but one must be extremely careful in suggesting that it has occurred, particularly when the alleged "self-deception" begins to occur in a wider and wider group of serious scientists—as has happened with cold fusion.

There are two ironies to the 1989 resurrection of Langmuir's thesis. One is that Langmuir was the fellow who gave hot fusion plasmas their name; he wasn't thinking of fusion at the time, but comparing ionized gases with blood plasma. Lesser known is that Langmuir engaged in quite a piece of dead-end science himself—that some would call "pathological." He was famous for his initiation and support of attempts to seed clouds to produce rain. This multiyear program that cost many millions of dollars and did not meet with discernible success in initiating precipitation, eventually petered out—just the way pathological science is supposed to!

What is particularly galling about charges of "pathological science" is that its levelers occasionally turn out by instinct, not by delving deeply into experiments, to have chosen the winning side. Felix Franks mentions this in his account of the polywater episode: "The 'experts' were once again proved right, after they had arrogantly asserted all along that polywater did not and could not exist. These assertions were based more on hunch than on reasoned argument or study of the evidence. Even those who had worked hard and persistently to disprove [polywater] claims were galled by the high-handed way in which members of the establishment dismissed the experimental evidence and yet turned out to be right in the end. The members of the elite are far from infallible, but they are conservative, and in science conservatism pays off more often than not."

I suggest that the study of "pathological science" makes an interesting parlor game, but that it is definitely not science and it should most certainly not be invoked to defame conscientious work by scientists. For all we know, Langmuir himself, if he were alive today, would be rolling up his sleeves and sweating—neutron detector in hand—over cold fusion cells.

✳ Ockham's Razor

The 14th-century medieval scholar William of Ockham (1285–1349) left an important legacy to a community of science that had not yet been born: the ideal of simplicity in scientific explanations. "Entities must not needlessly be multiplied," he wrote. His crisp dictum, which has attained wide currency among scientists as "Ockham's razor," urges the initial choice of a simple explanation to encompass complex evidence, rather than a host of ad hoc theories. His theory-shaving device (hence the term, razor) does not always work, of course, but it is worth thinking about when science is confronted with a bewildering array of seemingly contradictory experiments. The hazard is that Ockham's razor can sometimes be a two-edged sword—erroneously forcing simplicity on an inherently complex situation. On the other hand, if cold fusion pans out as a real energy-producing nuclear phenomenon, the virtue and fame of Ockham's razor may reach new heights. Scientists faced with a conundrum the likes of cold fusion would do well to heed Ockham.

✳ Theory Versus Experiment

A cartoon once posted on an MIT biologist's door was captioned, "Theory, like mist on eyeglasses, obscures vision." There is, indeed, an age-old tension in science between theory and experiment. Which should lead the wagon of science? Perhaps both should, but culturally and organizationally it isn't that simple. Some scientists specialize in one or the other approach, less frequently in both.

In cold fusion, we witness a full-fledged collision between experiment and theory—between two distinct approaches in the house of science. Some theorists refused to believe the results of experiments, which their theories could not explain—in particular, the finding of excess power and energy without readily apparent nuclear products to account for it. Experimentalists who found anomalies politely told these theorists to go back to their notebooks, but few did. It was easier for skeptical theorists to check out well-known physical mechanisms, satisfy themselves that these could not explain cold fusion, and believe confidently that the experimental results had to be in error.

Stanley Pons in his April 1989 testimony to Congress addressed the issue: "I think it's always dangerous to point at incorrect experimental data being based on theory. I think theory must be used to *explain* experimental data, not to *criticize* experimental data. If it's a well-established theory, then certainly you can raise questions. But I think that you need to consider first that the experimental data must be duplicated and explained, and then a theory put forth, rather than just saying your data must be wrong because the theory doesn't predict that."

John Bockris, speaking at the spring of 1990 American Chemical Society meeting in Boston echoed Pons: "One phenomenon has become so familiar to us that we've now forgotten that we don't understand it—the conductivity of organic compounds. It's now about 20 years since people began finding high-electronic conductivity in polymers, and to this day you cannot at all predict which compound will give it. There is no theory after 20 years. After three years, high-temperature superconductivity in ceramics is still not understood. So the fact that we don't yet understand how it's possible to get cold fusion—and there is a proliferation of theories, none of which are very convincing—shouldn't deter us. We should look only at the facts and we should keep basic scientific principles in our minds. Keep cool and gather good facts. Then when we've got the facts, and we've learned how to control it, then is the time to change our physics, if we have to to explain it."

Certainly not referring to cold fusion, condensed matter theoretical physicist Philip W. Anderson of Princeton similarly warned his colleagues with an essay in *Physics Today* (September 1990), "Solid-State Experimentalists: Theory Should Be on Tap, Not on Top": "We don't want to lose sight of the fundamental fact that the most important experimental results are precisely those that do *not* have a theoretical interpretation; the least important are often those that confirm theory to many significant figures. . . . The prejudice in favor of pat 'interpretation,' no matter how anomalous the observed phenomena, is particularly stifling when, as in journal refereeing and grant reviewing, it is essential to get consensus: Originality and independence of mind are least to be found in a committee."

One of the greatest lessons of cold fusion is that both experimenters and theorists should listen with great care to what each other has to offer, particularly in areas of deep controversy. Experiment is the hand of science, theory its mind.

✳ Peer Review

The peer review system in science does not operate monolithically to control information flow into each of the world's 40,000 or so scientific publications. In fact, no one can define exactly what peer review is, so uneven is its application! Some journals exercise loose control and researchers find it relatively easy to get their papers accepted for publication without inordinate checking and second-guessing by peers. But in some of the more prestigious journals, those that confer considerable honor and status for researchers published in their pages, the weeding-out and control process is much tighter and more stringent (not necessarily more fair). If there are obstinate reviewers and editors assigned to a particular review, or those with an axe to grind, an article will not see the light of day no matter how many objections are answered. If a paper presents a serious challenge to prevailing scientific opinion in presenting unexpected findings or radically new theories, it is certain to undergo greater scrutiny before being published in these prestigious journals than is a paper with less radical claims. Less extraordinary assertions—just as prone to error—usually have smoother sailing.

No one would deny that it is desirable in scientific publication to have a workable editing function—reviewers checking for obvious errors, apparent internal inconsistencies, or incomplete information. However, peer review has many times served to keep revolutionary data and theories from reaching a wider audience, for no other reason than the unwillingness of conservative scientists to allow such "obviously wrong" concepts to compete in a marketplace of scientific ideas. Since truth ultimately wins out in science, peer review cannot prevent a correct new idea from eventually being vindicated, but it can severely delay its acceptance.

This restriction function of peer review was clearly operating in certain publications during the cold fusion controversy. *Nature* magazine, for example, accepted only a very small number of experimental or theoretical papers that credited the claims of cold fusion proponents. On numerous occasions researchers with "positive" results sought to have their work published in *Nature*, to no avail. The reviewers and ultimate editorial judgment of *Nature* seemed to put a premium on showing how wrong the whole business was.

Even Nobel laureate in physics Julian Schwinger found it rough going to get his theory about cold fusion published. A prestigious journal of physics finally agreed to take his work, but only with a disclaimer. Schwinger's paper, "Nuclear Energy in an Atomic Lattice. Part 1," appeared in *Zeitschrift für Physik.** Attached was this prominent apologia, an Editorial Note:

> "Reports on cold fusion have stirred up a lot of activity and emotions in the whole scientific community as well as in political and financial circles. Enthusiasm about its potential usefulness was felt but also severe criticism has been raised. If in such a situation one of the pioneers of modern physics starts to attack the problem in a profound theoretical way we feel that it is our duty to give him the opportunity to explain his ideas and to present his case to a broad and critical audience. *We do, however, emphasize that we can take no responsibility for the correctness of either the basic assumptions and the validity of the conclusions nor of the details of the calculations. We leave the final judgement to our readers.*" [author's italics]

As though readers of scientific papers don't ordinarily exercise "final judgement!" This could be called the "Surgeon General's warning" approach to publication. Schwinger had earlier submitted another paper, "Cold Fusion: A Hypothesis," to *Physical Review Letters.* Of this episode he told me, "Although I anticipated rejection, I was staggered by the heights—or depths—to which the calumny reached. My only recourse was to resign from the American Physical Society." And he did.

Dr. Charles McCutchen, a creative physicist who worked for many years at the National Institutes of Health on biomechanical theories and experiments, characterized peer review as an "evolved conspiracy." Having himself suffered from peer review run amok, in 1976 he wrote, ". . . suppression of novelty by reviews is not a plot cooked up by referees and the establishment. But conspiracies can arise by evolution

**Zeitschrift für Physik*, D, Vol. 15, 1990: 221–225.

instead of by design, with the members falling into their roles by accident and finding them congenial." (*New Scientist*, April 29, 1976.)

McCutchen cited a litany of famous cases where peer review went astray: "Referees are supposed to despise error and cherish novelty. In fact they have suppressed important discoveries. F. W. Lanchester's circulation theory of aerodynamic lift was held up for ten years. J. J. Waterston's work on the kinetic theory of gases anticipated Maxwell by 12 or 13 years, and Boltzmann by 21 years. [His work is now vindicated, but Boltzmann committed suicide, in part because of attacks on his theories by holdouts who believed the world not to be made of atoms. This in 1906!] It was published 47 years after submission, only because Rayleigh found the manuscript in the archives of the Royal Society. It was 'nothing but nonsense, unfit even for reading before the society,' according to a referee. Publication of Krebs's citric acid cycle was delayed also."

The famous case of the laser also comes to mind. Theodore H. Maiman's paper announcing the construction of the first laser in 1960 at Hughes Research Laboratories was first rejected by *Physical Review Letters*, after which Maiman (correctly, it seems) resorted to the now-disparaged press conference route to announce his wondrous discovery and claim priority. Donald A. Glaser's effort to construct the first nuclear particle-detecting bubble chamber, which won him the 1960 Nobel Prize in Physics, was held up by funding agencies because his plan was considered "too speculative."

Science is a very organic process and invariably finds ways to work around restrictive policies at one journal that is impeding a new idea. For cold fusion, the *Journal of Electroanalytical Chemistry* became a hot-bed for cold fusion articles—obviously because its electrochemist editors, if not cold fusion proponents, were at least sympathetic to other electrochemists and their followers. An editor with vision, Dr. George Miley, opened up the pages of his publication *Fusion Technology* (normally reserved for hot fusion matters) to numerous cold fusion articles, many of which seemed to be rather loosely evaluated, but at least all could see and weigh them. Other journals too were open-minded about "positive" results cold fusion papers: the English language *Japanese Journal of Applied Physics*, the English language Italian journal, *Il Nuovo Cimento*, among others. Other mechanisms grew up for the conveyance of "subversive" cold fusion information—electronic networks, fax machines, and the telephone. Cold fusion may have been the first major scientific controversy that was seriously influenced by the two new technologies of information exchange—fax and electronic mail.

Though the peer review process should not be abandoned, some repair is called for. The cold fusion controversy suggests alternative philosophies of review that would be less inimical to unconventional research. What would be wrong, for example, with allowing editorial judgment to admit presumptively "wild" papers, but accompanying them with comments by skeptical reviewers who would point to serious possible defects? The new idea or experimental result would more readily be exposed to the fresh air of widespread scientific scrutiny. Unfortunately, the reason that some unconventional ideas are not fairly considered, sometimes has to do with an insidious notion that has crept into scientific publishing (whether it is fully admitted or not): that acceptance for publication confers a significant degree of legitimacy on the work. This should not be! Scientific papers are *made* to be found in error, and numerous ones are eventually found to be wrong, no matter how prestigious the journal in which they are published. In the very first year of the cold fusion controversy, a group of astronomers, to their chagrin (and credit), retracted a significant published finding about a neutron star that might be the remnant of the famous Supernova 1987A, because they found that an electronic glitch had completely invalidated their data.

Moreover, sometimes even peer-reviewed published work is intended not to be so widely understood right away! For example, all was not completely kosher with the Bednorz-Müller announcement of high-temperature superconductivity. Bednorz and Müller, who later deservedly won the Nobel prize for their seminal work, deliberately first published in a foreign language, obscure journal perhaps to slow down their competitors. And noted high-temperature superconductivity researcher Paul Chu in 1987 engaged, we are led to believe, in "non-standard" procedure when he announced to the press his group's new superconducting compound–*before* submitting his results to a scientific journal.

"Science by press conference" has been the disparaging remark of the cold fusion skeptics to denote the public announcement of work that has not passed through the pristine filter of peer review. The harping that peer review was outrageously violated by Fleischmann and Pons has been overdone to the extreme. In the first place, their initial printed work–sloppy and hurriedly done or not, thoroughly reviewed or not, was accepted for publication before that famous press conference! Yes, there were a few weeks after their announcement in which it was not in a journal, but thousands of scientists couldn't wait to get their hands on that faxed Fleischmann and Pons paper so they could begin their experiments with better information. Of course, they were rightly disappointed and angered when they found out that it was sparse in detail. When many good scientists did get the paper, they took it seriously enough because of its great potential–not because of its inherent completeness–to begin to conduct experiments immediately and many of these early experiments seem to have gotten useful results. Peer review in real time—an appropriate course for world-shaking science.

✱ The Fear of Error

The scientific process has become infected with an extraordinary virus, the fear of error. The phenomenon is not unique to the cold fusion episode. It has spread to other areas in science such as molecular biology, in which error and fraud have sometimes been egregiously mixed and confused by no less notable authorities on science than congressmen. If scientists increasingly feel threatened by accusations of fraud, when they may have only committed honest error, all science suffers. Now add to the list of what is to be feared in doing research the "fear of ridicule"—a quite warranted concern in view of what was done to cold fusion. Chemist Gregory A. Petsko, formerly of MIT, wrote in a letter to *Nature* before the cold fusion era in which he said, "What is needed is a decriminalization of error. Science often advances on the strength of theories that turn out to be incorrect, for a wrong hypothesis can produce many excellent experiments." (September 8, 1988.) This sentiment was a wise caution for what was to follow within six months.

✱ Vested Interests

There has been a perennial conflict in NASA over the virtues of manned versus unmanned spaceflights, a struggle that would not exist if every robot-favoring space scientist and every human space explorer were well taken care of in the budget. Similar conflicts exist between "big" and "small" sciences in a variety of fields—from superconducting supercolliders versus table-top physics, to human genome projects versus mom-and-pop biology. The common denominator is a too small budgetary pie being sliced too thinly.

When the cold fusion people entered the fusion arena and found it already well filled, hungering for resources, and always at the edge of oblivion, cold fusion acquired an instant natural enemy. Unfortunately, this enemy should have been a grateful ally, were there any reasonable margin in the national fusion program or physics programs in general for discretionary side trips into hinterlands of scientific novelty. As it was, threatened vested interests in the hot fusion effort drove it to sometimes irrational antagonism toward cold fusion. Not that there was not also a strong element of embarrassment and loss of pride among fusioneers who had overlooked the painfully simple concept—Fleischmann's "mad idea."

✳ Wishful Science or a Wish Come True?

Wishing that certain scientific observations might represent a new phenomenon is part and parcel of the drive of science, but so is the restraining force of conservatism and skepticism. Either too little or too much skepticism can be hazardous, on the one hand allowing too many wild notions to bewilder and occupy valuable research time or on the other permitting a new finding to be cast out with the bath water. For every astronomer like Percival Lowell at the turn of the century, whose intense wishing imaged in his mind the nonexistent canals of Mars, there is a scientist who does not believe what his data are telling him and passes up an amazing discovery. For example, the NASA scientists who initially dismissed apparently spurious data coming from satellite observations over Antarctica missed out on discovering the famous stratospheric ozone hole over that continent years before it was recognized.

Despite justifiable initial skepticism, most scientists wanted to believe that cold fusion could be real shortly after they first heard that two respected chemists had come up with the idea. In that sense, many *wished* the science to be correct even though they later became stern critics when the data turned out not to satisfy them. I met many a scientist at MIT—even outstanding physicists—who initially were willing to give the idea a chance, but who quickly became disenchanted when the going got tough. Today, many skeptics have come to believe that cold fusion is not a scientific wish come true but a case of *wishful science*—wishful on the part of experimentalists and theorists that cold fusion is real. Retired plasma physicist (formerly of Princeton Plasma Physics Laboratory) Milton Rothman put forth just such a view in his skeptical article on cold fusion, appropriately enough in *Skeptical Inquirer.**

Rothman, in fact, prefers the term *wishful science* to *pathological science*, believing that it is misguided intense wishing that is the driving force behind the pathology of what is ordinarily called pathological science. He wrote, "It was exaggerated belief in a theory that tilted the cold fusion work into disaster. Without that psychological factor the case would simply have been a matter of experimental error or misinterpretation of results, unfortunate circumstances that can happen to anybody."

But Rothman's highly critical article on cold fusion curiously betrays his own version of *wishful science.* Reading it is a good object lesson in how *not* to assess evidence of possible new phenomena. Sprin-

*"Cold Fusion: A Case History in 'Wishful Science'?", *Skeptical Inquirer,* Winter 1990: 161–170.

kled throughout are appeals to authority; for example, no "good" laboratory was verifying the alleged cold fusion anomalies. Witness: "After the initial period of stumbling about, the more cautious labs had their say. Caltech, MIT, Yale, Brookhaven, Oak Ridge, and others said their measurements had produced no evidence of either heat generation or emissions of neutrons." (In fact, not many months later, Oak Ridge researchers put forth very tantalizing evidence to support cold fusion.) Highly selective in his laboratories, don't you think? This he followed with the forementioned "polling" pitfall: "A meeting of the American Physical Society held in May unanimously rejected the Utah claims." Even if the meeting was strongly against cold fusion, how was it determined that there was a *unanimous* rejection? In fact, there was no such unanimous rejection.

With all these difficulties and imperfections evident in the scientific process, some have feared that the cold fusion controversy may have soured the public on science. In fact, people received much needed insight into how science does and does not go about its business. What cold fusion showed is that scientific illiteracy may be rampant, but not *interest* in science. The rise of public attention to a scientific matter was extremely gratifying to observe.

It is even possible that citizens may have been *turned on* to science through cold fusion! After all the touting of astronomically *big* science, they may relish the idea that a couple of dedicated basement enthusiasts—à la Edison or the Wright Brothers—had the courage to challenge accepted wisdom. It was nice to see people eagerly awaiting the fruits of laboratory research, rather than being indifferent or hostile. If the scientific community suffers at all from cold fusion, it will most likely be because of the many premature burials. If cold fusion is real and useful, Fleischmann and Pons and all who followed in their path will hardly be faulted by the public for deviating from "accepted procedure." Science almost always triumphs in the end.

18 *Whither Hot Fusion?*

Oh for a Muse of fire, that would ascend
The brightest heaven of invention. . . .

Prologue, Shakespeare's *The Life of Henry the Fifth*

Fusion pioneers of the 1950s saw the confinement of 100-million-degree plasmas as the one formidable obstacle to the release of fusion energy, and launched a brilliant and ultimately successful attack on it. They failed to guess that scientific success might have no direct consequences.

Harold P. Furth
Science, September 1990

Perhaps we should not be greatly troubled that our first attempt to develop such a marvelous thing will not be the success we had hoped. We can go on to seck a better alternative.

Lawrence Lidsky
"The Trouble with Fusion," *Technology Review*, 1983

✳ Antimagnetic Personality

IN AN IRONIC TWIST that was painful to hot fusion researchers, the cold fusion controversy erupted in the same year that magnetic confinement fusion had come excruciatingly close to a long sought goal—the close approach to energy breakeven with the Joint European Torus (JET) tokamak in Europe. While cold fusion merely "stole the fire" of the good news from JET, other trouble was brewing within the hot fusion camp itself.

With or without the stormy rise of cold fusion in 1989 and 1990, hot fusion research had reached a turning point. In August 1988, Dr. Robert O. Hunter, Jr., had become director of DOE's Office of Energy

Research. A friend of inertial confinement fusion (ICF) for many years, having worked for a company connected to the field, Hunter provoked a great outcry in the magnetic fusion community with his desire to force a "runoff" between the two approaches—a showdown between lasers and magnets. The contest would have ended around the turn of the century and decided the more promising route. Hunter wanted to foster competition within a field that already operated with a delicate self-imposed truce among factions.

Hunter attempted to lower the $345-million budget for magnetic fusion about $50 million and to raise the ante for ICF. This he tried in June 1989 amidst the initial wave of public excitement about cold fusion. How many "hits" could the hot fusion community take at one time! Among Hunter's initiatives was to put on hold plans to construct the $450- to $750-million Compact Ignition Tokamak (CIT) that was supposed to have begun in 1990 at the Princeton Plasma Physics Lab and in which MIT was collaborating.

If Fleischmann and Pons in the spring of 1989 thought that they had exhausted the wrath of the magnetic fusion community, they were wrong. There was much ammunition left for Hunter. The outrage of the magnetic community was in part why Hunter tendered his resignation from DOE to Energy Secretary Watkins in October 1989. His wild management style was to blame; among other actions, he interfered with the multibillion dollar Superconducting Supercollider Project (SSC) slated for Waxahatchie, Texas. Hunter angered the wrong congressmen and senators with on-again, off-again precipitous actions. He left DOE before he could take magnetic confinement fusion down a number of notches.

The ICF researchers were having problems too. Put on hold was the hoped-for Laboratory Microfusion Facility (LMF) that would cost hundreds of millions of dollars through the 1990s (perhaps as much as $2 billion), which Hunter had supported. Suspicions were from the very beginning that ICF was driven more by its potential military applications—simulation of thermonuclear weapon explosions—than its feasibility for civilian power reactors. A whiff of this leaked out during the cold fusion furor in May 1989. Lamar Coleman of LLNL, responding to a question in *Laser Focus World* (May 1989) about the effect of cold fusion said, "[I don't think] the ICF program would be seriously impaired, because it is needed for certain defense research applications."

Nonetheless, ICF work is proceeding apace. The multibeam NOVA laser built at the LLNL in the mid-1980s cost a few hundred million dollars. There is a 24-beam ICF laser at the University of Rochester, home of cold fusion critic John Huizenga. (While the magnetic fusion budget was being slashed by 15% late in 1990, the ICF laser at the University of Rochester received an $8.5-million boost.) And the ICF community is international too, with efforts underway in Japan, France, and England. In Japan and Europe, however, rather than being funded on par with magnetic fusion, ICF gets only about 10 percent of magnetic fusion's budget. On the other hand, the Japanese and Europeans are not as hamstrung as U.S. researchers by the ludicrous classification of work that has already surfaced in the open literature of those countries. Strange but true, open Japanese publications on laser fusion routinely publish designs that are classified in the United States. This has reportedly caused sometimes overzealous security people at places like Lawrence Livermore National Laboratory to attempt to "censor" copies of these journals received at the laboratory!

About a year before the cold fusion announcement, a bit of investigative reporting on the ICF program appeared on the front page of the *New York Times* (March 21, 1988). An underground testing program—dubbed Centurion-Halite—had reportedly been successful in bombarding fusion fuel pellets with X rays generated by nuclear explosions. The ICF laboratories in 1987 had promised enough hard data by 1991 to design a laser that would have 50 to 100 times the power of NOVA at LLNL and could ignite a fusion fuel pellet. Its estimated cost: somewhere between $1 billion and $20 billion. Even though $100 million lasers were the norm in ICF, this was just too much to swallow. Furthermore, the annual funding balance in favor of magnetic fusion—approximately $350 million versus $160 million—made sense if the goal was to get as soon as possible to a working power reactor (probably a tokamak).

Not that it was smooth sailing for magnetic fusion. In the spring of 1990, leaders of the U.S. House and Senate appropriations committees were talking about severe cuts in DOE's fiscal year '90 $320 million fusion budget, reductions that might lead to the demise of some of magnetic fusion's prized research devices—the tokamaks at Princeton and at MIT, for example, but also installations at Oak Ridge, Los Alamos, and General Atomics.

These actions, had they occurred, would have flown in the face of a May 1988 National Academy of Science study entitled, "Pacing the U.S. Magnetic Fusion Program," that recommended an immediate 20 percent increase in the the stagnant magnetic fusion budget and further major increases by the mid-1990s to keep the fusion option open for the 21st century. The study urged that the CIT be built and completed by the mid-1990s, followed by work with the Europeans, Japanese, and Soviets to build a prototype fusion power reactor to operate shortly after the turn of the century.

Magnetic fusion advocates again raised the specter of the United States falling behind the rest of the world in fusion technology (much as some cold fusion people have done). Ronald R. Parker of MIT warned, "If the Japanese or the Europeans develop fusion technology first, we will be buying it from them." With so many examples to point to in which America had already lost its competitive edge, who could deny that this could happen even with so demanding a technology as hot fusion?

✳ Going for Broke or Going Broke?

Near the first anniversary of the cold fusion announcement, DOE's Admiral Watkins set up a new Fusion Policy Advisory Committee to plan the future of fusion once again. One of three general paths were open: (1) continue with "business as usual"—research with existing machines, (2) go for broke with an accelerated program to build a working reactor; or (3) a scaled-down program with a focus on plasma physics research. The committee would also have to decide what should be the mix between ICF and magnetic fusion. Ironically, the committee held its first meeting on March 23, 1990—a famous anniversary. H. Guyford Stever, a former head of the National Science Foundation and science advisor to President Ford, who also had served on the DOE panel that rendered very negative conclusions on cold fusion, headed the committee.

To no one's surprise, in September 1990 the panel recommended a major boost in funding for magnetic fusion to $391 million in FY'91, $420 million in FY'92 , and increasing to $620 million by FY'96. However, the panel bowed to Secretary Watkins's desire for a more modest rise and offered an intermediate "constrained" budget plan as an option. But the panel urged that work on the Compact Ignition Tokamak should go forward, and that the United States should commit to working on the International Thermonuclear Experimental Reactor (ITER) in more than a design-level capacity.

Since the 1950s, the DOE and its forebears have already spent nearly $11 billion (in constant 1990 dollars) on hot fusion research—not very much considering the spectacular potential payoff should the effort achieve its long-sought goals. Compared to the cost of certain military hardware of unknown effectiveness, or measured against typical wastage in other sectors of government or industry, the hot fusion program and its scientific spin-offs have been an absolute bargain. Hot fusion is one possible and demonstrably very promising path to the control of nuclear fusion—an opening to the effectively infinite fuel reserves of mother ocean.

It is maddening. While the United States has pinched pennies with the $320-million budget of its magnetic fusion research program, it was seriously contemplating building 75 beautiful but expensive B-2 Stealth bombers, projected to cost about $900 million apiece. One of those spectacular marvels falling out of the sky on a training mission would be like losing three magnetic fusion programs in one pop. At this time also, the SDI research and development budget was still running at a several billion dollars per year clip while the Soviet empire was in near total chaotic collapse. While the energy security of the United States—not to forget that of the whole planet—hung in the balance, the hot fusion program remained the poor cousin of advanced military developments.

Perhaps Iraq's invasion of Kuwait in August 1990 and the resulting chaos in the world oil market might have prompted some truly longer range planning for energy independence. But no, the Stever Committee's recommendations are unlikely to be able to persuade a constrained DOE establishment to dramatically accelerate magnetic fusion. Yet there could be a turning point for hot fusion, if the national will were mustered. What the magnetic fusion community would like to see is a doubling of its budget within five to seven years. It would target 2025 to build a demonstration electricity-generating reactor with the obvious calling, "DEMO." It would then build the first commercial fusion plant by 2040—a little under a century after the controlled fusion quest was taken up. One estimate of the cost for the United States to get to DEMO in 2025 is another $18 billion (1990 dollars). Chicken feed for a gamble that could tap into infinite energy.

The magnetic fusion community is justifiably proud of its track record over the past 15 years. It points to the more than 10,000-fold increase in "power gain," meaning fusion power output relative to power input to heat the plasma, that it achieved in this period. If tritium had been added to the pure deuterium fuel in some of the recent near-breakeven experiments at JET and Princeton's Tokamak Fusion Test Reactor (TFTR), about 10 megawatts of fusion power would have been produced—albeit not power *in excess* of that required to sustain the plasma. As it is, with the pure deuterium fuel JET and TFTR have produced up to 50 kilowatts of fusion power. If the magnetic approach were allowed to run its course with the construction and testing of the CIT at Princeton, fusion researchers for the first time would be able to study "ignited" plasmas—self-heated ones producing some 100 to 500 megawatts of fusion power during burns of about five seconds duration.

There is a critical need for CIT, or so the hot fusioneers claim, because present experimental tokamak reactors do not allow all the physics of plasma self-heating by some fusion products (alpha particles from D-T fusion) to be examined. But the consensus—even though a not universally held one—is that enough is known about tokamaks to go forward with the first "burning plasma" experiment—CIT. It appears that deuterium-tritium fueled experiments planned at JET for 1992 will not be able to achieve ignition, though it will be close.

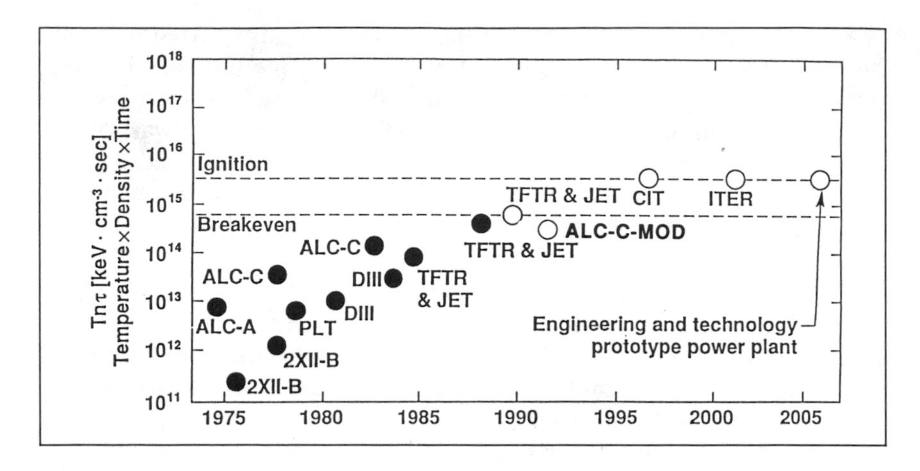

Progress in magnetic confinement fusion toward achieving the conditions required for fusion power—first breakeven, then ignition. Substantial progress has been made in the past 15 years, with current machines at approximate breakeven conditions (if tritium fuel were introduced into the plasma chambers of TFTR or JET). (Courtesy MIT Plasma Fusion Center)

The U.S. magnetic fusion community and its partners abroad are also looking beyond CIT to ITER, whose basic design was completed in 1990. The design collaboration has engaged the United States, the Soviet Union, Japan, and the European Community. In the spirit of the international cooperation that has long characterized the field, ITER is conceived as the "world's fusion engineering test reactor." Soon after the turn of the century, ITER would produce about 1,000 megawatts of fusion power during burns initially lasting some 200 seconds—ultimately to achieve continuous burning though at lower power levels. These powers would certainly bring hot fusion close to the power level range of contemporary electricity-generating stations, both fission and fossil. The practical details of a commercially feasible fusion power plant would still need to be worked out, which is why the anticipated wait to 2040. Among these "details" are the need to maintain plasma purity in the facc of sustained burning, materials issues to maintain the structural integrity of the reactor even in the face of high-intensity fluxes of energetic neutrons, and relatively benign (compared to fission) environmental issues having to do with disposing of spent reactor parts, and tritium fuel recovery and control.

ITER, DEMO, and their immediate successors would be deuterium-tritium burning reactors and would use a "standard" 50–50 mix of the two fuels. This has long been viewed the easier route to demonstrating hot fusion than possible fuels such as pure deuterium or a mixture of helium-3 and deuterium. The latter require higher operating temperatures, and the major problem that helium-3, while being abundant on the surface layers of the Moon (which entrap it from the solar wind), is virtually absent on Earth. The radioactive tritium part of the favored standard deuterium-tritium (D-T) fuel would be regenerated by the high-intensity flux of neutrons bombarding a lithium blanket—itself the means to make heat from the neutrons to generate steam and thereafter electricity. There are abundant domestic and terrestrial supplies of lithium. At least in principle, the basic foundation for an energy economy centered on D-T fusion exists.

One estimate of the capital cost of new U.S. electric power plant construction in the first half of the 21st century is about $4 trillion.* This would bring the installed U.S. electric generating capacity from the 670 billion watt level at present to 2,200 billion watts by 2050. To put hot fusion research expenditures in perspective, even if the $18 billion figure to get to DEMO were off by a factor of two, the money would still be less than one percent of the anticipated investment in fossil-fuel and/or fission nuclear plants to be constructed through the year 2050.

The hot fusion community believes it is on solid ground in its projections, but it has not been notably successful in the United States at convincing politicians to crank up the pursuit of its cherished goals. How the politics and science of fusion research will play out in the next few critical years when important decisions have to be made is difficult to foresee. It does not help that the Unites States, and indeed the whole world, appear to be in a period of great political and economic stress. Attention inevitably focuses on problems and solutions of the here and now. The battle between long and short planning horizons is engaged, and hot fusion's fate will ride on its outcome.

✳ Doubting Hot Fusion

Let there be no doubt: Any engineering undertaking that has consumed as much time as has hot fusion has *got* to be exceedingly difficult. This has led naturally to questioning its basic tenets, on occasion even by some of its original supporters. One of the best critical assessments of the direction of the hot fusion program appeared in MIT's *Technology Review* back in October 1983. In "The Trouble with Fusion," MIT Professor of Nuclear Engineering Lawrence Lidsky, then an associate director of the MIT Plasma Fusion Center and long involved in plasma physics, argued forcefully that even when the first D-T fusion reactor

*"A Plan for the Development of Magnetic Fusion Energy," March 1990, David E. Baldwin, E.C. Brolin, Stephen O. Dean, Alexander Glass, Rulon K. Linford, David O.Overskei, Ronald R. Parker, and John Sheffield.

were up and running some time in the next century, its very nature would make it uncompetitive with advanced types of fission reactors.

No longer at the Plasma Fusion Center, Lidsky is now trying to develop some of these so-called "passively safe" fission power plants—reactors that could be demonstrated to refuse to melt down or disintegrate even if all their coolant were to disappear or a "malicious operator" were to take control. Lidsky has many critics as well as admirers in the fission reactor community. Many in his field want to see a more incremental evolution to a safer next generation of light water nuclear power plant, rather than the radical design departure that Lidsky says is the only way to jump-start the moribund nuclear fission industry.

But Lidsky still keeps one of his twinkling eyes on fusion. The essence of his 1983 thesis (by which he still stands) was that the hot fusion program too early tied itself to D-T fusion, because this fuel cycle was by far the easiest route to demonstrating that the *scientific problem* of controlled fusion had at least one solution. Simply put, D-T is about a hundred times more reactive than pure D-D, the logical choice of fuel cycle if we had our druthers. Lidsky believes that so many engineering difficulties will attend maintaining economically functioning D-T reactors, that their practicality is doubtful. Hot fusion reactors are inherently more difficult beasts, he claims, because their power density is so much lower (less than one tenth) than that of fission reactors.

Fusion reactors will be finicky things with their need for ultra-cold superconducting magnets that confine plasmas at hundreds of millions of degrees, robotic devices to periodically maintain the tokamak walls, and so on. Thus, fusion reactors may be prone to many small accidents and interruptions—difficult to correct because of the radioactive operating environment within a tokamak. Watt for watt, fusion reactors will have to be larger and therefore more expensive than fission plants. The 14 MeV neutrons blasting out of the D-T plasma are many times as destructive to a fusion reactor's material integrity than are the short-range charged atomic fragments and lower energy neutrons in fission reactors.

The light water fission reactor may have been the victim of a too-easy early success that led to a design that the public (rightly or wrongly) no longer trusts. Fusion, by contrast, was not hobbled by early success but by the need for a direction that would work as quickly as possible and make believers of other scientists and politicians. Unfortunately, the extraordinary time this would take was never foreseen at the beginning of fusion research.

On the other hand, Lidsky and many other engineers believe that fusion reactors—even ones built for D-T fuel—would have problems with the disposal and handling of radioactive by-products that are very modest compared to those with fission reactors. Hot fusion power is often described qualitatively as being hundreds or even thousands of times less hazardous than fission power. By all accounts, this is almost certainly true. After a mere century of sleeping underground, a used stainless steel fusion vessel from a tokamak would be a millionfold less dangerous than the equivalent waste from a fission reactor that would remain potent for hundreds if not thousands of years (depending on its treatment). Among other good attributes, there is no possibility whatever of a "fusion meltdown." When plasma reactions terminate, they can't start up again independent of precise control. Nor is radioactive tritium notably difficult to deal with. And the material properties of fusion reactor walls can be adjusted to make disposal of used tokamak hulks less problematic.

Another course toward nearly limitless fuel reserves may be called for. Fusion may be "Mr. Clean," but it is not the only course. Will we perhaps have that new generation of passively safe nuclear power plants, coupled with a technology (pioneered in Japan) for extracting uranium from sea water? Perhaps we will invent a fusion reactor that would be neutronless or relatively so. One might say that if cold fusion ultimately proves to be practical to implement, virtually neutron-free controlled fusion may be approaching. Even though some fission cycles might provide hundreds or thousands of years of energy solace, one conclusion remains: Fission is unlikely to compete over eons with the fusion fuel in the deep blue sea.

✳ Neutronless "Hot" Fusion?

Too many neutrons! Both conventional magnetic fusion and ICF have this drawback that they would just as soon be without, because fast neutrons lead to structural damage in surrounding reactor parts and the buildup of radioactive materials. Though neutrons are the means by which energy is to be conveyed from hot plasmas to produce heat in a surrounding blanket of molten lithium metal, hot fusion would be far more appealing without neutrons. Some scientists think that neutronless hot fusion or fusion with precious few neutrons—so-called *aneutronic* fusion—might be practical. Certainly there are fusion reactions that produce no direct neutrons and few secondary ones.

Energy released in aneutronic reactions is always in the form of charged particles. Hence, the energy of these particles can go directly into the production of electrical energy, so power conversion efficiency is kept high. The "ultimate" aneutronic fusion reaction is $^3He + {}^3He$ reacting to form two protons and an alpha particle, with virtually zero neutron offspring or creation of radioactive materials. Unfortunately, there is little helium-3 on Earth, but other aneutronic reactions involving just lithium and hydrogen will do. (These are actually fission reactions.)

One American physicist, who for more than 15 years has pursued aneutronic fusion with the fervor of a crusade, is Yugoslavia-born and MIT-trained Bogdan Maglich. His company, Advanced Physics Corporation of Princeton, NJ, promotes the technical concept that he calls the "migma" reactor, an idea that he pioneered at Rutgers University in 1970. He thinks it is possible to create an incredibly small magnetic fusion reactor with an active volume of a mere single cubic centimeter! "Neutralized self-colliding beams" of particles are made to orbit in a rosette pattern within high magnetic fields in a migma reactor. Though Maglich has some prominent supporters for migma, such as Nobel laureates Glenn T. Seaborg and Murray Gell-Mann, he has struggled in vain to get funding support from DOE and its predecessors. Federal panels have spent many hundreds of hours reviewing proposals for migma funding and have always turned them down—another victim, some have said, of the "tokamak Mafia."

Maglich did apparently manage to get some funding from Saudi Arabian sources, and he has had Air Force support in more recent years. Maglich's struggle—even though migma has nothing technically in common with cold fusion circa 1989–90—bears a superficial resemblance to the trials of the electrochemical cold fusion proponents in the past two years. With not a little irony, the Second International Symposium on Aneutronic Power was held in Washington about one month after the cold fusion announcement, so news about migma was buried in the surrounding fusion furor.

Prior to the cold fusion upheaval, Maglich's claims were occasionally aired favorably in some of the popular science press. But as science historian Robert P. Crease outlined in an article in *The Scientist,* (November 27, 1989) Maglich's battles for migma have taken on a combative aspect that seems to transcend the technical issues of his reactor's feasibility.

Crease quotes MIT Physics Professor Martin Deutsch: "I know he has brought it on himself, but Bogdan is getting lynched. Nobody who comes up with a rational piece of physics ought to be treated the way he has. It's true that his claims are excessive and that finally there is nothing earthshaking in them. And it is true that he has a special talent for antagonizing people. Nevertheless, he does have some good ideas, and they ought to be funded. If Bogdan had been within the establishment, or at least not antagonized it, [his work] might well have a significant contribution to the mainstream fusion effort."

✳ The Great Blue Hope

Taking the long view, hot fusion is a far too valuable and cost-effective research program to let wither. The power source simply *must* be pursued because it holds the hope of releasing more energy from the deuterium in a single cubic kilometer of sea water than exists in all the world's known oil reserves. Even if cold fusion or some other unforeseen breakthrough were by good fortune to deliver in a few years a practical power-producing device, developing hot fusion should continue. The goal of abundant energy is far too important to global civilization for *all* advanced options not to be explored at the same time with a vengeance—including solar photovoltaic power cells, even though they are perhaps fated to be power producers of relatively low energy density.

Even if economical power production from hot fusion should look at some point like an increasingly tough proposition, plasma physics studies—the wonderful by-product of hot fusion reactor development—would be extremely valuable for their inherent scientific interest and other potential applications. Among these would be a better understanding of physical processes that govern more than 99 percent of the visible matter in the universe. On a practical level, the research could lead to advanced space propulsion systems—plasma fusion rockets—that could eventually open up the unending material resources of the solar system to exploration and exploitation. The vacuum of space, after all, is the natural playground of plasmas.

But by far the most shining promise of hot fusion—as with any route to controlling the fire of stars—is to get humankind permanently out from under the fossil fuel tyranny. Never again should desert demagogues hold the world in sway as they twiddle the oil spigot. To that end, few alternatives are as promising for the long haul as fusion power. As the Princeton Plasma Physics Lab's director Harold Furth so aptly characterized the matter at the American Chemical Society's April 1989 Cold Fusion session in Dallas: "In a mere 300 years, we'll blow the entire energy deposit in the fossil energy bank, which was laid down over 400 million years so that humanity could advance to a high level of civilization. We and our immediate descendants have the extraordinary privilege of blowing this entire bank account in one millionth of the time it took to accumulate. So I have a vision that our descendants in the year 2350 looking back and thinking about us and wondering what we had in mind—and whether there was anything redeeming that could be said about us. I think that one of the redeeming things that will be said about us was that we devoted some very small fraction of this bank to developing a nuclear energy source which could keep civilization going after we've blown all this stuff. If by chance we do succeed in developing fusion as an energy source, then people may think we weren't altogether bad."

But all is not well with the hot fusion program as it exists right now, and in late 1990 magnetic fusion suffered a severe 15 percent reduction, which was later partially reversed. There is, indeed, not enough money to support both the central thrust of the program—demonstration of D-T fusion in a power-producing reactor—and the parallel support of many other innovative fusion ideas, such as cold fusion. University of Illinois Professor George Miley's prepared statement for the House Committee on Science, Space, and Technology's cold fusion hearing on April 26, 1989, says it all: "One problem that is clear is that the fusion program has become so focussed on current major projects that innovative new work is curtailed due to lack of funding. For example, the National Science Foundation will not fund fusion-related research because it is the mission of DOE. Alternate approaches and innovative research receive less money each year from DOE's Office of Fusion Energy due to obligations to large projects. The Inertial Confinement Fusion Office doesn't even fund unsolicited research proposals, leaving that to the National Laboratories who obviously have other top priorities. This is no way to find innovative approaches in an area that should be a top national goal."

The hot fusion community, it seems, would increase its credibility and prospects if it abandoned its perennially defensive mentality. No need to "circle the wagons" every time some perceived threat comes along, whether it be cold fusion, migma fusion, or something still undreamed.

19 *Epilogue*

"Clarke's Third Law"

Any sufficiently advanced technology is indistinguishable from magic.

Arthur C. Clarke, *Profiles of the Future,* 1963

There is no reason why a new technology has to develop like fission and fusion on a thirty-year time scale. All it needs in order to go fast is small size of units, simple design, mass production, and a big market.

Freeman Dyson, *Infinite in All Directions,* 1988

Some say the world will end in fire,
Some say in ice.

Robert Frost, "Fire and Ice"

* Whence We Came, Where We Stand, and Where We Are Going

PEOPLE OF EARTH—the only planet in the Solar System now graced with oceans of water—are toying with fusion energy, arguably one of the most important inventions since their forebears learned to control fire a few million years ago. The similarity: Fire has an effectively infinite "fuel" supply—the oxygen and carbon of the biosphere—while fusion relies on the virtually inexhaustible sea that clings to the blue planet. Both chemical fire and fusion produce heat and other good things. The difference: fusion is 10 million times more potent per gram of fuel consumed.

This we know: Making the fusion Genie work in multimillion-degree plasmas is certainly possible. More than four decades of concentrated, heroic effort has made that abundantly clear. What is far less certain is whether humanity has the will and the courage to stay the difficult but not inordinately expensive course to make such a wonderful prospect commercially viable. Remember, the deuterium in one cubic kilometer of the ocean is vested with more potential fusion energy—hot or cold—than that of all known oil reserves in the world. If we are diligent, hot fusion could become practical sometime before the middle of the next century. One deeply hopes that this dream will be kept alive.

But now there seems to be another path open to tame the stars' fire. Following the discovery of what look extremely suspiciously like fusion phenomena at low temperature, a whole new world of possibilities has opened up. The hundreds of scientists the world over who continue to pursue the prospect of cold fusion power (even in the face of great ridicule and hostility) will inevitably get to the bottom of the matter. They will not stop investigating cold fusion until the truth is known. One way or other, the process of science will triumph as it always has. Give it time. If cold fusion turns out to be what many serious scientists think it is, great prospects may open for the burgeoning populations of the little world with the big dreams. Utopian visions may come to pass.

✳ Cold Fusion: Fact or Fiction?

Is cold fusion or the Pons-Fleischmann effect just a dream based on wishes? Perhaps, but this seems exceedingly unlikely right now, almost two years after Martin Fleischmann and B. Stanley Pons made their announcement. We know enough now to say two things:

1. The chance that the Pons-Fleischmann effect is a fantastic mass hallucination or error and not some *new physical effect*—either chemical or nuclear—is close to nil. That chance is close to the probability three years ago that high-temperature superconductivity was a mass delusion. Too many different effects associated with the phenomenon have been observed by researchers. This is not the infamous polywater, and cold fusion is most certainly not "pathological science."

2. The evidence is—at the very least—compelling that cold fusion is a *radically different and new kind of nuclear process* that has some prospect of leading to significant power generation and the possibility of transmuting isotopes of some elements into other elements. That is simply being fair and middle-of-the-road. But substitute the phrase *overwhelmingly compelling* for merely *compelling, but as yet incompletely explained*, and one might be closer to the truth.

At this time, the evidence for cold fusion is compelling but not yet *conclusive*, because no *certain* qualitative or quantitative theory yet exists about how a still unknown fuel (deuterium, hydrogen, or something else?) goes about producing excess heat and/or tritium and/or neutrons in cold fusion experiments. As soon as theory links intimately with experiment—and that could be soon—the reality of cold fusion will be established far beyond reasonable doubt.

Other developments may provc conclusively that cold fusion is real. One possibility: The evidence that tritium and/or neutrons are being produced in cold fusion reactions, though very strong now, is made even stronger by more compelling tests—particularly ones that are reproducible *on demand.* Experiments are proceeding in that direction right now. Another avenue: An excess heat-generating cold fusion device is developed that puts out such an obvious amount of surplus energy that no one can doubt its nuclear origin. Recent work on electrochemical systems using molten salts and deuterium/palladium/titanium are going in that direction. Still another possibility: Some other isotope emerges in an experiment that was clearly not present initially. Either its radioactive half-life is too short for it to have been there, or its abundance exceeds any reasonable possible original proportion.

✳ Cold Fusion: What It Isn't

Whatever cold fusion is, it can't possibly be based on *conventional* nuclear fusion reactions, which on the atomic level typically involve the combining of two atomic nuclei and the release of energetic particles or radiations. It is clear that the amount of excess heat production observed in some cold fusion experiments is billions of times larger than could be explained by conventional "two-body" deuterium-deuterium fusion with the same level of neutron end products. The corresponding copious production of neutrons, if it were conventional d-d fusion, would kill any unshielded experimenters and would constitute a much higher flux than has been observed.

Likewise for the tritium. The low levels of the isotope that researchers have measured are completely inconsistent with the amount of excess heat—far below one percent of what would be required. Moreover, even those low tritium levels are not observed regularly when heat is being produced. It is possible that neutrons and tritium are ancillary or auxiliary reactions to the main heat-producing reaction. Cold fusion may not be a single new phenomenon, but a class of related new nuclear processes. A whole new science may have opened up.

✳ Present Evidence

What is the compelling evidence right now that cold fusion is real? A number of apparent physical effects are involved, each supported by some degree of serious evidence. Depending on one's preference, any one of the following lines may make a strong case for cold fusion, but *taken together* the evidence is truly compelling. The most economical explanation is that a new phenomenon or a new class of related phenomena are occurring simultaneously. Though not all "cold fusion" experiments can be considered "correct"—including ones that have not given any *apparent* indication of new phenomena—it would be far more difficult to imagine that numerous systematic experimental errors are being made in a host of different experiments.

Cold fusion skeptic, physicist Richard Garwin, wrote in the early days of the cold fusion controversy (*Nature*, April 1989): "Large heat release from fusion at room temperature would be a multi-dimensional revolution. I bet against its confirmation." My own view, with the benefit of a further 18 months of evidence, is that the disappearance of cold fusion as a viable explanation of what is going on in electrochemical cells and allied systems would itself constitute a "multi-dimensional miracle" of staggering proportions. Dozens of different types of experiments and measurement methods would simultaneously each have to possess some fundamental flaw—and in regimes uncommon to some of the other experiments.

Yet more than a year after chemists Martin Fleischmann and B. Stanley Pons publicly claimed to have discovered cold nuclear fusion in table-top electrochemical cells, influential scientists and media were still writing obituaries for the beleaguered phenomenon. But undaunted by the barrage of invective, hundreds of serious scientists continue to pick at the thorny scientific knot of "anomalous phenomena observed in deuterated systems," which could conceivably transform the world's oceans into infinite fuel depots; so could thermonuclear or hot fusion when its technology is perfected in the coming decades.

If cold fusion reactions are taking place, they must be occurring in ways formerly not dreamed of. Fortunately, some theorists have proposed mechanisms, albeit difficult to embrace without reservations, that could explain cold fusion. For cold fusion to work, the energy of whatever fusion reactions may be happening must be distributed among atoms far and wide in the system, instead of coming out as high-energy particles.

At this stage, theories are not enough to put cold fusion on solid ground, but the experimental evidence is. A general summary of the most recent kinds:

1. a variety of experiments that show excess energy being produced by electrochemical cells, in amounts so large that no one has been able to explain it by chemical reactions or storage of chemical energy

2. power bursts from such cells that erratically turn on and off, but persist sometimes for hundreds of hours
3. the apparent generation of radioactive tritium and the inability to universally explain it away as mere "contamination"
4. observations of low-level bursts of neutrons from such cells and low-level random neutron emissions above background
5. other kinds of geometries and materials that also seem to produce radioactive tritium where no tritium should be
6. other kinds of isotopic anomalies in cold fusion experiments
7. physical-chemical grounds for believing that the reactions should be occurring erratically and not always reproducibly
8. convincing evidence that energetic tritium nuclei are being produced in some experiments
9. numerous control experiments in which no effects are seen in apparatus that has ordinary hydrogen substituted for deuterium, while effects are observed in equivalent deuterated systems

It is not that any one of these lines of evidence makes an ironclad case for cold fusion—as Fleischmann and Pons had originally claimed for their excess heat. Rather it is the totality of the anomalous phenomena, all associated with similar kinds of physical systems, that makes it difficult to understand how this could be a multidimensional mistake of vast proportions or an unprecedented mass hallucination by large numbers of scientists.

The canons of conventional nuclear physics, of course, are clear. Significant power-producing fusion reactions can't happen at room temperature. The combining of light atomic nuclei—fusion reactions—producing large amounts of heat with few if any neutrons being emitted cannot happen near room temperature, period! Yet the evidence that some new kind of nuclear process is at work in these strange and allegedly "impossible" cold fusion reactors seems to some observers to grow ever stronger; some are prepared to say that it is already beyond dispute. This, of course, is vehemently denied by many scientists. These skeptics should know how Leo Tolstoy might have viewed them:

> I know that most men, including those at ease with problems of the greatest complexity, can seldom accept even the simplest and most obvious truth if it be such as would oblige them to admit the falsity of conclusions they reached perhaps with great difficulty, conclusions which they have delighted in explaining to colleagues, which they have proudly taught to others, and which they have woven, thread by thread, into the fabric of their lives.

So don't expect the cold fusion controversy to disappear anytime soon. It may drag on for years before being resolved, unless good fortune shortly produces some very solid results that explain the numerous anomalies. The basic problem is that whatever the associated phenomena are, they are not always reproducible *on demand.* On the other hand, as repeatability becomes better and better in some experiments—it has risen to 80 or 90 percent in some cases—it may be easier for even skeptical scientists to believe that something new has really been discovered. They may then heed the eternal challenge of science not to follow where the worn path may lead, but to go instead where there is no path, and leave a trail.

A Fusion Resource Guide

The tide of information on fusion, both hot and cold, seems as overwhelming as the sea from which both draw their life. This guide merely introduces that huge body of literature. It was tempting to include a chronological listing of key popularized articles on fusion—particularly cold fusion—that have appeared in the past few years. That was quickly abandoned after their sheer volume became overwhelming. In its place is a concise list of publications that offer cold and hot fusion articles.

✳ Sources for Popular and Semitechnical Articles

Chemical & Engineering News
The Chronicle of Higher Education
Deseret News (A Salt Lake City newspaper)
Nature
The New Scientist
The New York Times
Physics Today (Published by the American Physical Society)
The Salt Lake Tribune (Salt Lake City)
Science
Science News
The Scientist (A newspaper published by the Institute for Scientific Information, Philadelphia)
21st Century Science and Technology
The Wall Street Journal

An invaluable source of information on cold fusion is the monthly newsletter, *Fusion Facts*, which is published by the Fusion Information Center, P. O. Box 58639, Salt Lake City, UT 84158.

To tap into the latest information on hot fusion, read the executive newsletter published monthly by Fusion Power Associates, 2 Professional Drive, Suite 248, Gaithersburg, MD 20879.

The Research Organization for Deuterium Systems (RODS) is an informal network of scientists, engineers, and interested citizens who are dedicated to understanding cold fusion phenomena: RODS, MIT Branch Post Office, Box 94, Cambridge, MA 02139.

In 1989, the National Science Foundation provided financial support to establish a Cold Fusion Archive at Cornell University, the John M. Olin Library, Ithaca, NY 14853–5301. Dr. Bruce V. Lewenstein of the Department of Communication, and Drs. Thomas F. Gieryn and William Dougan (the latter two no longer at Cornell) assembled and organized the archive. It is now open to researchers, who are encouraged to write to associate archivist Elaine Engst for information requests. The cold fusion archive includes:

- ✳ Original manuscripts and published materials by principals in the cold fusion controversy.
- ✳ Mass media articles on cold fusion (newspapers, magazines, books, radio and television broadcasts)—including original clippings, recordings, computer printouts, and transcripts.
- ✳ Taped interviews with researchers, research administrators, public information representatives, and journalists.
- ✳ Items of "material culture," such as sample electrodes, T-shirts, and "do-it-yourself kits."

Another outstanding resource for cold fusion articles, both technical and popular, is the massive *annotated* "Cold Nuclear Fusion Bibliography" assembled by Danish researcher Dieter Britz, a copy of which is in the Cornell Cold Fusion Archive.

✳ Primary Journals with Technical Articles on Cold Fusion

Electrochimica Acta
Europhysics Letters
Fusion Technology
Journal of Fusion Energy
Il Nuovo Cimento
Japanese Journal of Applied Physics
Journal of Electroanalytical Chemistry (and Interfacial Science)
Journal of Physics
Nature
Physical Review
Physical Review Letters
Science
Solid State Communications
Soviet Technical Physics Letters
Zeitschrift für Physik

✳ Books, Compendia, and Reports

Bromberg, Joan Lisa. *Fusion: Science, Politics, and the Invention of a New Energy Source.* Cambridge, MA: The MIT Press, 1982.

Close, Frank. *Too Hot to Handle: The Race for Cold Fusion.* London, W. H. Allen Publishing Co., 1990.

Energy Research Advisory Board. "Cold Fusion Research," A Report of the Energy Research Advisory Board to the U. S. Department of Energy, November 1989.

First Annual Conference on Cold Fusion (March 28–31, 1990): Conference Proceedings, by the National Cold Fusion Institute, Salt Lake City.

Journal of Fusion Energy. Special Issue: U. S. Department of Energy Workshop on Cold Fusion Phenomeneon. Part I, Vol. 9 (June 1990): 103–237; Part II, Vol. 9 (September 1990): 241–366.

Franks, Felix. *Polywater.* Cambridge, MA: MIT Press, 1981.

Heppenheimer, T. A. *The Man-Made Sun: The Quest for Fusion Power.* Boston and Toronto: Little, Brown and Company, 1984.

Herman, Robin. *Fusion: The Search for Endless Energy.* Cambridge, England: Cambridge University Press, 1990.

Iyengar, P. K., and M. Srinivasan. "BARC Studies in Cold Fusion (April–September 1989)," Bhabha Atomic Research Center, BARC–1500, December 1989.

Kuhn, Thomas S. *The Structure of Scientific Revolutions.* 2d ed. Chicago: University of Chicago Press, 1970.

Mamyrin, B. A., and I. M. Tolstikhin. *Helium Isotopes in Nature.* Amsterdam: Elsevier, 1984.

Morrison, Philip and Phylis. *The Ring of Truth: An Inquiry Into How We Know What We Know.* New York: Random House, 1987.

Mueller, W. M., J. P. Blackledge, and G. G. Libowitz. *Metal Hydrides.* New York: Academic Press, 1968.

Peat, F. David. *Cold Fusion: The Making of a Scientific Controversy.* Contemporary Books, 1989 (This book's coverage of the cold fusion saga ends very early).

Proceedings of NSF/EPRI Workshop on Anomalous Effects in Deuterated Materials. Washington, DC, October 16–18, 1989 (awaiting publication).

Proceedings of the Symposium on Cold Fusion. World Hydrogen Energy Conference #8, Hawaii, July 22–27, 1990.

Proceedings of an International Progress Review on Anomalous Nuclear Effects in Deuterium/Solid Systems. Brigham Young University, Provo, Utah. Sponsored by EPRI, DOE, and BYU, October 22–24, 1990, (publication in spring, 1991).

Rhodes, Richard. *The Making of the Atomic Bomb.* New York: Simon and Schuster, Inc., 1986.

Testimony on Cold Fusion Before the Committee on Science, Space, and Technology, U. S. House of Representatives. April 26, 1989, full transcript.

U.S. Congress, Office of Technology Assessment. *Starpower: The U. S. and the International Quest for Fusion Energy*. OTA-E-338 (Washington, DC: U. S. Government Printing Office, October 1987.)

Winterberg, Friedwardt. *The Physical Principles of Thermonuclear Explosive Devices*. New York: Fusion Energy Foundation, 1981.

Workshop on Cold Fusion Phenomena. Santa Fe, New Mexico, May 23–25, 1989, complete videotape of entire conference (11 tapes), available from the Los Alamos National Laboratory. Proceedings published in the *Journal of Fusion Energy*. Part I, Vol. 9, June 1990; Part II, September 1990.

✳ A Bibliography of Technical Cold Fusion Articles

[*Not comprehensive*, intended to reference only major works and articles of special interest.]

Adair, Robert, Stanley Bruckenstein, Loren Hepler, and Dale Stein. "Report of the Committee for the Review of the National Cold Fusion Institute." December 1990.

Albagli, D., R. Ballinger, V. Cammarata, X. Chen, R. Crooks, C. Fiore, M. Gaudreau, I. Hwang, C. K. Li, P. Linsay, S. Luckhardt, R. R. Parker, R. Petrasso, M. Schloh, K. Wenzel, and M. Wrighton. "Measurement and Analysis of Neutron and Gamma Ray Emission Rates, Other Fusion Products, and Power in Electrochemical Cells Having Pd Cathodes." MIT Plasma Fusion Center, PFC/JA-89-34 (July 1989).

Arata, Yoshiaki, and Yue-Chang Zhang. "Achievement of an Intense Cold Fusion Reaction." *Fusion Technology*, Vol. 18 (August 1990): 95–102.

Armstrong, R. D., "Editorial: The Cold Fusion Debate." *Electrochimica Acta*, Vol. 34, no. 9 (September 1989): 1287.

Armstrong, R. D., E. A. Charles, I. Fells, L. Molyneux, and M. Todd. "A Long-term Calorimetric Study of the Electrolysis of D_2O Using Palladium Cube Cathodes." *J. Electroanalytical Chemistry and Interfacial Electrochemistry,* Vol. 272 (10 November 1989): 293–297.

Armstrong, R. D., E. A. Charles, I. Fells, L.Molyneux, and M. Todd. "Some Aspects of Thermal Energy Generation During the Electrolysis of D_2O Using a Palladium Rod." *Electrochimica Acta*, Vol. 34, no. 9 (September 1989): 1319–1322.

Bennington, S. M., R. S. Sokhi, P. R. Stonadge, D. K. Ross, M. J. Benham, T. D. Beynon, P. Whithey, I. R. Harris, and J. P. G. Farr. "A Search for the Emission of X-Rays from Electrolytically Charged Palladium-Deuterium." *Electrochimica Acta*, Vol. 34, no. 9 (September 1989): 1323–1326.

Bertin, A., M. Bruschi, M. Capponi, S. De Castro, U. Marconi, C. Moroni, M. Piccinini, N. Semprini-Cesari, A. Trombini, A. Vitale, and A. Zoccoli. "Experimental Evidence of Cold Nuclear Fusion in a Measurement Under the Gran Sasso Massif." *Il Nuovo Cimento*, Vol. 101 A (June 1989): 997–1004.

Beuhler, R. J., G. Friedlander, and L. Friedman. "Cluster-Impact Fusion." *Physical Review Letters*, Vol. 63, no. 12 (18 September 1989): 1292–1295.

Bockris, John O'M., Guang H. Lin, and Nigel J. C. Packham. "A Review of the Investigations of the Fleischmann-Pons Phenomena." *Fusion Technology*, Vol. 18 (August 1990): 11–31.

Bressani, T., E. Del Giudice, and G. Preparata. "First Steps Toward an Understanding of Cold Nuclear Fusion." *Il Nuovo Cimento*, Vol. 101 A (May 1989): 845–849.

Broer, M. M., L. C. Feldman, A. C. W.P. James, J. S. Kraus, and R. S. Raghavan. "Search for Neutrons from Deuterium-Deuterium Nuclear Reactions in Electrochemically Charged Palladium." *Physical Review C*, Vol. 40, no. 4 (October 1989): R1559–R1562.

Bush, B. F., J. J. Lagowski, M. H. Miles, and G. S. Ostrom. "Helium Production During the Electrolysis of D_2O in Cold Fusion Experiments." *Journal of Electroanalytical Chemistry*, (to appear April or May 1991).

Bush, Robert T. "Cold 'Fusion': The Transmission Resonance Model Fits Data on Excess Heat, Predicts Optimal Trigger Points, and Suggests Nuclear Reaction Scenarios." *Fusion Technology*, Vol. 19 (March 1991): 313–356.

Butler, M. A., D. S. Ginley, J. E. Schriber, and R. I. Ewing. "High-Sensitivity Search for Neutrons During Electrochemical Reactions." *Fusion Technology*, Vol. 16 (November 1989): 388–390.

Celani, Francesco, et al. "Further Measurements on Electrolytic Cold Fusion with D_2O and Palladium at Gran Sasso Laboratory." *Fusion Technology*, Vol. 17 (July 1990): 718–724.

Chene, J., and A. M. Brass. "Tritium production during the cathodic discharge of deuterium on palladium." *J. Electroanalytical Chemistry*, Vol. 280 (28 February 1990): 199–205.

Claytor, T. N., D. G. Tuggle, P. Seeger, H. O. Menlove, R. K. Rohwer, and W. Doty. "Solid State Fusion." Los Alamos National Laboratory, internal memorandum (January 23, 1990).

Claytor, T. N., P. A. Seeger, R. K. Rohwer, D. G. Tuggle, and W. R. Doty. "Tritium and Neutron Measurements of a Solid State Cell." NSF/EPRI Workshop on Anomalous Effects in Deuterated Materials, Washington, DC (October 16–18, 1989); LANL Preprint UR-89-39-46.

Cohen, James S., and John D. Davies. "The Cold Fusion Family." *Nature*, 27 (April 1989) Vol. 338: 705–706.

"Cold Fusion: What's Going On?" (scientific correspondence: John M. Carpenter; Peroni Paolo; A. J. McCevoy and C. T. D. O'Sullivan; M. Gryzinski; Francesco Premuda) *Nature*, 27 (April 1989) Vol. 338: 711–712.

Cunnane, V. J., R. A. Scannell, and D. J. Schiffrin, "H_2 + O_2 Recombination in Non-Adiabatic Electrochemical Calorimetry of Water Electrolysis in an Undivided Cell." *J. Electroanalytical Chemistry and Interfacial Electrochemistry,* Vol. 269 (25 September 1989): 163–174.

Dalard, F., M. Ulmann, J. Augustynski, and P. Selvam. "Electrochemical Incorporation of Lithium into Palladium from Aprotic Electrolytes." *J. Electroanalytical Chemistry and Interfacial Electrochemistry,* Vol. 270 (10 October 1989): 445–450.

Dandapani, B., and Martin Fleischmann. "Electrolytic Separation Factors on Palladium." *J. Electroanalytical Chemistry and Interfacial Electrochemistry* 39 (1972): 323–332.

De Ninno, A., A. Frattolillo, G. Lollobattista, L. Martinis, M. Martone, L. Mori, S. Podda, and F. Scaramuzzi. "Evidence of Emission of Neutrons from a Titanium-Deuterium System." *Europhysics Letters*, Vol. 9 (1 June 1989): 221–224.

Dickinson, J. T., L. C. Jensen, S. C. Langford, R. Ryan, and E. Garcia. "Fracto-Emission from Deuterated Titanium: Supporting Evidence for a Fracto-Fusion Mechanism." *Journal of Materials Research*, Vol. 5 (January 1990): 109–120.

Ehrlich, A. C., D. J. Gillespie, and G. N. Kamm. "A Search for Neutrons in Single-Phase Palladium-Deuterium." *Fusion Technology*, Vol. 16 (December 1989): 529–531.

Ewing, R. I., M. A. Butler, and J. E. Schriber. "Negative Results and Positive Artifacts Observed in a Comprehensive Search for Neutrons from 'Cold Fusion' Using a Multidetector System Located Underground." *Fusion Technology*, Vol. 16 (November 1989): 404–407.

Fang, P. H. "Deuterium Fusion Through Nonequilibrium Induction." *Fusion Technology*, Vol. 19 (March 1991): 369–370.

Fleischmann, Martin, and Stanley Pons. "Electrochemically Induced Nuclear Fusion of Deuterium." *J. Electroanalytical Chemistry and Interfacial Electrochemistry,* Vol. 261 (10 April 1989): 301–308; Errata: Vol. 263 (10 May 1990): 187–188.

Fleischmann, Martin, Stanley Pons, Mark W. Anderson, Lian Jun Li, and Marvin Hawkins. "Calorimetry of the Palladium-Deuterium-Heavy Water System." *J. Electroanalytical Chemistry,* Vol. 287 (25 July 1990): 293–348.

Gai, M., S. L. Rugari, R. H. France, B. J. Lund, Z. Zhao, A. J. Davenport, H. S. Isaacs, and K. G. Lynn. "Upper Limits on Neutron and γ-Ray Emission from Cold Fusion." *Nature,* Vol. 340 (6 July 1989): 29–34.

Garwin, Richard L. "Consensus on Cold Fusion Still Elusive." *Nature,* Vol. 338 (20 April 1989): 616–617.

Gillespie, D. J., G. N. Kamm, A. C. Ehrlich, and P. L. Mart. "A Search for Anomalies in the Palladium-Deuterium System." *Fusion Technology,* Vol. 16 (December 1989): 526–528.

Goldanskii, A. I., and F. I. Dalidchik. "On the Possibilities of 'Cold Enhancement' of Nuclear Fusion." *Physics Letters B,* Vol. 234 (18 January 1990): 465–468.

Gozzi, D., P. L. Cignini, L.Petrucci, M. Tomellini, G. DeMaria, S. Frullani, F. Garibaldi, F. Ghio, and M. Jodice. "Evidences for Associated Heat Generation and Nuclear Products Release in Palladium Heavy-Water Electrolysis." *Il Nuovo Cimento,* Vol. 103 A (January 1990): 143–154.

Guinan, M. W., G. F. Chapline, and R. W. Moir. "Catalysis of Deuterium Fusion in Metal Hydrides by Cosmic Ray Muons." Lawrence Livermore National Laboratory, UCRL-100881 preprint, April 7, 1989.

Hagelstein, Peter. "Coherent Fusion Mechanisms." January 1991, submitted for Proceedings of an International Progress Review on Anomalous Nuclear Effects in Deuterium/Solid Systems, Brigham Young University, Provo, Utah. Sponsored by EPRI, DOE, and BYU, October 22–24, 1990 (publication in spring 1991).

Hagelstein, Peter L. "Coherent Fusion Theory." Paper presented at "Cold Fusion Session" of the winter annual meeting of the American Society of Mechanical Engineers, San Francisco (12 December 1989).

Iyengar, P. K., M. Srinivasan, et al. "Bhabha Atomic Research Center Studies in Cold Fusion." *Fusion Technology,* Vol. 18 (August 1990): 32–94.

Jones, Steven E. "Muon-Catalyzed Fusion Revisited." *Nature* 321, no. 6066 (May 8, 1987): 127–133.

Jones, S. E., E. P. Palmer, J. B. Czirr, D. L. Decker, G. L. Jensen, J. M. Thorne, S. F. Taylor, and J. Rafelski. "Observation of Cold Nuclear Fusion in Condensed Matter."*Nature*, 27 (April, 1989): 737–740.

Jorne, Jacob. "Neutron and Gamma-Ray Emission from Palladium Deuteride Under Supercritical Conditions." *Fusion Technology*, Vol. 19 (March 1991): 371–374.

Jorne, Jacob. "Unsteady Diffusion Reaction of Electrochemically Produced Deuterium in Palladium." *Journal of the Electrochemical Society*, Vol. 137 (January 1990): 369–370.

Kainthla, R. C., O. Velev, L. Kaba, G. H. Lin, N. J. C. Packham, M. Szklarczyk, J. Wass, and J.O'M. Bockris. "Sporadic Observation of the Fleischmann-Pons Heat Effect." *Electrochimica Acta*, Vol. 34, no. 9 (September 1989): 1315–1318.

Kamm, G. N., A. C. Ehrlich, D. J. Gillespie, and W. J. Powers. "Search for Neutrons from a Titanium-Deuterium System." *Fusion Technology*, Vol. 16 (November 1989): 401–403.

Keesing, R. G., R. C. Greenhow, M. D. Cohler, and A. J. McQuillan. "Thermal, Thermoelectric, and Cathode Poisoning Effects in Cold Fusion Experiments." *Fusion Technology*, Vol. 19 (March 1991): 375–379.

Klyuev, V. A., A. G. Lipson, Y. P. Toporov, B. V. Deryagin, V. I. Lushchikov, A. V. Strelkov, and E. P. Shabalin. "High-Energy Processes Accompanying the Fracture of Solids." *Soviet Technical Physics Letters* (12 November 1986): 551–552.

Koonin, S. E., and M. Nauenberg. "Calculated fusion rates in isotopic hydrogen molecules." *Nature*, (29 June 1989) Vol. 339: 690.

Kreysa, G., G. Marx, and W. Plieth. "A Critical Analysis of Electrochemical Nuclear Fusion Experiments," *J. Electroanalytical Chemistry and Interfacial Electrochemistry,* Vol. 266 (25 July 1989): 437–450.

Lewis, N. S., C. A. Barnes, M. J. Heben, A. Kumar, S. R. Lunt, G. E. McManis, G. M. Miskelly, R. M. Penner, M. J. Sailor, P. G. Santangelo, G. A. Shreve, B. J. Tufts, M. G. Youngquist, R. W. Kavanagh, S. E. Kellogg, R. B. Vogelaar, T. R.Wang, R. Kondrat, and R. New. *Nature* (17 August 1989): 525–530.

Liaw, B. Y., P-L Tao, P. Turner, and B. E. Liebert. "Elevated Temperature Excess Heat Production Using Molten-Salt Electrochemical Techniques." Proceedings of the Symposium on Cold Fusion, World Hydrogen Energy Conference #8, Hawaii (July 22–27, 1990).

Lin, G. H., R. C. Kainthla, N. J. C. Packham, and J.O'M. Bockris. "Electrochemical Fusion: A Mechanism Speculation." *J. Electroanalytical Chemistry*, Vol. 280 (28 February 1990): 207–211.

Lohr, Lawrence L., "Electronic Structure of Palladium Clusters: Implications for Cold Fusion." *The Journal of Physical Chemistry*, Vol. 93, no. 12, (15 June 1989): 4697–4698.

Matsumoto, Takaaki. " 'Nattoh' Model for Cold Fusion." *Fusion Technology*, Vol. 16 (December 1989): 532–534.

McBreen, James. "Absorption of Electrolytic Hydrogen and Deuterium by Pd: The Effect of Cyanide Adsorption." *J. Electroanalytical Chemistry*, Vol. 287 (25 July 1990): 279–291.

"Measurement of γ-rays from cold fusion." (scientific correspondence: Martin Fleischmann, Stanley Pons, and R. J. Hoffman; R. D. Petrasso, X. Chen, K. W. Wenzel, R. R. Parker, C. K. Li, and C. Fiore), *Nature*, Vol. 339 (29 June 1989): 667–669.

Mebrahutu, Thomas, Jose F. Rodriguez, Michael E. Bothwell, I. Francis Cheng, Del R. Lawson, John R. McBride, Charles R. Martin, and Manuel P. Soriaga. "Observations on the Surface Composition of Palladium Cathodes After D_2O Electrolysis in LiOD Solutions." *J. Electroanalytical Chemistry and Interfacial Electrochemistry,* Vol. 267 (10 August 1989): 351–357.

"Mechanism of solid-state fusion." (scientific correspondence: Vitalii I. Goldanskii and Fyodor I. Dalidchik; Mark Kirkpatrick and Cheryl D. Jenkins) *Nature* (16 November 1989): 231–232.

Miles, M. H., K. H. Park, and D. E. Stilwell. "Electrochemical Calorimetric Evidence for Cold Fusion in the Palladium-deuterium System." *J. Electroanalytical Chemistry*, Vol. 295 (26 November 1990): 241–254.

Miles, M. H., and R. E. Miles. "Theoretical Neutron Flux Levels, Dose Rates, and Metal Foil Activation in Electrochemical Cold Fusion Experiments." *J. Electroanalytical Chemistry*, Vol. 295 (26 November 1990): 409–414.

Morrey, John R., et al. "Measurements of Helium in Electrolyzed Palladium." *Fusion Technology*, Vol. 18 (December 1990): 659–668.

Nicol, Malcolm F., and Mostafa A. El-Sayed. "Editorial Comment." *The Journal of Physical Chemistry*, Vol. 93, no. 12 (15 June 1989): 4697.

Noninski, V. C., and C. I. Noninski. Comments concerning the calorimetric results in the MIT PFC/JA-89-34 report, submitted August 24, 1990, to the *Journal of Fusion Energy.*

Noninski, V. C., and C. I. Noninski. "Determination of the Excess Energy Obtained During the Electrolysis of Heavy Water." *Fusion Technology*, Vol. 19 (March 1991): 364–368.

Oriani, R. A., John C. Nelson, Sung-Kyu Lee, and J. H. Broadhurst. "Calorimetric Measurements of Excess Power Output During the Cathodic Charging of Deuterium into Palladium." *Fusion Technology*, Vol. 18 (December 1990): 652–658.

Oyama, Noboru, et al. "Electrochemical Calorimetery of D_2O Electrolysis Using a Palladium Cathode—An Undivided, Open Cell

System." *Bull. Chemical Society of Japan,* Vol. 63 (September 1990): 2659–2664.

Packham, N. J. C., K. L. Wolf, J. C. Wass, R. C. Kainthla, and J.O'M. Bockris. "Production of Tritium from D_2O Electrolysis at a Palladium Cathode." *J. Electroanalytical Chemistry and Interfacial Electrochemistry,* Vol. 270 (10 October 1989): 451–458.

Perfetti, P., F. Cilloco, R. Felici, M. Capozi, and A. Ippoliti, "Neutron Emission Under Particular Nonequilibrium Conditions from Pd and Ti Electrolytically Charged with Deuterium." *Il Nuovo Cimento,* Vol. 11 D (June 1989): 921–926.

Pons, Stanley, and Martin Fleischmann. "Calorimetric Measurements of the Palladium/Deuterium System: Fact and Fiction." *Fusion Technology,* Vol. 17 (July 1990): 669–679.

Preparata, Giuliano. "Theories of 'Cold' Nuclear Fusion: A Review." A private review paper prepared at the National Cold Fusion Institute (October 1990).

Price, P. B. "Search for High-Energy Ions from Fracture of LiD Crystals." *Nature,* Vol. 343 (8 February 1990): 542–544.

"Problems with the γ-ray Spectrum in the Fleischmann et al. Experiments." (scientific correspondence: R. D. Petrasso, X. Chen, K. W. Wenzel, R. R. Parker, C. K. Li, and C. Fiore) *Nature,* 18 (May 1989) Vol. 339: 183–184.

Proceedings of The First Annual Conference on Cold Fusion (March 28–31, 1990) National Cold Fusion Institute, Salt Lake City. (Highlights only):

✳ "A Study of Electrolytic Tritium Production," E. Storms and C. Talcott

✳ "An Overview of Cold Fusion Phenomena," M. Fleischmann

✳ "Anomalies in the Surface Analysis of Deuterated Palladium," D. Rolison, et al.

✳ "Anomalous Calorimetric Results During Long-term Evolution of Deuterium on Palladium from Alkaline Deuteroxide Electrolyte," A. J. Appleby, O. J. Murphy, et al.

✳ "Calorimetry and Electrochemistry in the D/Pd System," M. C. H. McKubre, et al.

✳ "Calorimetry of the Palladium-Deuterium System," S. Pons and M. Fleischmann

✳ "Does Tritium Form at Electrodes by Nuclear Reactions?" J. O'M Bockris, et al.

✳ "High-Sensitivity Measurements of Neutron Emission from Ti Metal in Pressurized D_2 Gas," H. Menlove

✳ "Isotopic Mass Shifts in Cathodically Driven Palladium via Neutron Transfer Suggested by a Transmission Resonance Model . . ." R. T. Bush

✳ "Nuclear Energy in an Atomic Lattice," J. Schwinger
✳ "Overview of BARC Studies in Cold Fusion," P. K. Iyengar and M. Srinivasan
✳ "Quantum Mechanics of Cold and Not-So-Cold Fusion," S. R. Chubb and T. A. Chubb
✳ "Recent Measurements of Excess Energy Production in Electrochemical Cells Containing Heavy Water and Palladium," M. Schreiber, R. A. Huggins, et al.
✳ "Search for Nuclear Phenomena by the Interaction Between Titanium and Deuterium," F. Scaramuzzi, et al.
✳ "Status of Coherent Fusion Theory," P. L. Hagelstein
✳ "Surface Reaction Mechanism and Lepton Screening for Cold Fusion with Electrolysis," Y. E. Kim
✳ "Technical Status of Cold Fusion Results," D. Worledge
✳ "The Effect of Velocity Distribution and Electron Screening on Cold Fusion," Y. E. Kim
✳ "The Initiation of Excess Power and Possible Products of Nuclear Interactions During the Electrolysis of Heavy Water," C. D. Scott, et al.
✳ "Theoretical Ideas on Cold Fusion," G. Preparata
✳ "Tritium Measurements and Deuterium Loading in D_2O Electrolysis with a Palladium Cathode," R. R. Adzic and E. Yeager, et al.

Rabinowitz, Mario, and David H. Worledge. "An Analysis of Cold and Lukewarm Fusion." *Fusion Technology*, Vol. 17 (March 1990): 344–349.

Rafelski, Johann, Mikolaj Sawicki, Mariusz Gajda, and David Harley. "How Cold Fusion Can be Catalyzed." *Fusion Technology*, Vol. 18 (August 1990): 136–142.

Rafelski, Johann, and Steven E. Jones. "Cold Nuclear Fusion." *Scientific American* 257, no. 1 (July, 1987): 84–89.

Rock, Peter A., William H. Fink, Donald A. McQuarrie, David H. Volman, and Yu-Fen Hung. "Energy Balance in the Electrolysis of Water with a Palladium Cathode." *Journal of Electroanalytical Chemistry*, Vol. 293 (October 25, 1990): 261–267.

Rolison, Debra R., and Patricia P. Trzaskoma. "Morphological Differences Between Hydrogen-loaded and Deuterium-loaded Palladium as Observed by Scanning Electron Microscopy." *J. Electroanalytical Chemistry*, Vol. 280 (25 July 1990): 375–383.

Rout, R. K., M. Srinivasan, A. Shyam, and V. Chitra. "Detection of High Tritium Activity on the Central Titanium Electrode of a Plasma Focus Device." *Fusion Technology*, Vol. 19 (March 1991): 391–394.

Salamon, M. H., et al. "Limits on the Emission of Neutrons, γ-Rays, Electrons, and Protons from Pons/Fleischmann Electrolytic Cells." *Nature*, Vol. 344 (29 March 1990): 401–405.

Sato, Tsutomo, Makoto Okamoto, Poong Kim, Yasuhiko Fujii, and Otohiko Aizawa. "Detection of Neutrons in Electrolysis of Heavy Water." *Fusion Technology*, Vol. 19 (March 1991): 357–363.

Schriber, J. E., M. A. Butler, D. S. Ginley, and R. I. Ewing. "Search for Cold Fusion in High-Pressure D_2-Loaded Titanium and Palladium Metal and Deuteride." *Fusion Technology*, Vol. 16 (November 1989): 397–400.

Schultze, J. W., U. König, A. Hochfeld, C. van Calker, and W. Kies. "Properties and Problems of Electrochemically Induced Cold Nuclear Fusion." *Electrochimica Acta*, Vol. 34, no. 9 (September 1989): 1289–1313.

Schwinger, Julian. "Cold Fusion: A Hypothesis." *Zeitschrift für Naturforschung*," Vol. 45 (May 1990): 756.

Schwinger, Julian. "Cold Fusion—Does It Have a Future?" Text of an address delivered in Tokyo (6 December 1990) the Yoshio Nishina Centennial Symposium. Also in *Evolutional Trends in Physical Science* (Springer Verlag, 1991).

Schwinger, Julian. "Nuclear energy in an Atomic Lattice. 1." *Zeitschrift für Physik D.*, Vol. 15 (1990): 221–225.

Schwinger, Julian. "Phonon Dynamics." *Proceedings of the National Academy of Sciences*, Vol. 87 (November 1990): 8370–8372.

Schwinger, Julian. "Phonon Representations." *Proceedings of the National Academy of Sciences*, Vol. 87 (September 1990): 6983–6984.

Scott, Charles D., et al. "Measurement of Excess Heat and Apparent Coincident Increases in the Neutron and Gamma Ray Count Rates During the Electrolysis of Heavy Water." *Fusion Technology*, Vol. 18: 103–114.

Scott, C. D., J. E. Mrochek, E. Newman, T. C. Scott, G. E. Michaels, and M. Petek. "A Preliminary Investigation of Cold Fusion by Electrolysis of Heavy Water." Oak Ridge National Laboratory, ORNL/TM-11322 (November 1989) also presented at "Cold Fusion Session" of the winter annual meeting of the American Society of Mechanical Engineers, San Francisco (12 December 1989).

Sona, Pier Giorgio, et al. "Preliminary Tests on Tritium and Neutrons in Cold Fusion Within Palladium Cathodes." *Fusion Technology*, Vol. 17 (July 1990): 713–717.

Storms, Edmund, and Carol Talcott. "Electrolytic Tritium Production." *Fusion Technology*, Vol. 17 (July 1990): 680–695.

Storms, E., and C. Talcott, "Electrolytic Tritium Production." NSF/ EPRI Workshop on Anomalous Effects in Deuterated Materials, Washington, DC (October 16–18, 1989).

Tabet, Eugenio, and Alexander Tenenbaum. "Nuclear Reactions from Lattice Collapse in a Cold Fusion Model." *Physics Letters A*, Vol. 144 (12 March 1990): 301–305.

Takahashi, Akito, Toshiyuki Iida, Fujio Maekawa, Hisashi Sugimoto, and Shigeo Yoshida. "Windows of Cold Nuclear Fusion and Pulsed Electrolysis Experiments." *Fusion Technology*, Vol. 19 (March 1991): 380–390.

Taniguchi, T., T. Yamamoto, and S. Irie. "Detection of Charged Particles Emitted by Electrolytically Induced Cold Nuclear Fusion." *Japanese Journal of Applied Physics*, Vol. 28 (November 1989): L2021–2023.

Van Siclen, C. DeW, and S. E. Jones. "Piezonuclear Fusion in Isotopic Hydrogen Molecules." *Journal of Physics G: Nuclear Physics*, Vol. 12 (1986): 213–221.

Wada, Nobuhiko, and Kunihide Nishizawa. "Nuclear Fusion in Solid." *Japanese Journal of Applied Physics*," Vol. 28 (November 1989): L2017–2020.

Wagner, Frederick T., Thomas E. Moylan, Michael E. Hayden, Ulrike Narger, and James L. Booth. "A Comparison of Calorimetric Methods Applied to the Electrolysis of Heavy Water on Palladium Cathodes." *J. Electroanalytical Chemistry*, Vol. 295 (26 November 1990): 393–402.

Walling, Cheves, and Jack Simons. "Two Innocent Chemists Look at Cold Fusion." *The Journal of Physical Chemistry*, Vol. 93, no. 12 (15 June 1989): 4693–4697.

Werle, H., G. Fieg, J. Lebkücher, and M. Möschke. "Trials to Induce Neutron Emission from a Titanium-Deuterium System." *Fusion Technology*, Vol. 16 (November 1989): 391–396.

Wiesmann, Harold. "Examination of Cathodically Charged Palladium Electrodes for Excess Heat, Neutron Emission, or Tritium Production." *Fusion Technology*, Vol. 17 (March 1990): 350–354.

Williams, D. E., D. J. S. Findlay, D. H. Craston, M. R. Sené, M. Bailey, S. Croft, B. W. Hooton, C. P. Jones, A. R. J. Kucernak, J. A. Mason, and R. I. Taylor. "Upper Bounds on 'Cold Fusion' in Electrolytic Cells." *Nature*, Vol. 342 (23 November 1989): 375–384.

Yamaguchi, Eiichi, and Takashi Nishioka. "Cold Nuclear Fusion Induced by Controlled Out-Diffusion of Deuterons in Palladium." *Japanese Journal of Applied Physics* (April 1990): L666–L669.

Zahm, Lance L., et al. "Experimental Investigations of the Electrolysis of D_2O Using Palladium Cathodes and Platinum Anodes."

Journal of Electroanalytical Chemistry, Vol. 281 (6 March 1990): 313–321.

Zahm, L. L., A. C. Klein, S. E. Binney, J. N. Reyes Jr., J. F. Higginbotham, A. H. Robinson, and M. Daniels. "Experimental Investigations of the Electrolysis of D_2O Using Palladium and Platinum Electrodes." (December 1989) Oregon State University, Department of Nuclear Engineering, OSU-NE-8914.

Ziegler, J. F., T. H. Zabel, J. J. Cuomo, V. A. Brusic, G. S. Cargill III, E. J. O'Sullivan, and A. D. Marwick. "Electrochemical Experiments in Cold Nuclear Fusion." *Physical Review Letters*, Vol. 62, no. 25 (19 June 1989): 2929–2932.

✳ Miscellaneous Articles

Barber, Bernard. "Resistance by Scientists to Scientific Discovery." *Science*, Vol. 134 (1 September 1961): 596–602.

Bockris, John, and Dalibor Hodko. "Is There Evidence for Cold Fusion?" *Chemistry and Industry* (5 November 1990): 688–692.

Broad, William, and Nicholas Wade. *Betrayers of the Truth: Fraud and Deceit in the Halls of Science.* New York: Simon and Schuster, 1982.

Brush, Stephen G. "Should the History of Science Be Rated X?" *Science*, Vol. 183 (22 March 1974): 1164–1172.

Dickman, Steven. "1920s Discovery, Retraction." *Nature*, Vol. 338 (27 April 1989): 692.

Douglas, John. "In Hot Pursuit of Cold Fusion." *EPRI Journal* (April/May 1989): 21–23.

Fry, Edward S., Jacob B. Natowitz, and John W. Poston. "Report of the Cold Fusion Review Panel at Texas A&M University." (October 15, 1990).

Furth, H. P. "Magnetic Confinement Fusion." *Science*, Vol. 249 (28 September 1990): 1522–1526.

"Fusion in 1926: plus ça change." *Nature* (27 April 1989): 706 [reprint of *Nature*, Vol. 118 (1926): 455, 556 and Vol. 119 (1927): 706].

Holdren, John P. "Fusion Energy in Context: Its Fitness for the Long Term." *Science*, Vol. 200 (14 April 1978): 168–180.

Hudson, Richard. "If You Read It First in Nature It's Big And (Usually) True." *Wall Street Journal* (May 15, 1989).

Jahnke, Art. "Weird Science." *Boston Magazine* (August 1989): 130–133, 146–149.

Langmuir, Irving. Transcribed and edited by Robert N. Hall. "Pathological Science." *Physics Today*, Vol. 42 (October 1989): 36–48.

Morrison, Douglas R. O. (CERN), "Cold Fusion News." Dispatched via international computer networks from March 31, 1989, through autumn 1990.

Rothman, Milton A. "Cold Fusion: A Case History in 'Wishful Science'?", *The Skeptical Inquirer*, Vol. 14 (Winter 1990): 161–170.

Segré, Emilio G. "The Discovery of Nuclear Fission." *Physics Today*, Vol. 42 (July 1989): 38–43.

Index